Patient-Specific Stem Cells

Patient-Specific Stem Cells

Edited by
Deepak A. Lamba

CRC Press
Taylor & Francis Group
Boca Raton London New York

CRC Press is an imprint of the
Taylor & Francis Group, an **informa** business

CRC Press
Taylor & Francis Group
6000 Broken Sound Parkway NW, Suite 300
Boca Raton, FL 33487-2742

First issued in paperback 2019

ISBN-13: 978-1-4665-8026-8 (hbk)
ISBN-13: 978-0-367-86801-7 (pbk)

Visit the Taylor & Francis Web site at
http://www.taylorandfrancis.com

and the CRC Press Web site at
http://www.crcpress.com

Contents

Preface

One of the biggest challenges faced in medical research has been the creation of accurate and relevant models of human disease. A number of good animal models have been developed to understand the pathophysiology. However, not all of them reflect the human disorder, a classic case being Usher's syndrome, where the mutant mouse does not have the same visual and auditory defects that patients face. There are others that have been even more difficult to model due to the multifactorial nature of the condition and due to the lack of discovery of a single causative gene such as age-related macular degeneration or Alzheimer's disease. Thus, a more relevant and accurate system will allow us to make better predictions on relevant therapeutic approaches.

The discovery of human pluripotent stem cells in 1998 followed by the technological advances to reprogram somatic cells to pluripotent stem cell-like cells in 2006 has completely revolutionized the way we can now think about modeling human development and disease. This now coupled with genome editing technologies such as transcription activator-like effector nucleases and clustered regularly interspaced short palindromic repeats and has set us up to develop *in vitro* models of both two-dimensional and three-dimensional organoids which can more precisely reflect the disease in the patients. These combinatorial technologies are already providing us with better tools and therapeutics in drug discovery or gene therapy.

This book summarizes both the technological advances in the field of the generation of patient-specific lines and the various gene editing approaches followed by its applicability in various systems. We hope that the book will serve as a reference for the current state of the field.

About the Editor

Dr. Deepak A. Lamba earned his medical degree from the University of Mumbai, Mumbai, India, and practiced as a physician there. He earned a master's degree in bioengineering from University of Illinois, in Chicago, where he worked on a chemically stimulating retinal prosthesis device, followed by a PhD degree from the University of Washington, in Seattle, where he focused on generating and transplanting retinal cells derived from human embryonic stem cells and induced pluripotent stem cells (iPSCs) in the lab of Dr. Thomas Reh. Dr. Lamba's research focuses on identifying new methods to treat degenerative vision disorders, including macular degeneration and retinitis pigmentosa, using stem cell technology. His laboratory is working on two broad areas: (a) feasibility of photoreceptor replacement therapy and hurdles to successful cellular integration and (b) modeling retinal degenerations *in vitro* using iPSCs as well as bioengineering and gene editing technologies.

Contributors

Elma Aflaki
Medical Genetics Branch
National Human Genome Research Institute
National Institutes of Health
Bethesda, Maryland

Kapil Bharti
Unit on Ocular and Stem Cell Translational Research
Bethesda, Maryland

Daniel K. Borger
Medical Genetics Branch
National Human Genome Research Institute
National Institutes of Health
Bethesda, Maryland

Shuibing Chen
Department of Surgery and Biochemistry
Weill Cornell Medical College
New York, New York

Emma L. Curry
Northern Institute for Cancer Research
Newcastle University
Newcastle, UK

Lisa M. Ellerby
Buck Institute for Research on Aging
Novato, California

Thelma Garcia
Buck Institute for Research on Aging
Novato, California

Rakesh Heer
Northern Institute for Cancer Research
Newcastle University
Newcastle, UK

Balendu Shekhar Jha
Unit on Ocular and Stem Cell Translational Research
National Eye Institute
National Institutes of Health
Bethesda, Maryland

Lindsay Lenaeus
Board of Governors Regenerative Medicine Institute
and
Induced Pluripotent Stem Cell Core
Cedars–Sinai Medical Center
Los Angeles, California

Mohammad Moad
Northern Institute for Cancer Research
Newcastle University
Newcastle, UK

Suranjit Mukherjee
Department of Surgery and Biochemistry
Weill Cornell Medical College
New York, New York

Robert O'Brien
Buck Institute for Research
on Aging
Novato, California

Loren Ornelas
Board of Governors Regenerative
Medicine Institute
and
Induced Pluripotent Stem Cell Core
Cedars–Sinai Medical Center
Los Angeles, California

In-Hyun Park
Yale Stem Cell Center
Yale School of Medicine
New Haven, Connecticut

Uthra Rajamani
Board of Governors Regenerative
Medicine Institute
and
Department of Biomedical Sciences
Cedars–Sinai Medical Center
Los Angeles, California

Karen Ring
Buck Institute for Research
on Aging
Novato, California

Craig N. Robson
Northern Institute for Cancer
Research
Newcastle University
Newcastle, UK

Dhruv Sareen
Board of Governors Regenerative
Medicine Institute
and
Induced Pluripotent Stem Cell Core
and
Department of Biomedical Sciences
Cedars–Sinai Medical Center
Los Angeles, California

Ruchi Sharma
Unit on Ocular and Stem Cell
Translational Research
National Eye Institute
National Institutes of Health
Bethesda, Maryland

Ellen Sidransky
Medical Genetics Branch
National Human Genome Research
Institute
National Institutes of Health
Bethesda, Maryland

Yoshiaki Tanaka
Yale Stem Cell Center
Yale School of Medicine
New Haven, Connecticut

Ningzhe Zhang
Buck Institute for Research on
Aging
Novato, California

1

Human-Induced Pluripotent Stem Cells: Derivation

Uthra Rajamani, Lindsay Lenaeus, Loren Ornelas, and Dhruv Sareen

CONTENTS

1.1 Introduction

The use of human embryonic stem cells (hESCs) in cellular therapies in regenerative medicine has, in the past decade, emerged as a promising strategy in the development of cell replacement therapies (1). However, several ethical concerns exist due to the embryonic source of hESCs. Further, hESCs could not be employed in the generation of a wide array of disease-specific cell lines, invaluable to disease modeling. Seminal work conducted by the laboratories of Shinya Yamanaka and James Thompson (2–4) showed that mouse and human fibroblasts can be directly reprogrammed to exhibit pluripotency, thus creating a breakthrough in stem cell research worldwide with the creation of induced pluripotent stem cells (iPSCs) (Figure 1.1). This method has since been demonstrated to be highly reproducible by numerous laboratories using a panoply of reprogramming methods (5,6). The discovery of iPSCs was a revolution in the field of stem cell research, which helped overcome the aforementioned drawbacks of hESCs, such as avoiding the use of human embryos for stem cells as well as the ability to produce patient-specific pluripotent cells, thereby bypassing immune compatibility

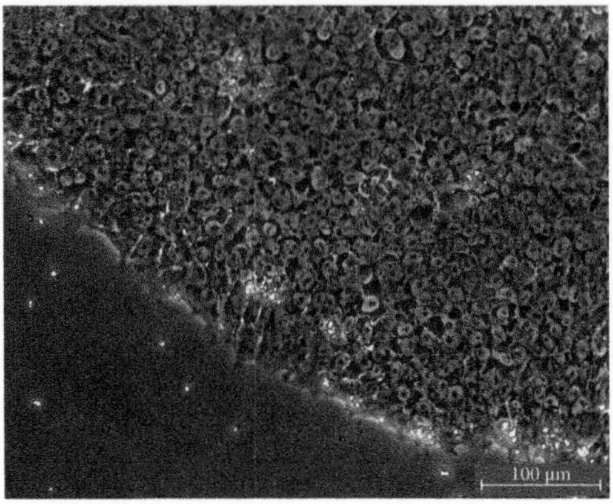

FIGURE 1.1
Bright-field image of human iPSC colony (×20 magnification) showing typical iPSC morphology.

challenges. Human iPSCs are able to differentiate into disease-relevant cell types, which serve as a key in patient-specific disease modeling and cell replacement therapies. This chapter will shed light on the various sources, methods of reprogramming, characterization and banking of iPSCs, their maintenance, and potential applications.

1.2 Derivation of iPS Cells

Human iPSCs are derived by the reprogramming of somatic cells to express pluripotency markers. Shinya Yamanaka's laboratory in their seminal work reprogrammed mouse and later human fibroblasts by retroviral transduction of pluripotency-inducing transcription factors, octamer-binding transcription factor 4 (Oct4), (sex-determining region Y)-box 2 (Sox2), Kruppel-like factor 4 (Klf4), and c-Myc (2,3). Concurrently, James Thompson's and Rudolf Jaenisch's laboratories also demonstrated that somatic adult human cells could be reprogrammed back to pluripotency by introducing Oct4 and Sox2 in combination with Nanog and Lin28 (5) or Klf4 and c-Myc (7). In the human system, Oct4 and Sox2 appear to be nonsubstitutable in inducing pluripotency, while Klf4, c-Myc, Nanog, and Lin28 are not indispensable. Several subsequent studies have added on to the list of reprogramming factors such as estrogen-related receptor beta (Esrrb) (8), spalt-like transcription factor (Sall4) (9), microribonucleic acid (miRNA) (10), simian vacuolating virus 40 large T antigen (SV40LT), and human telomerase reverse transcriptase (hTERT) (6,11). Several modes of delivery of the reprogramming factor expression vectors have been employed so far in attempts to maximize efficiency and minimize integration of reprogramming factor genes into the host iPSC genome.

1.2.1 Introduction to Reprogramming

The optimal delivery of reprogramming factors is crucial in achieving pluripotency, and as this is a rapidly developing field, several techniques have been employed for the generation of iPSCs. In addition, numerous starting cell types have been utilized in the derivation of iPSCs. As mentioned earlier, reprogramming involves the introduction of pluripotency transcription factors into somatic cells, rendering them with pluripotent abilities (Figure 1.2). The key reprogramming factors involved in inducing pluripotency are discussed in the following.

- Pou5f1 (Oct4): POU class 5 homeobox 1 (Pou5f1) was one of the first transcription factors to be identified as a regulator of pluripotency (12). Oct4-only transfections in the absence of other transcription factors have also successfully produced iPSCs (13). High expression of Oct4 has been found in primitive ectoderm, inner cell mass, and primordial germ cells in mouse development (14) as well as in embryonic stem cells (ESCs) *in vitro* (15) with a reduction in the expression upon differentiation (16). Studies have shown that the overexpression of Oct4 causes cells to move back to a pluripotent state (17), which confers upon cells a trilineage differentiation potential, giving them the ability to differentiate into endoderm, mesoderm (18), and neuroectoderm (19) lineages. On the contrary, the inhibition of Oct4 resulted in loss of pluripotency and differentiation (20). Oct4 introduction has so far been shown to be the only required factor for iPSC generation, thus proving the importance of Oct4 in somatic cell reprogramming.
- Sox2: Sox2 is yet another important pluripotency transcription factor, which is expressed in the epiblast, the inner cell mass, and the extra

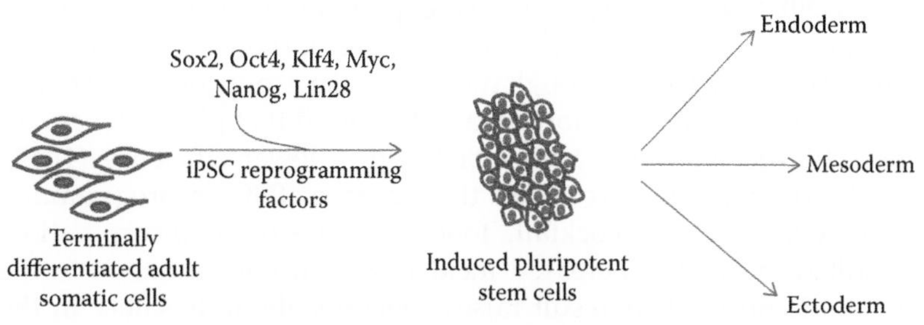

FIGURE 1.2
(See color insert.) A schematic representation of somatic cell reprogramming.

embryonic ectodermal cells of a developing mouse (21). Sox2 expression is maintained in neural stem cells, unlike Oct4 (22,23), and it interacts with several pluripotency transcription factors such as Oct4 to maintain pluripotency (24). It has been shown to be important in epiblast formation in mouse development, as Sox2 knockout mouse embryos did not form an epiblast or develop past implantation stage (25). Although other members of the Sox family, such as Sox1, Sox3, Sox15, and Sox18, have been suggested to compensate for Sox2 function, Sox2 remains a fundamental reprogramming factor widely used in iPSC generation protocols. However, Sox2 was dispensable when human neural stem cells were reprogrammed using only Oct4 (26,27), and in the mouse system, Sox2 was replaced by a small molecule inhibitor of transforming growth factor beta signaling, called *RepSox* (28).

- Nanog: Nanog is known to be a master regulator of stem cell pluripotency (29), which is expressed in the inner cell mass, the primordial germ cell, and the epiblast in early embryos (30). The name of the Nanog gene was chosen by Professor Ian Chambers, taking a cue from Irish mythology legend, "Tír na nÓg," which signifies immortality conferred by Nanog upon the stem cells. Nanog plays a key role in suppressing Cdx2, which in turn leads to the specification of the fate of the inner cell mass (31). Nanog also works in synchrony with Oct4 and Sox2 in maintaining the pluripotency of stem cells (32,33). Although Dr. Yamanaka and his group proved that the initial derivation of iPSCs was possible without the ectopic expression of Nanog (2,3), Dr. Thomson's group showed that Nanog is a key component of the reprogramming factor cocktail. It is, nonetheless, an important marker for the fully reprogrammed germ line-competent human iPSC state (34–36).
- c-Myc and L-Myc: The helix-loop–helix/leucine zipper transcription factor c-Myc is well documented for its role in control of cell cycle, apoptosis, protein biosynthesis, and metabolism (37,38). c-Myc overexpression has been shown to resist differentiation in mouse ESCs (39). Interestingly, c-Myc overexpression in hESCs drove them to apoptosis and differentiation to extra-embryonic endoderm and trophectoderm (40). Although the Yamanaka group identified c-Myc as a reprogramming factor, it was later found that pluripotency can be achieved without c-Myc, albeit with a lower efficiency (41,42). Furthermore, it was revealed that murine iPSCs reprogrammed with c-Myc in the cocktail, following blastocyst incorporation, resulted in greatly increased incidence of tumor formation in the chimeric mice . This result raises concerns about its safety in the use for iPSCs generated for clinical purposes while also potentially confounding disease modeling phenotypes if randomly integrated

into the host genome. It has since been shown to regulate a different set of target genes to induce pluripotency compared to Oct4, Sox2, and Klf4 (43). c-Myc is proposed to repress the expression of somatic genes in early reprogramming, promote the activation of pluripotency genes (43), promote cell proliferation, and repress cell cycle arrest genes (44). It is also proposed that c-Myc may activate the self-renewal gene program by modifying the chromatin structure, thus allowing somatic cells to attain pluripotency (45). With recent advances in reprogramming technologies, a transformation-deficient form of c-Myc, called *L-Myc*, is routinely employed as a reprogramming factor. Mice generated by substitution of L-Myc results in drastically decreased tumor formation in comparison to c-Myc (46).

- Klf4: The Klf4 family of reprogramming factors, belonging to the Kruppel-like factor family of zinc finger transcription factors, has emerged as a key regulator of pluripotency. Klf4 is expressed in moderately regenerative adult tissues such as gut, skin, testis, and intestine (47) and was one of the first reprogramming factors employed alongside Oct4, Sox2, and c-Myc (2,3). Klf4 is also an important component of core transcription network which regulates the expression of Oct4, Sox2, Nanog, c-Myc, and itself (33,48), primarily required for increasing reprogramming efficiencies (49,50). Furthermore, Klf4 has also been shown to interact and act along with Oct4 and Sox2, resulting in the activation of ESC-specific genes such as Nanog and left–right determination factor 1 (49,50). Additionally, it also suppresses p53-mediated apoptosis (51), thereby assisting in reprogramming somatic cells to pluripotency.
- Lin28: Lin28 encodes for a translation-enhancing RNA-binding protein (52) and was first shown to regulate developmental timing in *Caenorhabditis elegans* (53). In the induction of pluripotency, Lin28 was first used in combination with Nanog, Oct4, and Sox2 to generate human iPSCs (5). It can block the processing of the lethal 7 (let7) family of noncoding miRNAs, which are a group of prodifferentiation miRNAs (54) that also act as tumor suppressors by decreasing the expression of oncogenes such as c-Myc and Ras (55). As a result, the Lin28-mediated suppression of let7 results in cell proliferation and drives cell transformation. Lin28 can also reside in polysomal ribosome fractions, which translate messenger RNAs (mRNAs) (56). It also possibly binds directly to Oct4 mRNA in hESCs and interacts with RNA helicase A to facilitate translation (57). Lin28 can also reside in processing bodies, which degrade mRNA (56). Taken together, a possible mechanism of Lin28-mediated induction of pluripotency could be via the regulation of posttranscriptional gene expression by selective activation of antidifferentiation and

repression of prodifferentiation mRNAs. Further studies are needed to fully decipher this mechanism, but it is well understood that the presence of Lin28 enhances reprogramming efficiencies (58).

Several more reprogramming factors have been identified since the seminal work by Yamanaka (2), such as Sall4 (9), Esrrb (59), miRNA (10,60), SV40LT, hTERT (6,11), and proreprogramming factor, GLIS1 (61,62).

1.2.2 Source of Tissue Samples

Several sources of somatic cells have been employed to be utilized for reprogramming. This section will briefly discuss a few cell types that have been successfully reprogrammed thus far.

1.2.2.1 Primary Mesenchymal Cells

1. Blood
   a. Fresh blood: The generation of iPSCs from CD34+-mobilized human peripheral blood cells using retroviral transduction of Oct4/Sox2/Klf4/c-Myc was first accomplished by Loh et al (63). They showed that blood-derived human iPSCs are very similar to hESCs in terms of morphology, expression of surface antigens and pluripotency-associated transcription factors, deoxyribonucleic acid (DNA) methylation status at pluripotent cell-specific genes, and their ability to differentiate *in vitro* and *in teratomas* (63). This finding also paved way for generating iPSCs for somatic genetic disorders specific to the hematopoietic system, which cannot be mimicked in the fibroblast reprogramming technique. Furthermore, the use of peripheral blood to generate iPSCs obviates the need for skin biopsies and provides cells that require minimal maintenance in culture.
   b. Lymphoblastoid cell lines (LCLs): Human iPSCs can be generated using Epstein–Barr virus (EBV)-immortalized B cells from the blood known as *lymphoblastoid cells* (64). The EBV infections in the immortalized cells are mostly latent and hence express only a limited number of viral proteins. LCLs are phenotypically characterized by activated B cells expressing activation markers such as CD23, CD39, and adhesion molecules such as lymphocyte function-associated antigen 1 (LFA1), LFA3, and intercellular adhesion molecule (65). LCLs can be explanted from the peripheral blood of EBV-seropositive individuals without exogenous infection (66) or by *in vitro* infection of peripheral blood mononuclear cells (PBMCs) with EBV (67). Advents in recent methods allow for much more reliable reprogramming

of iPSC lines using episomal plasmids (67). The vast potential of this source of tissue allows greater utilization of numerous repositories worldwide that already store well-annotated and whole-genome-sequenced patient LCLs. These LCLs can be readily converted to iPSCs and subsequently to any differentiated cell type for research studies.

2. Fibroblasts—Initial studies by Takahashi and Yamanaka (2) utilized mouse embryonic fibroblasts and tail tip fibroblasts to generate iPSC. These cells were similar in morphology and growth properties to ESCs and were also seen to express all ESC marker genes. Following this seminal work, several other groups also generated iPSCs from fibroblasts by using human fibroblasts (3,5). Dermal fibroblasts derived from skin punch biopsies, until recently, have been the most commonly used source of tissue for iPSC generation and disease-modeling studies. Skin punch biopsies are enzymatically digested, and the resultant cell pellet is plated in flasks with fibroblast growth medium to acquire fibroblasts (68). Alternatively, skin biopsies cut into small fragments supplied with fibroblast growth media can also be used to acquire fibroblasts (69). It usually takes around seven days for fibroblasts to grow from the skin biopsy fragments along with other cell types such as keratinocytes. These cells are enzymatically detached and plated onto fresh dishes cultured in fibroblast growth media to favor the growth of fibroblasts over keratinocytes, and after two to three passages, a relatively homogenous culture of fibroblast may be obtained. The process of obtaining, expanding, and banking fibroblasts takes around four to eight weeks (69) before they can be utilized for reprogramming.
3. Adipose tissue—Human adipose tissue has also been shown to be a source of iPSCs. Adipose tissue can be obtained from various depots such as bone marrow; brown, mammary, and white adipose tissues; and mechanical tissues. These sources show differences in stem cell content, with arms showing the greatest stem cell recovery in humans (70). Human lipoaspirates have been reported to be a source for multipotent stem cells (71). Due to the ease in obtaining adipose tissue, using them as a source of iPSCs holds promise. Studies have shown that the reprogramming of the adipose tissue-derived cells with Oct4, Sox2, and Klf4 resulted in iPSCs forming hESC-like colonies and expressing pluripotency markers (72). Interestingly, despite not employing c-Myc to induce pluripotency due to its oncogenic nature, this study showed successful induction of pluripotency in adipose tissue cells and exhibited the ability to differentiate into all three germ layers.
4. Oral progenitors—Another potential source of iPSC is the oromaxillofacial region, where cells can be derived from dental pulp (73),

gingiva (74), buccal mucosa (75), and periodontal ligament (76). Deriving iPSCs from buccal mucosa is a safe method as the collection of buccal mucosa is a simple procedure performed on a dental chair which does not cause functional damage and allows quick healing (75). Periodontal ligament fibroblasts have rapid turnover and good remodeling, renewal, and repair ability (76). These cells have been reprogrammed by the introduction of Oct4, Sox2, Klf4, and C-Myc using retroviral gene transfer.

1.2.2.2 Primary Epithelial Cells

1. Organ-specific epithelial cells—iPSCs have also been derived from isolated primary stomach and liver cells from patients that come in the clinic for respective gastroenterological procedures. This was achieved by the introduction of four transcription factors (Oct4, Sox2, Klf4, and c-Myc) by retroviral vectors into hepatocytes or gastric epithelial cells (77). This study suggested that c-Myc had a smaller role to play in the generation of iPSCs from the stomach/the liver compared to fibroblasts.
2. Mammary—The immortalized mammary epithelial cell line (MCF-10A) has also been utilized for producing iPSCs from mammary tissues (78). MCF-10A cells do not possess tumor initiation abilities but have shown promise in oncogenic transformation (79). These cells were reprogrammed using retroviral vectors to introduce Oct4, Sox2, Klf4, and c-Myc (78). Similarly, isolated primary mammary epithelial cells can also be utilized for reprogramming to generate iPSC.
3. Kidney and urine—Kidney mesangial (80) and renal proximal tubular epithelial cells (81) isolated from biopsies during clinical procedures have also been converted to iPSCs. Recently, exfoliated renal epithelial cells present in urine have been used to generate iPSCs (82). This source tissue is a noninvasive and cost-effective source of cells for the generation of iPSCs. However, this was achieved by utilizing the integrating retrovirus to introduce factors. Urinary iPSCs show high differentiation potential comparable to that of ESCs and show potential of differentiation into lineages derived from the three germ layers (82).
4. Eye—Limbal epithelial cells have also been utilized for the generation of iPSCs (83,84). The introduction of Oct4, Sox2, Klf4, Lin28, and L-Myc has shown successful reprogramming of limbal epithelial cells (84). The differentiation of iPSCs to corneal epithelium has been a challenge (85,86), although there have been a few studies that have produced lens, retinal pigment epithelial cells, and photoreceptor cells (87,88). However, limbal epithelial cell-derived iPSCs have

shown more promise in the generation of limbal-like cells better than fibroblast-derived iPSCs (84). The use of the same tissue of origin to derive iPSCs for the differentiation of a particular tissue can contribute to improved success. This is possibly due to the partial retention of tissue-specific epigenetic signatures including methylation patterns (84).

1.2.3 Methods of Reprogramming

Since the first reprogramming of somatic cells to pluripotency by using the Yamanaka factors with retroviral delivery, this field has seen several other reprogramming techniques with improved efficiencies and clinical safety. Some of the commonly used reprogramming methods and their efficiencies based on available literature are compiled here (Table 1.1).

1.2.3.1 Integrating Vectors

1. Retroviral vectors—Retroviruses possess the ability to integrate with the genome of the host cell and form a continuous transgene expression and thus serve as an efficient gene delivery method. Retroviral vectors were the first used method of reprogramming fibroblasts to a pluripotent state (2). Platinum-E cells were used to transduce viral vectors by using FuGENE 6 transfection reagent. The viral supernatants derived from transduction were then filtered and used to infect the target cells. Retroviral vectors were also successfully employed by other groups in reprogramming human adult dermal fibroblasts to pluripotency (6,89). This method was shown to have an efficiency of 0.001–0.5%. Although this was the first method to induce pluripotency in somatic cells, actively dividing cells are required, and this method is not as efficient in nondividing (senescent) or slow-dividing cells. The major disadvantage of this method is the random sites of transgene integration, which could result in unpredictable mutation load in the host genetic background, thus complicating disease modeling of patient-specific phenotypes.
2. Lentiviral vectors—Inducible lentiviral vectors have been shown to improve temporal control over reprogramming factor expression levels and have been widely used for studying molecular changes and timing during reprogramming. Although initial attempts with lentivirus only yielded 0.01% efficiency, which was far less than that of retrovirus (5), subsequent improvements to vector delivery methods and reprogramming cocktails helped increase efficiency by about a 100-fold (11,90). Lentiviral vectors are also known to randomly integrate into the host genome and hence may not be clinically

TABLE 1.1

Commonly Used Reprogramming Techniques and Efficiencies

	Method	Cell Type	Efficiency (%)	Pros	Cons
Integrating	Retroviral	Fibroblasts, neural stem cells, mesenchymal cells, keratinocytes, blood cells, adipose cells	~0.001–0.5	Efficient	Genomic integration
	Lentiviral	Fibroblasts, keratinocytes	~0.01	Efficient, transduces dividing and nondividing cells	Genomic integration
	Inducible lentiviral	Fibroblasts, β cells, keratinocytes, blood cells, and monocytes	~0.1–1	Efficient, allows for controlled expression of factors	Genomic integration
Excisable	Transposons (PiggyBac)	Fibroblast	~2.5	No genomic integration	
	LoxP-flanked lentiviral plasmid	Fibroblast	~0.1–1	No genomic integration	
Nonintegrating	Adenoviral	Fibroblasts and liver cells	~0.001	No genomic integration	Low efficiency
	Plasmid	Fibroblast, neural stem cells, mesenchymal cells, keratinocytes, blood cells, adipose cells, lymphoblastoid cell lines	~0.001–0.1	Occasional genomic integration	Occasional vector genomic integration
DNA-free	Sendai virus	Fibroblast, CD34+, PBMCs, keratinocytes, myoblasts, T-cells	~1	No genomic integration	Expensive
	Modified mRNA	Fibroblast	~1–4.4	No genomic integration	Requirement for multiple rounds of transfection
	MicroRNA	Fibroblast	~0.02	No genomic integration	Low efficiency

applicable. Further, like retroviral vectors, the incomplete silencing of reprogramming transgenes by using this delivery method leads to greater variability, disease phenotypes, and cell differentiation efficiencies. Lentiviral vectors can be used to reprogram both dividing and nondividing cells with stable transgene expression and low immunogenicity (91). Including an additional vector expressing reverse tetracycline transactivator, which drives the expression of reprogramming factors only in the presence of doxycycline, serves as a helpful method to select cells that have completely reprogrammed to pluripotency because the removal of doxycycline from the system after initial transduction leads to nonsurvival of partially reprogrammed cells (92–94). However, a drawback to this method is the use of more than one viral vector, as only some cells will be infected by all the viral vectors, thus reducing reprogramming efficiency. This concern was addressed by the use of single cassette reprogramming vectors with each of the reprogramming factors separated by a self-cleaving peptide signal (95–97). The introduction of locus of crossover in P1 (loxP) sites to these vectors helped the excision of any integrated sequence by the overexpression of Cre-recombinase (96,97). Recently, the STEMCCA vector has been designed and is widely used with around 0.1–1.5% reprogramming efficiency but still leaves behind a small genetic footprint in the host genome even after excision.

1.2.3.2 Nonintegrating Vectors

1. Adenovectors—Adenoviral vectors do not integrate with the host genome and hence have been regarded to be more advantageous than integrating viral vectors with high expression of exogenous genes and minimal integration of viral transgenes. The virus level dilutes with every cell division and allows only a transient expression of transgenes. Initially, adenoviral vectors were employed for somatic cell reprogramming with a very low efficiency range of 0.0001–0.001% (98,99). Interestingly, about 20% of the transduced cells were tetraploid (98), which is probably a result of fusion of cells. Given these findings, further research is needed to optimize and refine the use of adenoviral vectors in somatic cell reprogramming to pluripotency.
2. PiggyBac transposon—PiggyBac transposons allow for the efficient delivery of reprogramming factors by means of a cut–paste mechanism. A transposase can integrate to the chromosomal TTAA site that can later be excised from the genome upon reexpression of transposase. This system has been shown to be efficient in excisable gene delivery of up to 10 kb DNA fragments (100). The PiggyBac transposon system can be completely removed from the host genome

without changing sequences at the integration sites (101). The PiggyBac reprogramming technique thus allowed the generation of mouse and human iPSCs by using tetracycline-inducible or polycistronic expression of reprogramming factors (102–104). This method reported a high efficiency rate of 2.5%. This system provides an efficient nonintegrative reprogramming approach with minimal changes to host genome. However, transposase excision of transgenes could still be accompanied by microdeletions of genomic DNA (101), which questions its clinical applications. Also, a small footprint of the plasmid may be left behind.

3. mRNA (±miRNA)—mRNA is another nonintegrating approach for the introduction of reprogramming factors. The use of mRNA allows for the reprogramming factors to be transcribed by the cells themselves. This method was first tested on human fibroblasts with 1.4% efficiency (105). Interestingly, the addition of Lin28 to the Yamanaka factors during reprogramming increased efficiency to 4.4%. Even though mRNAs for reprogramming factors are commercially available, this method has not been tested on other cell types other than fibroblasts, besides being labor intensive and not very reliable. HESCs have been shown to express certain miRNA clusters, and hence when synthetic equivalents of mature miR-302b or miR-372 alongside Yamanaka factors are introduced to fibroblasts, reprogramming efficiency close to 10–15% has been reported (106). Certain miRNAs can reprogram cells at high efficiencies even in the absence of Yamanaka factors (107). Additionally, the transfection of human cells with certain miRNAs such as miR-200c, miR-302s, and miR-369s also resulted in their reprogramming at an efficiency of 0.002% in 20 days after the first transfection (108). In 2010, *in vitro* synthesized RNA transfection of Oct4, Lin28, Sox2, and Nanog successfully produced iPSCs from human fibroblasts (109). RNA transfection reprogramming had an efficiency of 0.05% and also required numerous transfections before successful reprogramming as their half-life is limited.

4. Protein-based reprogramming—In recent years, protein-based reprogramming has shown promise in the production of clinical grade iPSCs. Protein tagged with a C-terminus polyarginine domain, which has been previously shown to allow efficient protein transduction across the membrane (110), can be employed as a direct protein delivery method. Protein transduction has been employed in both mouse and human iPSC generation with efficiencies of 0.006% and 0.001%, respectively (111,112). Moreover, protein transduction, like RNA transfection, required multiple rounds of treatment for successful reprogramming; although both these studies have a very low

reprogramming efficiency, the cells produced were free of genetic modification and hence protein transduction offers promise in producing iPSCs for clinical application.

5. Sendai virus—Sendai virus-based reprogramming has been considered advantageous since Sendai viruses are RNA viruses and hence do not enter the nucleus. As a result, they are diluted and eventually lost by the cells in about 10 passages from infection. They are also capable of producing large amounts of reprogramming proteins. Sendai viruses can be used to reprogram fibroblasts as well as blood cells (113–115). This method has shown 0.1% efficiency with blood and 1% efficiency with fibroblasts. Sendai viruses are also available commercially as ready-to-use viral extracts with the Yamanaka factors, the cost being relatively high. A disadvantage of the use of Sendai virus is that it can take more than 15 passages to be lost from the cells.
6. DNA-based plasmids—This is a facile nonintegrating approach for the transient expression of reprogramming factors. Initially employed for the generation of nonintegrated mouse iPSCs using polycistronic plasmids (116), a number of iPSC colonies showed integration of transgenes. This technique requires multiple rounds of transfection for a sustained transgene expression required for reprogramming and has a significantly low efficiency compared to integrating vector methods (116,117). Subsequently, the use of minicircle nonviral polycistronic plasmids was employed to reprogram human adipose tissue cells (118). These plasmids provided a higher transfection rate and a slower dilution rate, resulting in fewer rounds of transfection with an efficiency of 0.005%. Origin of viral replication (OriP)/Epstein-Barr nuclear antigen 1 (EBNA1) vectors, derived from EBV, is a commonly used episomal plasmid since these do not need viral packaging and can be eventually removed from cells by culturing them in the absence of drug selection (4). OriP/EBNA1 vectors stably replicate extrachromosomally and only require a cis-acting OriP (119) and a trans-acting EBNA gene (120). OriP/EBNA vectors replicate in a cell cycle just once and, with drug selection, can bring about stable transfection in almost 1% of the transfected cells (121,122). As mentioned earlier, in the absence of drug selection, the episomes are lost at the rate of 5% per generation and cells are rendered plasmid-free (123).
7. Small molecules—The use of small molecules has provided a new approach to reprogramming with reduced safety concerns, avoiding the use of viral vectors and having the ability for large-scale production. They are fast acting with reversible effects, allowing for dose-dependent control for fine-tuning outcomes. Small molecules are also easy to handle and can be applied *in vitro* and *in vivo*. Several small molecules have been employed to improve reprogramming efficiency,

such as valproic acid, a widely used histone deacetylase (HDAC) inhibitor that facilitates Oct4-only-mediated reprogramming (13). BIX-01294 is a G9 histone methyltransferase which has been employed as a substitute for Oct4 in neural progenitor cell reprogramming (124). CHIR99021, a GSK3 inhibitor, has been shown to improve Oct4-mediated reprogramming in combination with A-83-01, an ALK4/5/7 [ALK4 (ACVR1B) = activin receptor-like kinase 4 (activin receptor type-1B)] inhibitor; sodium butyrate, an HDAC inhibitor; and pyruvate dehydrogenase kinase 1 activator PS48 (13). Although an increasing number of small molecules are being employed in improving reprogramming efficiencies, care must be taken while using precious samples because small molecules have some disadvantages, including having multiple off-target and cytotoxicity effects.

A survey conducted recently on the most used and most adopted techniques for reprogramming revealed that lentivirus, Sendai virus, and episomal DNA plasmid-based methods have been most used by laboratories (125). When it comes to the adoption of a successful technique, Sendai virus and episomal DNA-based plasmid based reprogramming had the highest adoption rate compared to lentiviral and retroviral methods, suggesting that most laboratories were inclined toward nonintegrating reprogramming methods. Episomal DNA-based (oriP/EBNA1) methods have a significant cost advantage over all other techniques because they require minimal laboratory investments and simple plasmid amplification systems. Episomal plasmid and Sendai virus reprogramming showed high efficiency and reliability with low workload and an almost complete loss of transgene sequences in most iPSC lines toward later passages (125). Episomal plasmid and mRNA methods have advantages such as a complete absence of integration or relatively fast loss of reprogramming agent, respectively. With these techniques constantly evolving, means to improve efficiency of these methods will help researchers choose the best-suited method for their requirements. Although successful in mouse embryonic fibroblasts, completely defined chemical-based small molecule-only methods of reprogramming in the human system have remained elusive.

Clone selection and isolation is key to obtaining a clean homogenous culture of iPSCs. For selection, the colonies are visually inspected for ESC-like morphology and manually isolated (7). Live staining and expression of pluripotency-associated cell surface glycoproteins, such as stage-specific embryonic antigen 3 (SSEA3) and SSEA4, T-cell receptor alpha locus 1-60 (TRA-1-60), and TRA-1-81, can be assessed (1,89,126,127), as these pluripotency markers are linked to fluorescent proteins to assist in colony selection (36,128,129). The selected colonies are picked using a pipette and transferred on to a fresh extracellular matrix-coated dish and expanded (130). This is, however, a highly skilled and time-consuming process since the choice of clones picked for propagation determines the homogeneity of the iPSC line.

Acknowledgments

Dhruv Sareen is supported by funding from National Institutes of Health (NIH)/National Center for Advancing Translational Sciences (UL1TR00014), NIH/National Institute of Neurological Disorders and Stroke/Library of Integrated Network-Based Cellular Signatures (1U54NS091046-01), NIH/National Eye Institute (R01 EY023429-01) funds and institutional start-up funds from Cedars-Sinai Medical Center.

References

1. Pera, M. F., B. Reubinoff, and A. Trounson, Human embryonic stem cells. *J Cell Sci*, 2000. 113 (Pt 1): pp. 5–10.
2. Takahashi, K. and S. Yamanaka, Induction of pluripotent stem cells from mouse embryonic and adult fibroblast cultures by defined factors. *Cell*, 2006. 126(4): pp. 663–76.
3. Takahashi, K. et al., Induction of pluripotent stem cells from adult human fibroblasts by defined factors. *Cell*, 2007. 131(5): pp. 861–72.
4. Yu, J. et al., Human induced pluripotent stem cells free of vector and transgene sequences. *Science*, 2009. 324(5928): pp. 797–801.
5. Yu, J. et al., Induced pluripotent stem cell lines derived from human somatic cells. *Science*, 2007. 318(5858): pp. 1917–20.
6. Park, I. H. et al., Reprogramming of human somatic cells to pluripotency with defined factors. *Nature*, 2008. 451(7175): pp. 141–6.
7. Meissner, A., M. Wernig, and R. Jaenisch, Direct reprogramming of genetically unmodified fibroblasts into pluripotent stem cells. *Nat Biotechnol*, 2007. 25(10): pp. 1177–81.
8. Feng, B. et al., Reprogramming of fibroblasts into induced pluripotent stem cells with orphan nuclear receptor Esrrb. *Nat Cell Biol*, 2009. 11(2): pp. 197–203.
9. Wong, C. C. et al., High-efficiency stem cell fusion-mediated assay reveals Sall4 as an enhancer of reprogramming. *PLoS One*, 2008. 3(4): p. e1955.
10. Marson, A. et al., Connecting microRNA genes to the core transcriptional regulatory circuitry of embryonic stem cells. *Cell*, 2008. 134(3): pp. 521–33.
11. Mali, P. et al., Improved efficiency and pace of generating induced pluripotent stem cells from human adult and fetal fibroblasts. *Stem Cells*, 2008. 26(8): pp. 1998–2005.

12. Scholer, H. R. et al., Octamer binding proteins confer transcriptional activity in early mouse embryogenesis. *EMBO J*, 1989. 8(9): pp. 2551–7.
13. Zhu, S. et al., Reprogramming of human primary somatic cells by OCT4 and chemical compounds. *Cell Stem Cell*, 2010. 7(6): pp. 651–5.
14. Pesce, M. and H. R. Scholer, Oct-4: Gatekeeper in the beginnings of mammalian development. *Stem Cells*, 2001. 19(4): pp. 271–8.
15. Rosner, M. H. et al., A POU-domain transcription factor in early stem cells and germ cells of the mammalian embryo. *Nature*, 1990. 345(6277): pp. 686–92.
16. Assou, S. et al., A meta-analysis of human embryonic stem cells transcriptome integrated into a web-based expression atlas. *Stem Cells*, 2007. 25(4): pp. 961–73.
17. Sareen, D. and C. N. Svendsen, Stem cell biologists sure play a mean pinball. *Nat Biotechnol*, 2010. 28(4): pp. 333–5.
18. Niwa, H., J. Miyazaki, and A. G. Smith, Quantitative expression of Oct-3/4 defines differentiation, dedifferentiation or self-renewal of ES cells. *Nat Genet*, 2000. 24(4): pp. 372–6.
19. Shimozaki, K. et al., Involvement of Oct3/4 in the enhancement of neuronal differentiation of ES cells in neurogenesis-inducing cultures. *Development*, 2003. 130(11): pp. 2505–12.
20. Matin, M. M. et al., Specific knockdown of Oct4 and beta2-microglobulin expression by RNA interference in human embryonic stem cells and embryonic carcinoma cells. *Stem Cells*, 2004. 22(5): pp. 659–68.
21. Miyagi, S. et al., The Sox-2 regulatory regions display their activities in two distinct types of multipotent stem cells. *Mol Cell Biol*, 2004. 24(10): pp. 4207–20.
22. Ellis, P. et al., SOX2, a persistent marker for multipotential neural stem cells derived from embryonic stem cells, the embryo or the adult. *Dev Neurosci*, 2004. 26(2–4): pp. 148–65.
23. Graham, V. et al., SOX2 functions to maintain neural progenitor identity. *Neuron*, 2003. 39(5): pp. 749–65.
24. Yuan, H. et al., Developmental-specific activity of the FGF-4 enhancer requires the synergistic action of Sox2 and Oct-3. *Genes Dev*, 1995. 9(21): pp. 2635–45.
25. Avilion, A. A. et al., Multipotent cell lineages in early mouse development depend on SOX2 function. *Genes Dev*, 2003. 17(1): pp. 126–40.
26. Kim, J. B. et al., Direct reprogramming of human neural stem cells by OCT4. *Nature*, 2009. 461(7264): pp. 649–53.
27. Kim, J. B. et al., Oct4-induced pluripotency in adult neural stem cells. *Cell*, 2009. 136(3): pp. 411–9.
28. Ichida, J. K. et al., A small-molecule inhibitor of Tgf-β signaling replaces *Sox2* in reprogramming by inducing *Nanog*. *Cell Stem Cell*, 2009. 5(5): pp. 491–503.
29. Mitsui, K. et al., The homeoprotein Nanog is required for maintenance of pluripotency in mouse epiblast and ES cells. *Cell*, 2003. 113(5): pp. 631–42.
30. Chambers, I. et al., Functional expression cloning of Nanog, a pluripotency sustaining factor in embryonic stem cells. *Cell*, 2003. 113(5): pp. 643–55.
31. Chen, L. et al., Cross-regulation of the Nanog and Cdx2 promoters. *Cell Res*, 2009. 19(9): pp. 1052–61.
32. Boyer, L. A. et al., Core transcriptional regulatory circuitry in human embryonic stem cells. *Cell*, 2005. 122(6): pp. 947–56.
33. Kim, J. et al., An extended transcriptional network for pluripotency of embryonic stem cells. *Cell*, 2008. 132(6): pp. 1049–61.

34. Maherali, N. et al., Directly reprogrammed fibroblasts show global epigenetic remodeling and widespread tissue contribution. *Cell Stem Cell*, 2007. 1(1): pp. 55–70.
35. Okita, K., T. Ichisaka, and S. Yamanaka, Generation of germline-competent induced pluripotent stem cells. *Nature*, 2007. 448(7151): pp. 313–7.
36. Wernig, M. et al., In vitro reprogramming of fibroblasts into a pluripotent ES-cell-like state. *Nature*, 2007. 448(7151): pp. 318–24.
37. Kendall, S. D., S. J. Adam, and C. M. Counter, Genetically engineered human cancer models utilizing mammalian transgene expression. *Cell Cycle*, 2006. 5(10): pp. 1074–9.
38. Patel, J. H. et al., Analysis of genomic targets reveals complex functions of MYC. *Nat Rev Cancer*, 2004. 4(7): pp. 562–8.
39. Cartwright, P. et al., LIF/STAT3 controls ES cell self-renewal and pluripotency by a Myc-dependent mechanism. *Development*, 2005. 132(5): pp. 885–96.
40. Sumi, T. et al., Apoptosis and differentiation of human embryonic stem cells induced by sustained activation of c-Myc. *Oncogene*, 2007. 26(38): pp. 5564–76.
41. Wernig, M. et al., c-Myc is dispensable for direct reprogramming of mouse fibroblasts. *Cell Stem Cell*, 2008. 2(1): pp. 10–2.
42. Nakagawa, M. et al., Generation of induced pluripotent stem cells without Myc from mouse and human fibroblasts. *Nat Biotechnol*, 2008. 26(1): pp. 101–6.
43. Sridharan, R. et al., Role of the murine reprogramming factors in the induction of pluripotency. *Cell*, 2009. 136(2): pp. 364–77.
44. Vermeulen, K., Z. N. Berneman, and D. R. Van Bockstaele, Cell cycle and apoptosis. *Cell Prolif*, 2003. 36(3): pp. 165–75.
45. Knoepfler, P. S. et al., Myc influences global chromatin structure. *EMBO J*, 2006. 25(12): pp. 2723–34.
46. Okita, K. et al., A more efficient method to generate integration-free human iPS cells. *Nat Methods*, 2011. 8(5): pp. 409–12.
47. Nandan, M. O. and V. W. Yang, The role of Kruppel-like factors in the reprogramming of somatic cells to induced pluripotent stem cells. *Histol Histopathol*, 2009. 24(10): pp. 1343–55.
48. Chen, X. et al., Integration of external signaling pathways with the core transcriptional network in embryonic stem cells. *Cell*, 2008. 133(6): pp. 1106–17.
49. Nakatake, Y. et al., Klf4 cooperates with Oct3/4 and Sox2 to activate the Lefty1 core promoter in embryonic stem cells. *Mol Cell Biol*, 2006. 26(20): pp. 7772–82.
50. Wei, Z. et al., Klf4 interacts directly with Oct4 and Sox2 to promote reprogramming. *Stem Cells*, 2009. 27(12): pp. 2969–78.
51. Rowland, B. D., R. Bernards, and D. S. Peeper, The KLF4 tumour suppressor is a transcriptional repressor of p53 that acts as a context-dependent oncogene. *Nat Cell Biol*, 2005. 7(11): pp. 1074–82.
52. Polesskaya, A. et al., Lin-28 binds IGF-2 mRNA and participates in skeletal myogenesis by increasing translation efficiency. *Genes Dev*, 2007. 21(9): pp. 1125–38.
53. Moss, E. G., R. C. Lee, and V. Ambros, The cold shock domain protein LIN-28 controls developmental timing in *C. elegans* and is regulated by the lin-4 RNA. *Cell*, 1997. 88(5): pp. 637–46.
54. Melton, C., R. L. Judson, and R. Blelloch, Opposing microRNA families regulate self-renewal in mouse embryonic stem cells. *Nature*, 2010. 463(7281): pp. 621–6.

55. Viswanathan, S. R., G. Q. Daley, and R. I. Gregory, Selective blockade of microRNA processing by Lin28. *Science*, 2008. 320(5872): pp. 97–100.
56. Balzer, E. and E. G. Moss, Localization of the developmental timing regulator Lin28 to mRNP complexes, P-bodies and stress granules. *RNA Biol*, 2007. 4(1): pp. 16–25.
57. Qiu, C. et al., Lin28-mediated post-transcriptional regulation of Oct4 expression in human embryonic stem cells. *Nucleic Acids Res*, 2010. 38(4): pp. 1240–8.
58. Liao, J. et al., Enhanced efficiency of generating induced pluripotent stem (iPS) cells from human somatic cells by a combination of six transcription factors. *Cell Res*, 2008. 18(5): pp. 600–3.
59. Zhang, X. et al., Esrrb activates Oct4 transcription and sustains self-renewal and pluripotency in embryonic stem cells. *J Biol Chem*, 2008. 283(51): pp. 35825–33.
60. Barroso-delJesus, A. et al., Embryonic stem cell-specific miR302-367 cluster: Human gene structure and functional characterization of its core promoter. *Mol Cell Biol*, 2008. 28(21): pp. 6609–19.
61. Yoshioka, N. et al., Efficient generation of human iPSCs by a synthetic self-replicative RNA. *Cell Stem Cell*, 2013. 13(2): pp. 246–54.
62. Maekawa, M. and S. Yamanaka, Glisl, a unique pro-reprogramming factor, may facilitate clinical applications of iPSC technology. *Cell Cycle*, 2011. 10(21): pp. 3613–4.
63. Loh, Y. H. et al., Generation of induced pluripotent stem cells from human blood. *Blood*, 2009. 113(22): pp. 5476–9.
64. Pope, J. H., M. K. Horne, and W. Scott, Transformation of foetal human leukocytes in vitro by filtrates of a human leukaemic cell line containing herpes-like virus. *Int J Cancer*, 1968. 3(6): pp. 857–66.
65. Rickinson A. M. et al., *A New Look at Tumor Immunology*. CSHL Press, New York, 1992: pp. 53–80.
66. Nilsson, K. et al., The establishment of lymphoblastoid lines from adult and fetal human lymphoid tissue and its dependence on EBV. *Int J Cancer*, 1971. 8(3): pp. 443–50.
67. Barrett, R. et al., Reliable generation of induced pluripotent stem cells from human lymphoblastoid cell lines. *Stem Cells Transl Med*, 2014. 3(12): pp. 1429–34.
68. Wang, H. et al., Improved enzymatic isolation of fibroblasts for the creation of autologous skin substitutes. *In Vitro Cell Dev Biol Anim*, 2004. 40(8–9): pp. 268–77.
69. Vangipuram, M. et al., Skin punch biopsy explant culture for derivation of primary human fibroblasts. *J Vis Exp*, 2013(77): p. e3779.
70. Schipper, B. M. et al., Regional anatomic and age effects on cell function of human adipose-derived stem cells. *Ann Plast Surg*, 2008. 60(5): pp. 538–44.
71. Zuk, P. A. et al., Multilineage cells from human adipose tissue: Implications for cell-based therapies. *Tissue Eng*, 2001. 7(2): pp. 211–28.
72. Aoki, T. et al., Generation of induced pluripotent stem cells from human adipose-derived stem cells without c-MYC. *Tissue Eng Part A*, 2010. 16(7): pp. 2197–206.
73. Lizier, N. F., I. Kerkis, and C. V. Wenceslau, *Generation of Induced Pluripotent Stem Cells from Dental Pulp Somatic Cells: Pluripotent Stem Cells*, D. Bhartiya (Ed.), InTech. DOI: 10.5772/55856. Available from http://www.intechopen.com/books/pluripotent-stem-cells/generation-of-induced-pluripotent-stem-cells-from-dental-pulp-somatic-cells. 2013.

74. Egusa, H. et al., Gingival fibroblasts as a promising source of induced pluripotent stem cells. *PLoS One*, 2010. 5(9): p. e12743.
75. Miyoshi, K. et al., Generation of human induced pluripotent stem cells from oral mucosa. *J Biosci Bioeng*, 2010. 110(3): pp. 345–50.
76. Nomura, Y. et al., Human periodontal ligament fibroblasts are the optimal cell source for induced pluripotent stem cells. *Histochem Cell Biol*, 2012. 137(6): pp. 719–32.
77. Aoi, T. et al., Generation of pluripotent stem cells from adult mouse liver and stomach cells. *Science*, 2008. 321(5889): pp. 699–702.
78. Nishi, M. et al., Induction of cells with cancer stem cell properties from nontumorigenic human mammary epithelial cells by defined reprogramming factors. *Oncogene*, 2014. 33(5): pp. 643–52.
79. Debnath, J., S. K. Muthuswamy, and J. S. Brugge, Morphogenesis and oncogenesis of MCF-10A mammary epithelial acini grown in three-dimensional basement membrane cultures. *Methods*, 2003. 30(3): pp. 256–68.
80. Song, B. et al., Generation of induced pluripotent stem cells from human kidney mesangial cells. *J Am Soc Nephrol*, 2011. 22(7): pp. 1213–20.
81. Montserrat, N. et al., Generation of induced pluripotent stem cells from human renal proximal tubular cells with only two transcription factors, OCT4 and SOX2. *J Biol Chem*, 2012. 287(29): pp. 24131–8.
82. Zhou, T. et al., Generation of human induced pluripotent stem cells from urine samples. *Nat Protoc*, 2012. 7(12): pp. 2080–9.
83. Hayashi, R. et al., Generation of corneal epithelial cells from induced pluripotent stem cells derived from human dermal fibroblast and corneal limbal epithelium. *PLoS One*, 2012. 7(9): p. e45435.
84. Sareen, D. et al., Differentiation of human limbal-derived induced pluripotent stem cells into limbal-like epithelium. *Stem Cells Transl Med*, 2014. 3(9): pp. 1002–12.
85. Yu, D. et al., Differentiation of mouse induced pluripotent stem cells into corneal epithelial-like cells. *Cell Biol Int*, 2013. 37(1): pp. 87–94.
86. Hanson, C. et al., Transplantation of human embryonic stem cells onto a partially wounded human cornea in vitro. *Acta Ophthalmol*, 2013. 91(2): pp. 127–30.
87. Tibbetts, M. D. et al., Stem cell therapy for retinal disease. *Curr Opin Ophthalmol*, 2012. 23(3): pp. 226–34.
88. Qiu, X. et al., Efficient generation of lens progenitor cells from cataract patient-specific induced pluripotent stem cells. *PLoS One*, 2012. 7(3): p. e32612.
89. Lowry, W. E. et al., Generation of human induced pluripotent stem cells from dermal fibroblasts. *Proc Natl Acad Sci USA*, 2008. 105(8): pp. 2883–8.
90. Zhao, Y. et al., Two supporting factors greatly improve the efficiency of human iPSC generation. *Cell Stem Cell*, 2008. 3(5): pp. 475–9.
91. Wong, R. C. B., E. L. Smith, and P. J. Donovan, *New techniques in the generation of pluripotent stem cells—Differentiation and pluripotent alternatives*, M. S. Kallos (Ed.), ISBN: 978-953-307-632-4, InTech. Available from http://www.intechopen.com/books/embryonic-stem-cells-differentiation-and-pluripotent-alternatives/newtechniques-in-the-generation-of-induced-pluripotent-stem-cells.
92. Brambrink, T. et al., Sequential expression of pluripotency markers during direct reprogramming of mouse somatic cells. *Cell Stem Cell*, 2008. 2(2): pp. 151–9.
93. Maherali, N. et al., A high-efficiency system for the generation and study of human induced pluripotent stem cells. *Cell Stem Cell*, 2008. 3(3): pp. 340–5.

94. Stadtfeld, M. et al., Defining molecular cornerstones during fibroblast to iPS cell reprogramming in mouse. *Cell Stem Cell*, 2008. 2(3): pp. 230–40.
95. Carey, B. W. et al., Reprogramming of murine and human somatic cells using a single polycistronic vector. *Proc Natl Acad Sci USA*, 2009. 106(1): pp. 157–62.
96. Chang, C. W. et al., Polycistronic lentiviral vector for "hit and run" reprogramming of adult skin fibroblasts to induced pluripotent stem cells. *Stem Cells*, 2009. 27(5): pp. 1042–9.
97. Sommer, C. A. et al., Induced pluripotent stem cell generation using a single lentiviral stem cell cassette. *Stem Cells*, 2009. 27(3): pp. 543–9.
98. Stadtfeld, M. et al., Induced pluripotent stem cells generated without viral integration. *Science*, 2008. 322(5903): pp. 945–9.
99. Zhou, W. and C. R. Freed, Adenoviral gene delivery can reprogram human fibroblasts to induced pluripotent stem cells. *Stem Cells*, 2009. 27(11): pp. 2667–74.
100. Ding, S. et al., Efficient transposition of the piggyBac (PB) transposon in mammalian cells and mice. *Cell*, 2005. 122(3): pp. 473–83.
101. Wang, W. et al., Chromosomal transposition of PiggyBac in mouse embryonic stem cells. *Proc Natl Acad Sci USA*, 2008. 105(27): pp. 9290–5.
102. Kaji, K. et al., Virus-free induction of pluripotency and subsequent excision of reprogramming factors. *Nature*, 2009. 458(7239): pp. 771–5.
103. Woltjen, K. et al., piggyBac transposition reprograms fibroblasts to induced pluripotent stem cells. *Nature*, 2009. 458(7239): pp. 766–70.
104. Yusa, K. et al., Generation of transgene-free induced pluripotent mouse stem cells by the piggyBac transposon. *Nat Methods*, 2009. 6(5): pp. 363–9.
105. Warren, L. et al., Highly efficient reprogramming to pluripotency and directed differentiation of human cells with synthetic modified mRNA. *Cell Stem Cell*, 2010. 7(5): pp. 618–30.
106. Subramanyam, D. et al., Multiple targets of miR-302 and miR-372 promote reprogramming of human fibroblasts to induced pluripotent stem cells. *Nat Biotechnol*, 2011. 29(5): pp. 443–8.
107. Onder, T. T. and G. Q. Daley, microRNAs become macro players in somatic cell reprogramming. *Genome Med*, 2011. 3(6): p. 40.
108. Anokye-Danso, F. et al., Highly efficient miRNA-mediated reprogramming of mouse and human somatic cells to pluripotency. *Cell Stem Cell*, 2011. 8(4): pp. 376–88.
109. Yakubov, E. et al., Reprogramming of human fibroblasts to pluripotent stem cells using mRNA of four transcription factors. *Biochem Biophys Res Commun*, 2010. 394(1): pp. 189–93.
110. Matsushita, M. et al., A high-efficiency protein transduction system demonstrating the role of PKA in long-lasting long-term potentiation. *J Neurosci*, 2001. 21(16): pp. 6000–7.
111. Zhou, H. et al., Generation of induced pluripotent stem cells using recombinant proteins. *Cell Stem Cell*, 2009. 4(5): pp. 381–4.
112. Kim, D. et al., Generation of human induced pluripotent stem cells by direct delivery of reprogramming proteins. *Cell Stem Cell*, 2009. 4(6): pp. 472–6.
113. Fusaki, N. et al., Efficient induction of transgene-free human pluripotent stem cells using a vector based on Sendai virus, an RNA virus that does not integrate into the host genome. *Proc Jpn Acad Ser B Phys Biol Sci*, 2009. 85(8): pp. 348–62.
114. Seki, T. et al., Generation of induced pluripotent stem cells from human terminally differentiated circulating T cells. *Cell Stem Cell*, 2010. 7(1): pp. 11–4.

115. Ban, H. et al., Efficient generation of transgene-free human induced pluripotent stem cells (iPSCs) by temperature-sensitive Sendai virus vectors. *Proc Natl Acad Sci USA*, 2011. 108(34): pp. 14234–9.
116. Gonzalez, F. et al., Generation of mouse-induced pluripotent stem cells by transient expression of a single nonviral polycistronic vector. *Proc Natl Acad Sci USA*, 2009. 106(22): pp. 8918–22.
117. Okita, K. et al., Generation of mouse induced pluripotent stem cells without viral vectors. *Science*, 2008. 322(5903): pp. 949–53.
118. Jia, F. et al., A nonviral minicircle vector for deriving human iPS cells. *Nat Methods*, 2010. 7(3): pp. 197–9.
119. Yates, J. et al., A cis-acting element from the Epstein–Barr viral genome that permits stable replication of recombinant plasmids in latently infected cells. *Proc Natl Acad Sci USA*, 1984. 81(12): pp. 3806–10.
120. Yates, J. L., N. Warren, and B. Sugden, Stable replication of plasmids derived from Epstein–Barr virus in various mammalian cells. *Nature*, 1985. 313(6005): pp. 812–5.
121. Leight, E. R. and B. Sugden, Establishment of an oriP replicon is dependent upon an infrequent, epigenetic event. *Mol Cell Biol*, 2001. 21(13): pp. 4149–61.
122. Yates, J. L. and N. Guan, Epstein-Barr virus-derived plasmids replicate only once per cell cycle and are not amplified after entry into cells. *J Virol*, 1991. 65(1): pp. 483–8.
123. Nanbo, A., A. Sugden, and B. Sugden, The coupling of synthesis and partitioning of EBV's plasmid replicon is revealed in live cells. *EMBO J*, 2007. 26(19): pp. 4252–62.
124. Shi, Y. et al., A combined chemical and genetic approach for the generation of induced pluripotent stem cells. *Cell Stem Cell*, 2008. 2(6): pp. 525–8.
125. Schlaeger, T. M. et al., A comparison of non-integrating reprogramming methods. *Nat Biotechnol*, 2015. 33(1): pp. 58–63.
126. Thomson, J. A. et al., Embryonic stem cell lines derived from human blastocysts. *Science*, 1998. 282(5391): pp. 1145–7.
127. Chan, E. M. et al., Live cell imaging distinguishes bona fide human iPS cells from partially reprogrammed cells. *Nat Biotechnol*, 2009. 27(11): pp. 1033–7.
128. Pfannkuche, K. et al., Initial colony morphology-based selection for iPS cells derived from adult fibroblasts is substantially improved by temporary UTF1-based selection. *PLoS One*, 2010. 5(3): p. e9580.
129. Tan, S. M. et al., A UTF1-based selection system for stable homogeneously pluripotent human embryonic stem cell cultures. *Nucleic Acids Res*, 2007. 35(18): p. e118.
130. Ohnuki, M., K. Takahashi, and S. Yamanaka, Generation and characterization of human induced pluripotent stem cells. *Curr Protoc Stem Cell Biol*, 2009. Chapter 4: pp. Unit 4A 2.

2

Human-Induced Pluripotent Stem Cells: Banking and Characterization

Uthra Rajamani, Lindsay Lenaeus, Loren Ornelas, and Dhruv Sareen

CONTENTS

2.1 iPSC Culture

Meticulous, thorough, and sterile stem cell culture techniques are necessary to sustain the quality and utility of iPSCs from the point of clone generation to expansion and iPSC banking. Culture conditions such as environment, cleaning, passaging methods, and use of suitable substrates and feeders are important in maintaining pluripotency and minimizing spontaneous differentiation. This section delineates the salient aspects of iPSC culturing.

2.1.1 Environment

When culturing iPSCs, it is vital that the conditions at which they are kept remain stable. The environment for these cells is maintained by using a temperature- and CO_2-regulated incubator. The cells should be cultured at 37°C with 95% relative humidity and an O_2 level of 20–21%, which represents atmospheric, albeit nonphysiological, concentrations. It is also important to maintain the CO_2 levels at 5% in order to maintain physiological neutral pH at around 7.4. However, some studies suggest that culturing iPSCs under normoxic or hypoxic conditions (≤5% O_2) showed improved efficiency of iPSC generation (1). In comparison to normoxic conditions, hypoxia showed better pluripotency markers and improved efficiency of iPSC generation with nonviral vectors such as plasmid vectors and PiggyBac transposons (1). However, hypoxia is also said to induce cytotoxicity, and that is determined by the degree of hypoxia. Maintaining a stable culture environment is crucial in determining the health of iPSCs.

2.1.2 Cell Morphology

All pluripotent stem cells (PSCs) (hESCs and human-induced pluripotent stem cells [hiPSCs]) often prefer to grow in colonies of approximately >1000 cells. Undifferentiated iPSC colonies have clear borders distinct from differentiated cells. The individual cells are round and small with large nuclei and prominent nucleoli. They also typically have a high nuclear-to-cytoplasm ratio. When differentiated, cells morph into a cobblestone fibroblast-type appearance that allows for easier distinction between undifferentiated pluripotent cells and differentiated cells under a bright-field microscope. This typically occurs at the edges of the colony and, although infrequently, can also occur in the center.

2.1.3 Cleaning (Grooming)

A defining characteristic of iPSCs is their ability to be directed to almost any cell type in the human body; however, in order to keep an unlimited supply of pluripotent cells that can be expanded into billions of cells, their quality has to be controlled by maintaining them in an undifferentiated state. A small percentage of cells in iPSC culture spontaneously differentiate into random cell types that exert undesirable effects on the surrounding pluripotent cells in a culture dish. As a result, maintaining these cells in an undifferentiated pluripotent state requires daily media changes and their upkeep. This is often referred to as *cleaning* unwanted cells. Differentiated cells are cleaned out or groomed using manual removal processes in order to retain a pure population of iPSCs (Figure 2.1). Currently, there is a variety of manual tools used to fulfill these cleaning requirements. Reliable automation of these processes is still under development. In most cases, the spontaneously differentiated cells are removed by physically scraping the large/flat cells off of the plate under a phase/bright-field microscope (Figure 2.1), without removing the iPSCs that should retain a high nuclear–cytoplasmic ratio. One way to signify where these differentiated cells are on the plate is to use a self-inking object marker attached to a microscope objective turret wheel. A circle of ink is placed around colonies that are differentiated, and the designated colonies can be either manually scratched or vacuum-aspirated from within each ink circle. This method is most useful when maintaining a high volume of iPSCs that do not require vigorous cleaning. Another method is live grooming and following similar methods while viewing under a microscope. Two different style tools can be utilized: (a) P1000 filtered pipette tip inserted into a P200 nonfiltered pipette tip or (b) pulled cotton plugged Pasteur pipette made into a curved, pointed tip. For general maintenance and removal of large areas of differentiation, the P1000 tip inserted into a P200 tip is more effective. It removes large areas of differentiation more efficiently than the latter tool

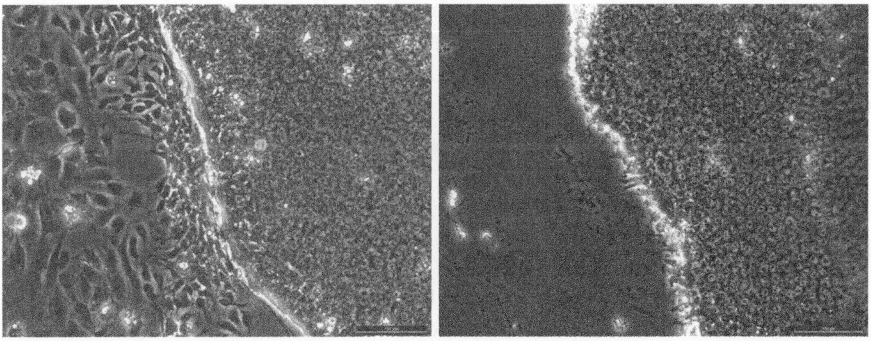

(a) (b)

FIGURE 2.1
Comparison of iPSC colonies (a) before and (b) after cleaning. Note the distinctly visible differentiated cells before cleaning which have been removed and cleaned out.

style. The pulled Pasteur pipette is more effective for fine precision cleaning in cultures where there are not many iPSC colonies growing. This technique is most efficient when the volume of iPSC lines being maintained are not high because it can take a substantial amount of time to get through a plate. Regardless of which cleaning method is used, postcleaning fresh media needs to be supplied to cells and typical culturing methods require daily replenishment of fresh pluripotent maintenance media. Everyday media change is necessary since the activity of the essential pluripotency growth factor basic fibroblast growth factor (bFGF) rapidly depletes with an *in vitro* half-life of only 12 hours (2). Also, the daily media change removes colonies that have been lifted from the culture.

2.1.4 Passaging

Passaging of iPSCs refers to moving them from their current substrate and dish in a new dish coated with fresh substrate (typically, a mix of an extracellular matrix [ECM]) to allow for continual expansion, growth, and banking. This is performed every four to seven days by using a few well-established methods. There is a variety of commercial techniques available, such as using a StemPro® EZ Passage Tool. This tool allows for the cutting of the larger colonies in a well of 80–90% confluence into smaller uniform-sized colonies by simply mowing through the well in a uniform motion. This form of passaging is nonenzymatic and not chemical based. Therefore, it is considered a gentle technique for newly established lines because it maintains a more stable low-stress environment by keeping the cells in contact with their neighboring cells in clumps (~100–1000 cells). Another technique commonly utilized is a chemical-based ethylenediaminetetraacetic acid (EDTA) solution, which chemically dissociates the large colonies of iPSCs into much smaller clump sizes (~10–25 cells). This method is better to use on well-established lines that are considered low-maintenance and do not require much manual cleaning and removal of differentiated cells. Both techniques allow for easy passaging in feeder-free conditions; however, passaging cells that are cultured on a feeder layer, such as inactivated mouse embryonic fibroblasts (MEFs), requires lifting the colonies using an enzymatic approach. The most common technique is to use collagenase type IV to lift the iPSC colony edges away from the feeder layer. Once the cells have been lifted from the plate, they are moved to a new MEF feeder plate.

2.1.5 Growth Substrates

HiPSCs can be cultured on a variety of feeder-based or feeder-free substrates based on the researcher's needs. These substrates assist in efficient attachment of iPSCs as well as in propagation and maintenance of pluripotency in comparison to plastic tissue culture dishes. It has also been shown that the iPSCs have mechanical memory of their substrates, which influences their growth and cellular fate (3).

2.1.5.1 Feeder-Based

The first substrate that was used in PSC culture was MEFs as feeders. The maintenance of PSCs on MEFs kept the cells in an undifferentiated state for prolonged culturing (4) by making PSC cultures more stable by releasing not well-characterized nutrients and cytokines into the media. While more defined and convenient ECM substrates have been developed for PSC culturing, MEFs, however, remain a gold standard as a reliable substrate for basic PSC research. MEFs are also commonly used as a support to aid the recovery and the survival of iPSCs when genetic manipulations are being performed due to the stress imposed during such processes. As the utility of iPSCs transitions from fundamental basic science research to greater clinical and translational uses, the use of mouse feeders to grow human PSCs is undesirable, especially as long-term culturing of PSCs on MEFs has been shown to express immunogenic nonhuman proteins, such as sialic acid (5). Therefore, novel and efficient methods for xeno-free culturing of PSCs have to be developed, and there has been some recent progress in this regard.

2.1.5.2 Undefined ECMs

PSCs require some form of an ECM substrate to remain in an undifferentiated state. A commonly used version is a mix of ECM proteins that are removed from Engelbreth–Holm–Swarm tumors in mice (6). The common variety of this form of ECM in most laboratories is the growth factor-reduced version of Matrigel™ (BD), Geltrex® (Life Technologies), and Millipore ECMs. These predominantly comprise of basement proteins including, laminin, collagen IV, and enactin, which are all components of the basement membrane (6). These substrates have all been shown to successfully support the self-renewal and the pluripotency of PSCs *in vitro,* but they suffer from not being completely defined and amenable for translational uses as they are derived from mouse tumors (7). Standard Matrigel also consists of various growth factors that can alter results because of the cellular interactions, which can be avoided by using growth factor-reduced versions. However, there can be significant batch variations and lots need to be routinely tested as traces of unknown cytokines may be present (6). The use of such complex and undefined matrices is still widely employed for convenience in most research laboratories; however, studies that need clinical-grade PSCs need to move to xenogenic-free substrates (7). Further, mechanisms by which PSCs proliferate and maintain pluripotency on these complex substrates are poorly understood.

2.1.5.3 Defined ECMs and Recombinant Laminin

During early embryonic development, the first major proteins to arise for the ECM are laminins (8). Laminins are heterotrimeric glycoproteins that are made up of α, β, and γ chains (8). The alpha chain has five different isoforms, while the beta and gamma chains each exist in three genetically variant forms (8). In humans, 15 different isoforms of laminins are present, and these various combinations support various cellular functions (9). Of the multiple combinations of laminin α, β, and γ chains, two, LN-511 and LN-521, have been proven to support the growth of human PSCs (10). The successful growth and maintenance of PSCs on such recombinant human laminins is a great advance for moving PSCs into clinical-grade applications that are more amenable for regenerative therapies in humans.

2.1.5.4 Vitronectin

Vitronectin is a serum glycoprotein that is a component of ECM basement membranes and promotes cell adhesion and spreading (11). It has been used widely as an ECM for hESC and iPSC culture. Specifically, two variant forms of vitronectin have been identified to support stem cell attachment and survival when grown in defined E8 media, namely, VTN-N and VTN-NC (12). It has been shown that vitronectin better supports initial attachment and survival of cells when passaged in small clumps using EDTA as a passaging method (12).

2.1.6 Media

2.1.6.1 Serum-Based or Serum Replacement Media

The use of serum as a supplement in iPSC culturing media is not routinely recommended as it contains undefined and inconsistent amounts of growth factors, hormones, and other cytokines and xenobiotics, which inhibit pluripotency and promote stem cell differentiation. As a result, standardizing culture conditions and having a clear idea of the role of various factors in iPSC culture become a challenge. Although fetal bovine serum is a universal supplement utilized for cell culture (in primary and cancer cell lines), its use hinders researchers' ability to relate PSC responses to exogenous stimuli (13). Additionally, the serum also has the risk of containing hemoglobin and endotoxins. Knockout serum-replacement (KOSR) media have been employed to control the disadvantages posed by serum-based media. KOSR is a defined and serum-free supplement, which has been successfully employed to replace serum in the culture of undifferentiated iPSCs. KOSR provides consistent growth conditions and results in less differentiation of iPSCs compared to serum-based media.

2.1.6.2 Serum-Free Media: mTeSR1

After the initial protocols describing PSC culturing on mouse feeder cells, researchers have been concerned about whether the animal feeders and the xenogenic media components had nonphysiological effects on signaling pathways in the human system in addition to their immunogenic nature for preparing cellular therapeutics. This inquiry created efforts to discovering xeno- and feeder-free culture medium leading to the discovery of a now widely used media in stem cell laboratories, TeSR1 (14). This is the first xeno- and feeder-free maintenance medium that was widely accepted and eventually made commercially available (14). TeSR1 media is composed of 18 additional components that are not included in Dulbecco's modified Eagle medium (DMEM)/F12, of which five are indispensable: bFGF, lithium chloride, γ-aminobutyric acid, pipecolic acid, and transforming growth factor-beta (TGFβ). These components were all tested, and it was discovered that the removal of any of these factors would diminish either cell growth or expression of PSC markers; however, bFGF was the only element that affected both parameters (14). A complete list of all components of TeSR1 is available in the original manuscript supplementary information of Ludwig et al. (14). Even though this media is beneficial to translational studies for *in vivo* work, it is not very functional for everyday maintenance and research because of the cost of human protein products (15). This led to the development of mTeSR1, which is one of the most widely used feeder-free PSC culture media currently used (15). mTeSR1, which stands for "modified TeSR1," still consists of the same components as TeSR1, except for the use of human holo-transferrin, human serum albumin, and human insulin. These were replaced by their bovine counterparts (15). Also, to bring down media costs, human bFGF was replaced with purified zebra fish bFGF (15). mTeSR1 is suitable for feeder-free conditions where the use of animal products is not strictly prohibited. However, it needs to be noted that one of the main components of both TeSR1 and mTeSR1 is bovine-derived albumin.

2.1.6.3 Essential Defined Media

The use of albumin creates variability in the production of media because, similar to serum, albumin is not clearly defined, and it is not consistent from lot to lot. Recently, more defined and minimal-component media have been developed such as E8 (12). This medium is composed of eight essential components, including the basal media, and is chemically defined. The final, optimized version of E8 contains DMEM/F12, insulin, selenium, transferrin, L-ascorbic acid, bFGF, TGFβ (or nodal), and sodium bicarbonate ($NaHCO_3$) (12). DMEM/F12 was chosen as the basal medium because it performed better over 11 other types of basal media compared in the study. bFGF and insulin increase cell survival and cell proliferation; L-ascorbic acid is vital to cell cloning, and selenium, for sustainability of long-term cultures. These five components, with the addition of $NaHCO_3$ for pH adjustment, could be used to grow PSCs for short term, but it was found that after long-term culturing, they would be more prone to sporadic differentiation. This was mitigated through the addition of TGFβ that elevated the expression of the pluripotency maintaining gene, Nanog. Finally, transferrin was included for enhanced survival and PSC propagation (12). Many versions of E8 media are now marketed by commercial vendors; however, this media is relatively easy to formulate in any individual stem cell laboratory. The simplicity of manufacturing chemically defined media such as E8 allows for more reproducible results across laboratories.

However, it should be pointed out that the different passaging methods and feeder environments yield variable culturing efficiencies, cell morphologies, and phenotypes in iPSCs. After a recent study that compared the effects of enzymatic and mechanical passaging and MEFs and feeder-free substrates on the genetic and the epigenetic stability of PSCs (16), enzymatic passaging on a feeder-free substrate were associated with higher buildup of genetic aberrations when compared to mechanical passaging on feeders (16). Enzymatic passaging was thought to be the biggest contributor of genetic aberrations. Notably, the duration of culture plays a large role in the genetic stability of PSCs. Although enzymatic passaging on feeder-free substrates has cytogenetic aberrations at lower passages, mechanical passaging on feeder layers also results in genetic aberrations upon long-term culturing. This emphasizes the importance of developing excellent practices and routine quality control (QC) procedures for iPSC culturing in stem cell laboratories. The best suitable passage method and substrate should be chosen where the laboratory is confident in the quality of the iPSC lines—primarily, normal cytogenetics and maintenance of pluripotency—while carefully banking multiple cryovials at lower passages.

2.1.7 iPSC Characterization

Performing genetic manipulation of somatic cells and reprogramming to pluripotency requires QC measures at multiple stages of the iPSC generation and expansion processes. Evidence that the isolated clone is a bona fide iPSC line is required. Various assays are used to say, unequivocally, that the cells generated are indeed legitimate iPSCs; these assays test gene expression profiles, cell biomarkers, and differentiation potential.

2.1.8 Genetic Assays

2.1.8.1 G-Band Karyotype and Comparative Genomic Hybridization

The process of reprogramming and long-term culture of iPSCs raises questions about culture-adapted chromosomal abnormalities. ESCs, derived from the developing blastocyst, have diploid karyotypes that can remain stable after many passages *in vitro* (17,18). However, there have been many reports that these cells may acquire specific recurrent chromosomal abnormalities after prolonged culture. These include aneuploidy with gain of chromosomes (trisomy) 12, 17, and X in human ESCs (18–21) and trisomy 8 and

11 in mouse ESCs (22,23). These abnormalities lead to differential growth rates (18–20), thus lessening the reproducibility and the reliability. The conventional technique for PSC culture QC and identifying these chromosomal alterations has been G-band karyotyping (24). Entire chromosomes are visualized with this technique to discover any acquired abnormalities over time such as aneuploidy, deletions, insertions, and translocations (Figure 2.2) (25). Karyotypes can identify aberrations at the resolution of a few megabases, whereas newer techniques are able to recognize deviations at much higher resolution level (26), such as comparative genomic hybridization (CGH). CGH is used to examine the entire genome of a sample to detect any imbalances in a genomic complement (27). CGH used as an array (aCGH) can be used to recognize genome-wide variations in chromosomal structure and number (28). Both G-band karyotypes and aCGH can be used to detect aneuploidies, but because aCGH has a higher resolution of detection, it can detect subchromosomal aberrations, making it more precise for mapping the entire chromosome (28). On the other hand, aCGH is more expensive and, unlike G-band karyotype, is unable to detect any balanced chromosomal variations such as inversions and translocations. Because of this, G-band karyotypes are still considered the standard, but advances in technology are paving the way for aCGH to be more cost-effective for routine QC of iPSC cultures, especially for large public iPSC initiatives and repositories (28).

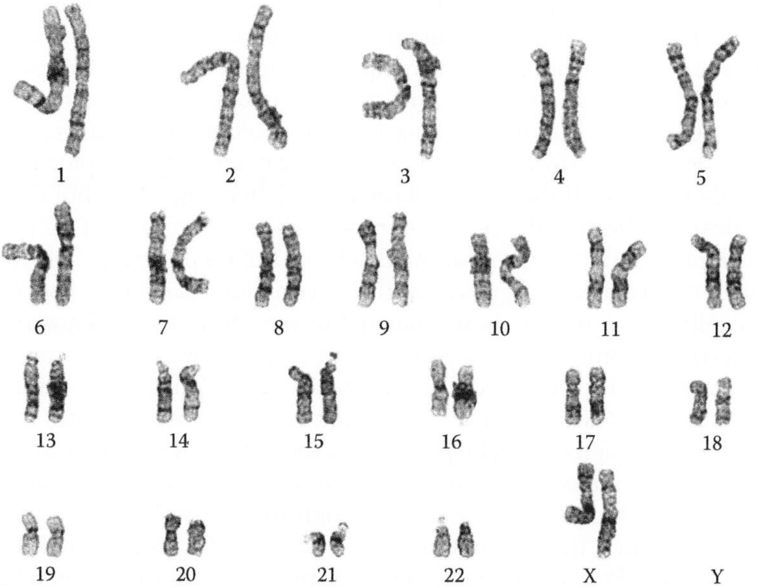

FIGURE 2.2
G-band karyotyping of an iPSC line.

2.1.8.2 Identity Testing by Short Tandem Repeat

The main prerequisite for newly generated iPSC lines is whether the cell line is authentic and matches the identity of the original donor. Microsatellites in the human genome harboring short tandem repeat (STR) DNA markers allow the identification of individual cell lines at the DNA level. Polymerase chain reaction (PCR) amplification of multiple highly polymorphic microsatellite STR loci and gender determination have been proven to be the best tools for screening the uniqueness of DNA profiles in publicly available STR database. Matching profiles between the source tissue and the reprogrammed iPSC line should be conducted routinely, especially before distributing the cell lines and publishing in a manuscript.

2.1.8.3 PluriTest

A stable genetic profile is vital for the success of iPSCs; therefore, along with detecting chromosomal alterations, the expression of genetic markers for pluripotency should be investigated after iPSCs have been generated. PluriTest (http://pluritest.org) (29) is an open-source platform that is freely accessible and provides assessment of pluripotency of human PSCs (29). PluriTest combines bioinformatics algorithms in an easy to use web interface with microarray-based gene expression assays that determine how well a sample measures up to a set of known standards of PSCs (hESCs and hiPSCs). Pluripotency is computed based on two scores: (a) pluripotency score and (b) novelty score. The pluripotency score utilizes a database, referred to as the *Stem Cell Matrix* (SCM), to compare genes expressed in the samples with genes from both pluripotent and nonpluripotent cells (29). A sample with a high pluripotent score (>20) is interpreted as being pluripotent (hESCs, iPSCs, or germ cell tumors), while a low pluripotent score (<20) signifies a somatic, primary or differentiated cell type (29). Notably, the pluripotency score only indicates if the cells contain a pluripotent signature but does not categorically indicate if the cell preparation is a normal diploid PSC culture. For example, partially reprogrammed cells or hESCs and germ cell tumors that have abnormal karyotypes may also have a high pluripotency score (30). The novelty score reveals how related the sample is to the known well-characterized profiles of PSCs in the SCM (29). A low novelty score (<1.6) demonstrates that the queried sample's gene expression is similar to that of the PSCs in the SCM; a sample with a high novelty score indicates that the pattern of expression cannot be justified by the PSC patterns in the SCM (29). Partially differentiated iPSCs or cells with abnormal karyotypes are most likely to have a high novelty score, although they may show high pluripotency score. Hence, the pluripotency score combined with the novelty score accounts for the PluriTest's sensitivity and specificity, thus necessitating this as an assay to determine pluripotency.

2.1.8.4 TaqMan® hPSC™ Scorecard for Pluripotency

The TaqMan hPSC Scorecard Panel provided by Life Technologies assesses pluripotency and trilineage differentiation potential using real-time TaqMan quantitative PCR arrays on a focused set of genes accompanied by analysis software. This assay can be useful in determining pluripotency, as a focused set of well-validated pluripotency genes are utilized for generation of quantitative transcriptome profile and a pluripotency score. Comparing expression patterns with these profiles allows for the inclusion of fully reprogrammed iPSCs. Researchers can choose between five to six compatible real-time PCR systems that may be available in a genomics core laboratory. Results are analyzed with relative ease using accompanying software provided by the vendor. On a price point/sample, the TaqMan hPSC Scorecard and the PluriTest are comparable; however, large-scale microarrays on which PluriTest is based also provide whole-transcriptome data that could be utilized for studying other relevant iPSC biology.

2.1.9 Pluripotency Antigens and Immunophenotyping

A rapid test for pluripotency is an alkaline phosphatase (AP) stain. AP is an enzyme that will hydrolyze various molecules containing phosphates when conditions become alkaline (31). AP activity has been shown to be elevated in undifferentiated PSCs such as ESCs, embryonic germ cells, as well as iPSCs. This stain utilizes AP by marking the undifferentiated pluripotent cells using a chromogen (typically, red or purple reaction product) and leaves differentiated cells and feeder cells unstained (Figure 2.3). Another commonly used method to assess the status of pluripotency antigens in PSCs is immunocytochemistry (ICC). ICC utilizes a primary antibody that reacts with a known intracellular or surface antigen. This antibody is either directly conjugated to a fluorophore or binds a secondary antibody conjugated to a fluorophore. Stage-specific embryonic antigens 3 and 4 (SSEA3 and SSEA4), high-molecular weight glycoproteins, and tumor rejection antigens (TRA 1-60 and TRA 1-81) are cell surface antigens that are employed for the detection of PSCs but get downregulated as pluripotent cells begin to terminally differentiate (32). Originally, these antigens were recognized as pluripotency markers in embryonal carcinoma and germ cell tumors (33–36). Importantly, a number of nuclear transcription factor markers are key regulators of pluripotency. Oct4 and Nanog are critical for the self-renewal ability of PSCs (37,38), while Sox2 also plays a vital role in PSCs (39), neuroectoderm, and endodermal stem cells (Figure 2.3). Oct4, Sox2, and Nanog form a feedback self-regulatory circuit to maintain pluripotency in both mouse and human ESCs. All three transcription factors regulate themselves as well as one another by binding to their own promoters as well as mutually co-occupying the same gene region. Oct4 binds to the Nanog promoter and regulates its expression in a negative feedback

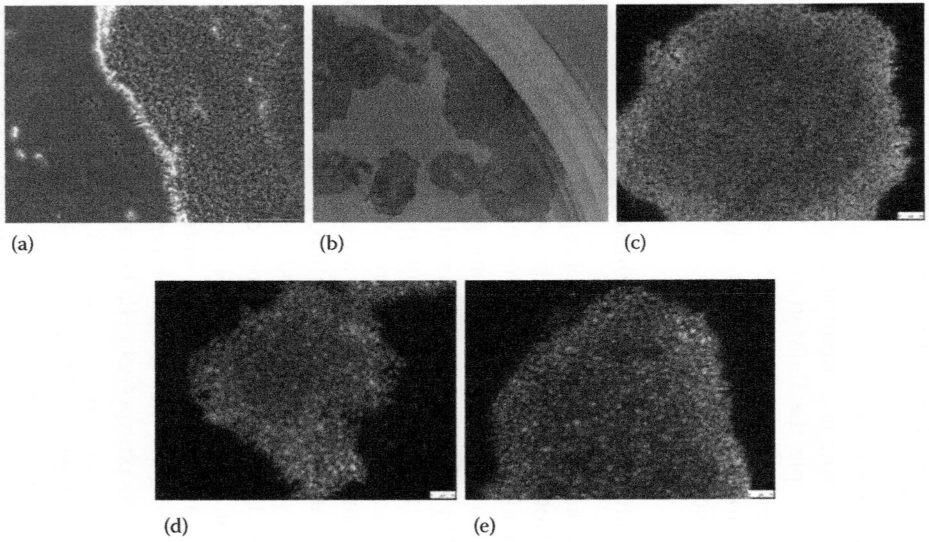

FIGURE 2.3
(See color insert.) Characterization of iPSCs by ICC: (a) bright-field image of a healthy iPSC colony and (b) alkaline phosphatase-stained iPSC colony showing positive AP staining of undifferentiated pluripotent cells. Immunofluorescence staining images showing pluripotency markers such as (c) SSEA4 (green) and OCT3/4 (red), (d) TRA 1-81 (green) and SOX2 (red), and (e) TRA 1-60 (green) and NANOG (red).

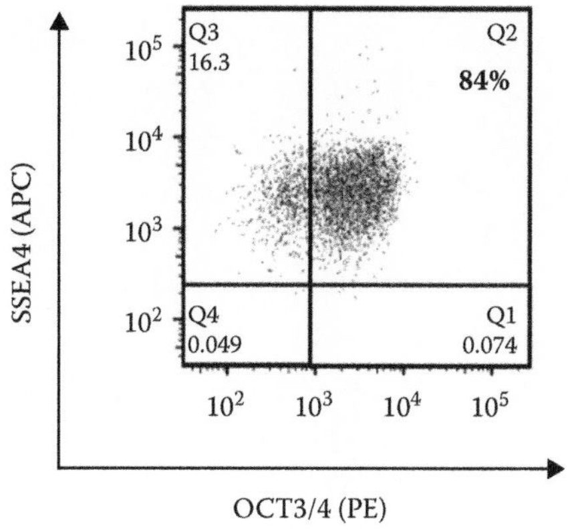

FIGURE 2.4
(See color insert.) Flow cytometric analysis of iPSCs for the presence of pluripotency markers SSEA4 and OCT4.

manner whereby Nanog is repressed when Oct4 levels are above normal. This regulatory loop helps maintain ESC properties and identity (40,41).

Lastly, flow cytometry is a quantitative technique used to verify the levels of pluripotency antigens of an iPSC culture by immunostaining and processing with similar combinations of markers. A combination immunostain for SSEA4 and Oct4 is frequently utilized to quantify the undifferentiated status and cell health of an iPSC culture, typically considered to be greater than 80% double positive for two antigens (cell surface and nuclear) in the cell population (Figure 2.4).

2.1.10 Differentiation Potential

PSCs possess the ability to give rise to cells derived from all three germ layers—endoderm, mesoderm, and ectoderm. These three germ layers have the capacity to develop into all other cell types of the body (42). Several means to determine the differentiation potential of iPSCs have been established, of which embryoid body (EB) formation is a commonly employed method. EB formation is a facile method for characterizing hiPSCs because it has been a well-established technique for measuring pluripotency and trilineage differentiation potential in hESCs. Spontaneous formation of EBs involves mechanical or enzymatic harvesting of adherent iPSC colonies as clumps into three-dimensional aggregates in suspension cultures (Figure 2.5). Notably, to allow for the differentiation program to begin, the medium is changed from pluripotency maintenance (mTeSR1/E8) to a differentiation medium by primarily removing the exogenous pluripotency maintenance factor bFGF. This allows for the cells to aggregate and spontaneously differentiate within the aggregates while beginning the trilineage differentiation

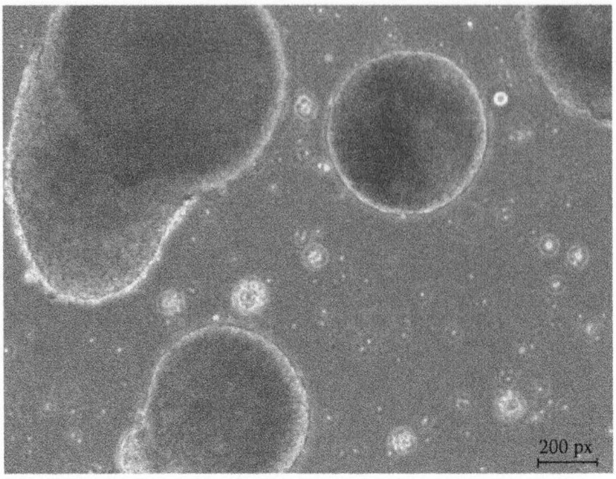

FIGURE 2.5
(See color insert.) Embryoid bodies are clumps of spherical aggregates that contain differentiated derivatives representing all three germ layers (ectoderm, endoderm, mesoderm).

program, without being directed toward a specific cell fate by addition of exogenous morphogens.

2.1.10.1 Reverse Transcription PCR

Usually between 10–30 days of differentiation, the EBs are harvested and mRNA is extracted, reverse transcribed into complementary DNA, and subjected to PCR for germ layer-specific genes (43–45). The results reveal the ability of the EBs to spontaneously differentiate into ectoderm, mesoderm, and endoderm. At the Cedars-Sinai iPSC Core, the three germ layers are identified through the expression of various known markers for endoderm (alpha-fetoprotein, SOX17, hepatocyte nuclear factor 4 alpha [HNF4A], forkhead box protein A1, or GATA1), mesoderm (heart and neural crest derivatives expressed 1 and Msh homeobox 1), and ectoderm (neural cell adhesion molecule, orthodenticle homeobox 2, and paired box protein 6) (44). If all three markers are detected at some point during the EB differentiation, then the cells are qualified as retaining a trilineage germ layer differentiation potential.

2.1.10.2 TaqMan hPSC Scorecard for Differentiation Potential

In addition to assessing pluripotency, the TaqMan hPSC Scorecard assay also allows for quick and comprehensive characterization of trilineage potential of hiPSCs using highly quantitative gene expression profiling (46). Gene enrichment analysis for lineage marker genes is employed and combined in one bioinformatics scorecard with a software analysis suite available on the Life Technologies website. This well-validated assay panel provides an ideal utility for assessing differentiation potential of any newly generated iPSC

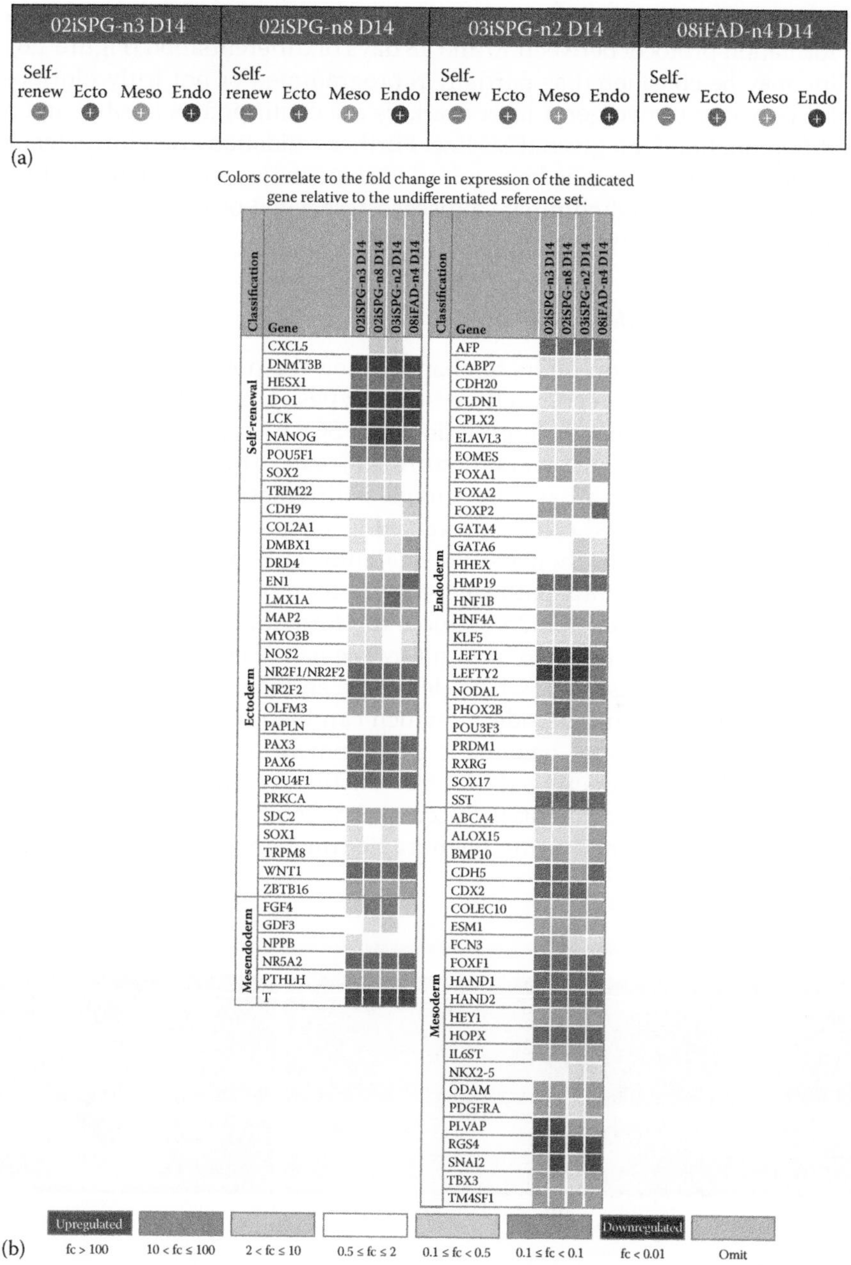

FIGURE 2.6
(See color insert.) TaqMan hPSC Scorecard Assay. Panel (a) shows in a nutshell the pluripotency of EB indicating their ability to form the various germ layers. Panel (b) shows the expression levels of the genes that are involved in the formation of the three germ layers. Upregulated genes are shown in red and the downregulated genes in blue.

line. This assay is typically performed on the described simple spontaneous EB formation protocol between 10 and 18 days of differentiation (Figure 2.6). A line may be considered as partially reprogrammed or not truly pluripotent if only one to two germ layer markers are confirmed as positive upon EB differentiation for a given iPSC line. Further validation and pluripotency characterization assays may be required to characterize a questionable iPSC line as pluripotent and eliminate a disease-associated or lack of germ layer differentiation phenotype.

2.1.10.3 Directed Differentiation

Directed differentiation refers to channelizing the potential of stem cells to differentiate into specific cell or tissue types. This utilizes the use of morphogens that direct differentiation of iPSCs to a germ layer of interest (from the three germ layers) (Figure 2.7). Cues from early embryonic development serve as a basis for directed differentiation. Subsequently, the germ layer-specified cells are directed to specific cell types by the use of a combination of morphogens and small molecules known to activate or inhibit key developmental pathways specific for a cell type (47). For example, activin A alongside Wnt3A induces the formation of definitive endoderm. This can be directed to hindgut by the use of FGF/Wnt (48–50). On the other hand, the use of bFGF on definitive endoderm directs them into anterior ventral foregut endoderm, which can then be directed to airway epithelia by suppressing pancreatic development using sonic hedgehog (51). The ability of directing iPSCs into all three germ layers also certifies for their trilineage potential. Successful differentiation of the cell type of interest is determined using phenotypic functional analyses, qRT-PCR and IF/ICC.

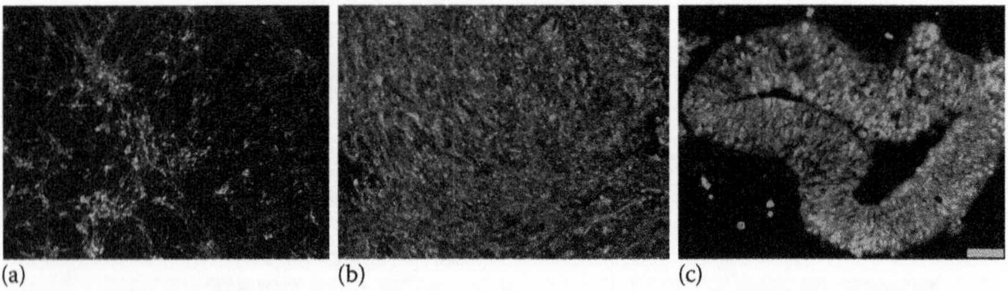

(a) (b) (c)

FIGURE 2.7
(See color insert.) iPSCs differentiated into specific cell types: (a) differentiation of iPSCs to motor neurons (neurectoderm) expressing TuJ1 (red) and Nkx6.1 (green); (b) iPSCs differentiated into chondrocytes showing collagen type I (green) through mesoderm specification; (c) generation of foregut organoids from iPSCs through endoderm differentiation. Organoid expresses Sox2 (green), a posterior foregut marker on day 20.

2.1.11 Teratoma Assay

The teratoma assay is an *in vivo* method for demonstrating pluripotency and differentiation capabilities of iPSCs. Teratomas are rapidly growing benign tumors and often comprised of all three germ layers. Teratoma assay involves the injection of iPSCs into immunodeficient mice (52), and these eventually develop into tumors that can be excised and pathologically examined (52). The ability of iPSCs to form tumors and tissues consisting of all three germ layers *in vivo* serves as an indicator of pluripotency (Figure 2.8). This assay involves the sacrifice of mice and can be time consuming. Approximately 1–5 × 10^6 iPSCs are injected into an immunodeficient mouse, and the growth of tumor is monitored. The injection site of iPSCs varies from one research group to another. Some common sites include subcutaneous (53), intra-myocardial, and skeletal muscles (54); kidney capsules (45,55); and testes (56). The injected animals have to be monitored for about two to three months as teratomas develop, and this process is laborious. Immunohistochemical

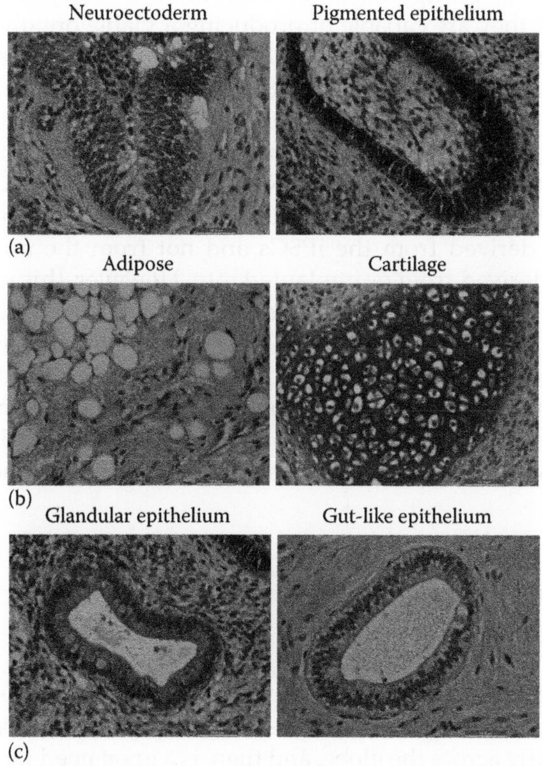

FIGURE 2.8
(See color insert.) Panels showing teratoma staining exhibiting propensity to form all three germ layers, namely, (a) the ectoderm, (b) the mesoderm, and (c) the endoderm.

and pathological analyses of the teratoma are performed and often provide only a qualitative measure of a few cell types representative from each of the three germ layers. Teratoma formation assays are time consuming and expensive, and with recent technological advances in iPSC characterization, they have become redundant. Given these limitations and the push to reduce the burden of unnecessary animal experimentation, teratoma formation assay is now less frequently used for assessing iPSC pluripotency.

2.1.12 Tetraploid Complementation Assay

The tetraploid complementation assay is a technique employed to test for pluripotency alongside applications in making genetically modified organisms and to study the effects of mutations in embryonic development. However, this measure of pluripotency is only technically feasible and ethical in non-human-derived iPSCs, primarily mice. This technique involves combining two mammalian embryos to form a tetraploid embryo (57). The mouse iPSCs are considered to be pluripotent and similar to their embryonic counterparts (mouse ESCs) if they are capable of producing a viable organism. In 2009, it was shown for the first time that iPSCs can support full-term development of embryos from tetraploid blastocysts (58) when tetraploid embryos were first generated and iPSCs were microinjected into the cavity of the blastocysts. These complemented blastocysts, after being transferred into pseudopregnant Institute of Cancer Research (ICR) mice, were allowed to develop full term resulting in the successful formation of viable progeny, indicating that the pups were derived from the iPSCs and not from the mice, which the embryos were derived from or implanted into. However, this process is very inefficient with only ~25% chance of success in developing viable progeny.

Every newly generated human iPSC line should be tested using a combination of most of these assays and accompanied by a Certificate of Analysis document listing the characterization of its pluripotent state. These measures establish standardization and rigorous QC for subsequent applications, safeguarding against genetically aberrant or partially reprogrammed iPSC lines.

2.2 iPSC Banking

As the demand for quality-controlled, research-grade, and disease-relevant iPSC lines increases, related data and stem cell characterization services are expanding rapidly across the globe, and there is a great need for the standardization of iPSC banking procedures. The implications and the potential of the iPSCs are vast. They have the potential to produce research tools for modeling many debilitating diseases and create a myriad of therapeutically relevant target cells. Therefore, the recipe for creating a successful and large-scale iPSC repository would require building a reliable supply chain for the generation of these iPSC lines from patients with multiple disease indications, producing them over specification QC measures, developing operationally responsible banking procedures, a robust distribution service, innovating and developing reliable differentiation protocols, and reliable customer support.

2.2.1 Cryopreservation

The standardization of cryopreservation procedures is vital to the creation of large-scale iPSC banks. Cryopreservation is the long-term storage of cryogenically preserved PSCs. The two most common methods to freeze iPSCs are either by vitrification or by a slow freezing method. Each method requires the collection of cells from the substrate. This can be enzymatic or mechanical, with some sort of cryoprotectant reagent. Vitrification is the process by which cells are suspended in cryomedia, placed in cryovials, submerged in liquid nitrogen, and stored for long term in the gas phase of liquid nitrogen (59). When a colony of cells is vitrified, the intracellular medium becomes very viscous, even glass-like. This consistency prevents the formation of ice crystals within the cells, which stops the cells from bursting from within upon thawing (59). The method of vitrification has a high recovery rate postthaw but is not ideal for large quantity banking because the technique is very labor intensive and requires a high concentration of cryoprotectant chemicals, which can be very toxic to the cells when the procedure is not performed with alacrity (59). The second common form of cryopreservation is through slow freezing by way of a control rate freezer or a nonmechanical cryocontainer, often referred to as a Mr. Frosty™. Similar to vitrification, the cells are first harvested and placed in cryomedia then into cryovials, moved into either a control rate freezer or a cryocontainer, and finally stored for long term in the vapor phase of liquid nitrogen (60). The use of the control rate freezer or the cryocontainer is necessary because it cools the cells by 1°C per minute, which is vital to allow for water to leave the cells before the freezing is complete. This is important for the slow freezing process because it allows for extracellular ice formation, which increases osmotic concentrations resulting in the dehydration of the cells (61). This hyperosmotic environment prefreezing decreases the potential for intracellular ice formation thus preventing cell damage and death upon thawing (61). The method of freezing is just as important as the cryopreservation media. Numerous commercially available media optimized for PSCs (serum-containing and serum-free) are available with a varying concentration of a cryoprotective agent and a specific agent to neutralize the toxicity of the cryoprotectant. Some common serum-free media are Cryostor CS10 and mFreSR from STEMCELL Technologies and Synth-a-Freeze® sold by Life Technologies. As the market continues to grow and expand, more media will be available with even less toxicity than dimethyl sulfoxide (DMSO); for example FREEZEstem™ from BioLamina is a defined DMSO- and xeno-free cryopreservation medium that can be used for iPSCs (62).

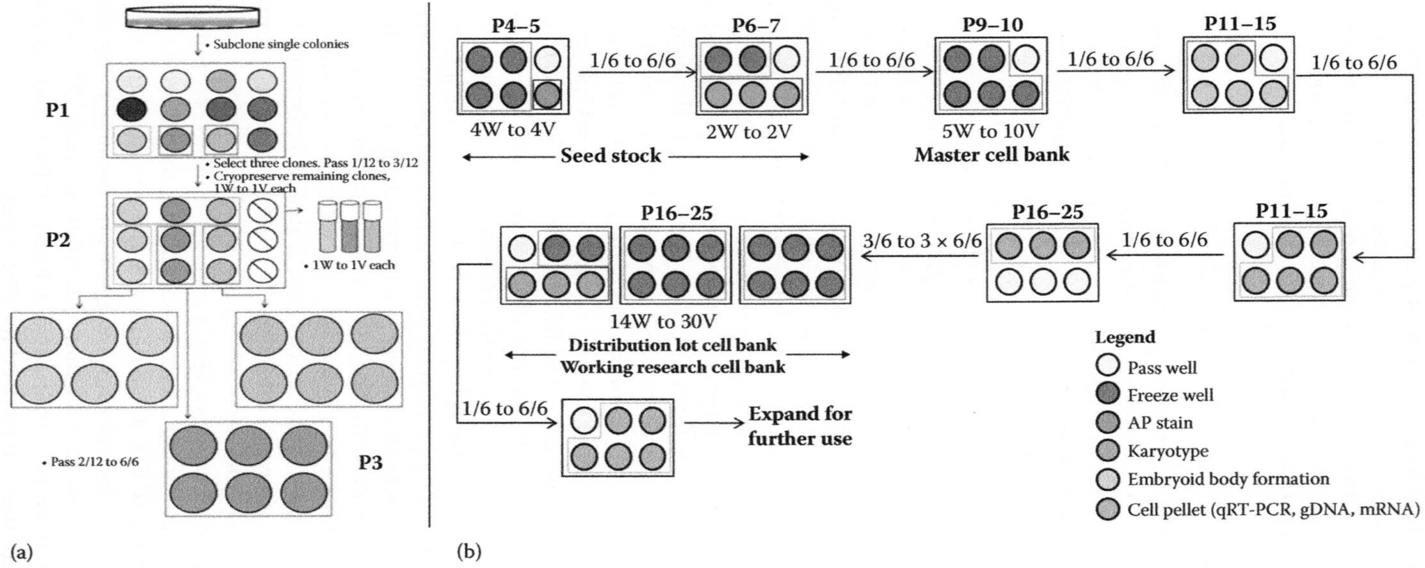

FIGURE 2.9
(See color insert.) (a) Schematic representation of single clone isolation and selection. The 12 colors at passage 1 (P1) indicates 12 single clones. Three clones of the initial 12 are selected for further expansion and characterization. One well of a 12-well plate (1/12) is expanded to 3 wells of a 12-well plate (3/12), after which 2/12 is expanded into 6 wells of a 6-well plate (6/6). The remaining clones are cryopreserved with 1 well (1W) distributed into one vial (1V) each. Clones are selected based on cell morphology, growth rate, and rate of differentiation. (b) Schematic representation of single clone expansion, characterization, and banking. This process is repeated for each clone selected at P2 (a). Typically, 1 well of a 6-well plate (1/6) is passaged into 6 wells of a 6-well plate (6/6).

2.2.2 Cell Expansion and Banking

The banking of quality iPSC lines is crucial for basic and translational research. This involves careful QC and thorough characterization of several iPSC lines. There is a need for great competency in cell line production and handling and an infrastructure to support maintenance and sharing of cell stocks with an international research community in order to maintain reproducibility and consistency in hiPSC research. Typically, the plan for optimal cell banking would require isolation of a clonal iPSC line and creation of an early passage seed stock of six vials between passages (P) 4 and 7 (Figure 2.9). One cryovial of alternate clonal lines is frozen as backup at P2. The seed stock is expanded to create a master cell bank of 10 cryovials between P9 and P10. Between P16 and P25, a working research cell bank and distribution lot iPSC bank is created for respective purposes. The early passage seed stock can also provide for the creation of a Current Good Manufacturing Practices (cGMP) master cell bank and working cell bank if future applications demand for iPSC-derived cells to be used in transplantation therapies. Therefore, maintaining a seed and master bank of low passage reference iPSCs is critical for future replacement.

Forums such as the International Stem Cell Forum are coordinating efforts with existing cell repositories, such as WiCell, Coriell, and RUDCR Infinite Biologics; stem cell cores within universities and academic centers; and new stem cell banks for consensual management and distribution plans (63). Multiple National Institutes of Health (NIH)-funded initiatives at the institutional level have also promoted such measures. Setting high standards for international banking and supply of iPSCs is an integral part of this approach. Infrastructures, proper governance structures, and streamlining protocols for institutional review board, stem cell research oversight committee approvals, and universal material transfer agreements in creating and obtaining iPSC lines are some of the necessary processes. Stringent selection criteria to ensure the cells are of excellent quality and compliant with ethical norms is a primary requirement alongside thorough documentation of the source and the applications of the iPSCs generated.

2.3 iPSC Technology: Bench Top to Bedside

The tremendous evolution of iPSC research in a short period stands testimony for the promise this technology holds in clinical applications. The prospect of having a renewable source of patient-specific pluripotent cells can prove to be of great significance in medicine. Several groups around the world have successfully developed, characterized, and differentiated iPSCs into specific cell types. Interestingly, the cell type-specific differentiation of iPSCs, when induced sequentially, closely resembles the development *in vivo* and is also helpful in better understanding the stages of organ development. Researchers have reported success in differentiating retinal cells from iPSCs generating both photoreceptor-like and retinal pigment epithelial-like phenotypes (64,65). The development of a human optic cup from iPSCs, which can be used for transplants, is an exciting prospect with great future clinical applications such as retinal grafts for cell replacement therapies of inherited retinal diseases (66). In addition to monogenetic diseases (44,67), even complex neurological diseases like schizophrenia (68,69) and Alzheimer's disease (70,71) can be modeled using patient iPSC lines. Gaining molecular insights into complex metabolic diseases, such as obesity, using iPSC disease modeling has great prospects offering a better understanding of neural metabolic regulatory mechanisms. Studies have reported the use of patient-specific HNF1A mature onset diabetes of the young three reprogrammed cells to investigate molecular mechanisms for diabetic pathophysiology and to model clinical phenotypes of diabetes (72). This group also showed that diabetic patient-derived iPSCs helped better understand mechanisms of defective insulin production and vascular dysfunction. iPSC-derived gastric organoids have also been reported (73), whereby they show formation of posterior foregut particularly antral organoids that form a robust *in vitro* system to elucidate mechanisms of stomach development and disease. Therefore, iPSCs are a great setup to study disease-in-a-dish models and gain insights into disease progression. This concept offers great promise in personalized therapy development paving the way for unhindered testing of novel therapeutic options (74).

Pharmacological studies on differentiated iPSCs also promise to form an important part of nonclinical drug evaluation prior to human clinical trials. Well-defined patient-specific models have helped understand disease mechanisms that are propelling the identification of relevant targets for drug discovery. Additionally, with the idea of precision medicine gaining steam, since aspects of patient iPSCs in a dish mimic their human counterpart's cell, the efficiency and the accuracy of drug screening should be significantly increased. Drug screening tests performed on human PSC-derived cardiomyocytes have helped identify compounds that cause predictable responses in the cells similar to pharmacological responses in the human heart (75). With regard to disease modeling and drug screening, studies on iPSC-based models of diabetic cardiomyopathy (76), neurological diseases (77), and liver diseases (78), to name a few, have shown promise in identifying relevant drugs for treatment of the disease.

A key question in the clinical use of the iPSC technology has been the reliability of these cells in maintaining genetic stability. Retroviral or lentiviral induction may be sufficient for *in vitro* applications, but alteration of genome during transgene integration is a major concern when considering applications in medicine. The possibility of iPSCs turning tumorigenic cannot be overlooked as the use of iPSCs in clinical use is still in its infant stage. A case report involving the use of donor stem cells for the treatment of ataxia talengiectasia outside the United States reported multifocal tumors in the brain and the spinal cord of a young patient after long-term follow-up (79). However, this patient had received a frighteningly large number of fetal neural stem cells transplanted at multiple sites than would ever be feasible in United States Food and Drug Administration-approved clinical trial. On the other hand, the first report that employed hESC-derived retinal pigment epithelium for treatment of age-related macular degeneration showed no tumorigenicity or ectopic tissues four months after transplantation (80). These reports highlight the care that needs to be taken in choosing iPSCs for therapy. The successful and stable generation of iPSCs for use in medicine still requires modifications in culture conditions, safer pluripotency induction methods, and selection of reprogramming factors that avoid c-Myc and other proto-oncogenic factors. However, the promise this technology offers in clinical applications and therapeutics is undeniable, and future research is sure to lead us to the ultimate goal of employing iPSCs in the bedside.

Acknowledgments

Dhruv Sareen is supported by funding from NIH/National Center for Advancing Translational Sciences (UL1TR00014), NIH/National Institute of Neurological Disorders and Stroke/Library of Network-Based Cellular Signature (1U54NS091046-01), NIH/National Eye Institute (R01 EY023429-01) funds and institutional start-up funds from Cedars-Sinai Medical Center.

References

1. Yoshida, Y. et al., Hypoxia enhances the generation of induced pluripotent stem cells. *Cell Stem Cell*, 2009. 5(3): pp. 237–41.
2. Westall, F. C., R. Rubin, and D. Gospodarowicz, Brain-derived fibroblast growth factor: A study of its inactivation. *Life Sci*, 1983. 33(24): pp. 2425–9.
3. Yang, C. et al., Mechanical memory and dosing influence stem cell fate. *Nat Mater*, 2014. 13(6): pp. 645–52.
4. Thomson, J. A. et al., Embryonic stem cell lines derived from human blastocysts. *Science*, 1998. 282(5391): pp. 1145–7.
5. Martin, M. J. et al., Human embryonic stem cells express an immunogenic nonhuman sialic acid. *Nat Med*, 2005. 11(2): pp. 228–32.
6. Hughes, C. S., L. M. Postovit, and G. A. Lajoie, Matrigel: A complex protein mixture required for optimal growth of cell culture. *Proteomics*, 2010. 10(9): pp. 1886–90.

7. Chen, K. G. et al., Human pluripotent stem cell culture: Considerations for maintenance, expansion, and therapeutics. *Cell Stem Cell*, 2014. 14(1): pp. 13–26.
8. Rodin, S. et al., Long-term self-renewal of human pluripotent stem cells on human recombinant laminin-511. *Nat Biotechnol*, 2010. 28(6): pp. 611–5.
9. Lu, H. F. et al., A defined xeno-free and feeder-free culture system for the derivation, expansion and direct differentiation of transgene-free patient-specific induced pluripotent stem cells. *Biomaterials*, 2014. 35(9): pp. 2816–26.
10. Wang, Y., L. Cheng, and S. Gerecht, Efficient and scalable expansion of human pluripotent stem cells under clinically compliant settings: A view in 2013. *Ann Biomed Eng*, 2014. 42(7): pp. 1357–72.
11. Boron, W. F. and E. L. Boulpaep, *Medical Physiology*. Philadelphia, PA: Saunders, 2012.
12. Chen, G. et al., Chemically defined conditions for human iPSC derivation and culture. *Nat Methods*, 2011. 8(5): pp. 424–9.
13. Kusuda Furue, M. et al., Advantages and difficulties in culturing human pluripotent stem cells in growth factor-defined serum-free medium. *In Vitro Cell Dev Biol Anim*, 2010. 46(7): pp. 573–6.
14. Ludwig, T. E. et al., Derivation of human embryonic stem cells in defined conditions. *Nat Biotechnol*, 2006. 24(2): pp. 185–7.
15. Ludwig, T. E. et al., Feeder-independent culture of human embryonic stem cells. *Nat Methods*, 2006. 3(8): pp. 637–46.
16. Garitaonandia, I. et al., Increased risk of genetic and epigenetic instability in human embryonic stem cells associated with specific culture conditions. *PLoS One*, 2015. 10(2): p. e0118307.
17. Hoffman, L. M. and M. K. Carpenter, Characterization and culture of human embryonic stem cells. *Nat Biotechnol*, 2005. 23(6): pp. 699–708.
18. Baker, D. E. et al., Adaptation to culture of human embryonic stem cells and oncogenesis in vivo. *Nat Biotechnol*, 2007. 25(2): pp. 207–15.
19. Draper, J. S. et al., Recurrent gain of chromosomes 17q and 12 in cultured human embryonic stem cells. *Nat Biotechnol*, 2004. 22(1): pp. 53–4.
20. Spits, C. et al., Recurrent chromosomal abnormalities in human embryonic stem cells. *Nat Biotechnol*, 2008. 26(12): pp. 1361–3.
21. Meisner, L. F. and J. A. Johnson, Protocols for cytogenetic studies of human embryonic stem cells. *Methods*, 2008. 45(2): pp. 133–41.
22. Morshead, C. M. et al., Hematopoietic competence is a rare property of neural stem cells that may depend on genetic and epigenetic alterations. *Nat Med*, 2002. 8(3): pp. 268–73.
23. Sugawara, A. et al., Current status of chromosomal abnormalities in mouse embryonic stem cell lines used in Japan. *Comp Med*, 2006. 56(1): pp. 31–4.
24. Speicher, M. R. and N. P. P. Carter, The new cytogenetics: Blurring the boundaries with molecular biology. *Nat Rev Genet*, 2005. 6(10): pp. 782–92.
25. O'Connor, C., Karyotyping for chromosomal abnormalities. *Nat Educ*, 2008. 1(1): p. 27
26. Theisen, A., Microarray-based comparative genomic hybridization (aCGH). *Nat Educ*, 2008. 1(1).
27. Venkataraman, K. S., J. Ramkrishna, and N. Raghavan, Human leptospirosis: A recent study in Madras, India. *Trans R Soc Trop Med Hyg*, 1991. 85(2): p. 304.

28. Elliott, A. M., K. A. Elliott, and A. Kammesheidt, High resolution array-CGH characterization of human stem cells using a stem cell focused microarray. *Mol Biotechnol*, 2010. 46(3): pp. 234–42.
29. Muller, F. J. et al., A bioinformatic assay for pluripotency in human cells. *Nat Methods*, 2011. 8(4): pp. 315–7.
30. Müller, F.-J., B. Brändl, and J. F. Loring, Assessment of human pluripotent stem cells with PluriTest StemBook. In *The Stem Cell Research Community*. StemBook. Cambridge, MA, 2012.
31. Singh, U. et al., Novel live alkaline phosphatase substrate for identification of pluripotent stem cells. *Stem Cell Rev*, 2012. 8(3): pp. 1021–9.
32. Wright, A. J. and P. W. Andrews, Surface marker antigens in the characterization of human embryonic stem cells. *Stem Cell Res*, 2009. 3(1): pp. 3–11.
33. Schopperle, W. M. and W. C. DeWolf, The TRA-1-60 and TRA-1-81 human pluripotent stem cell markers are expressed on podocalyxin in embryonal carcinoma. *Stem Cells*, 2007. 25(3): pp. 723–30.
34. Kannagi, R. et al., Stage-specific embryonic antigens (SSEA-3 and -4) are epitopes of a unique globo-series ganglioside isolated from human teratocarcinoma cells. *EMBO J*, 1983. 2(12): pp. 2355–61.
35. Damjanov, I. et al., Immunohistochemical localization of murine stage-specific embryonic antigens in human testicular germ cell tumors. *Am J Pathol*, 1982. 108(2): pp. 225–30.
36. Marrink, J. et al., TRA-1-60: A new serum marker in patients with germ-cell tumors. *Int J Cancer*, 1991. 49(3): pp. 368–72.
37. Pesce, M. and H. R. Scholer, Oct-4: Gatekeeper in the beginnings of mammalian development. *Stem Cells*, 2001. 19(4): pp. 271–8.
38. Chambers, I. et al., Nanog safeguards pluripotency and mediates germline development. *Nature*, 2007. 450(7173): pp. 1230–4.
39. Rizzino, A., Sox2 and Oct-3/4: A versatile pair of master regulators that orchestrate the self-renewal and pluripotency of embryonic stem cells. *Wiley Interdiscip Rev Syst Biol Med*, 2009. 1(2): pp. 228–36.
40. Boyer, L. A. et al., Core transcriptional regulatory circuitry in human embryonic stem cells. *Cell*, 2005. 122(6): pp. 947–56.
41. Loh, Y. H. et al., The Oct4 and Nanog transcription network regulates pluripotency in mouse embryonic stem cells. *Nat Genet*, 2006. 38(4): pp. 431–40.
42. U.S. Department of Health and Human Services, The stem cell. In *Stem Cell Information*. Bethesda, MD: National Institutes of Health, U.S. Department of Health and Human Services, 2009. https://stemcells.nih.gov/info/basics/6htm, accessed May 19, 2015.
43. Itskovitz-Eldor, J. et al., Differentiation of human embryonic stem cells into embryoid bodies compromising the three embryonic germ layers. *Mol Med*, 2000. 6(2): pp. 88–95.
44. Sareen, D. et al., Targeting RNA foci in iPSC-derived motor neurons from ALS patients with a C9ORF72 repeat expansion. *Sci Transl Med*, 2013. 5(208): p. 208ra149.
45. Sareen, D. et al., Differentiation of human limbal-derived induced pluripotent stem cells into limbal-like epithelium. *Stem Cells Transl Med*, 2014. 3(9): pp. 1002–12.

46. Bock, C. et al., Reference maps of human ES and iPS cell variation enable high-throughput characterization of pluripotent cell lines. *Cell*, 2011. 144(3): pp. 439–52.
47. Williams, L. A., B. N. Davis-Dusenbery, and K. C. Eggan, SnapShot: Directed differentiation of pluripotent stem cells. *Cell*, 2012. 149(5): pp. 1174–1174 e1.
48. Dessimoz, J. et al., FGF signaling is necessary for establishing gut tube domains along the anterior-posterior axis in vivo. *Mech Dev*, 2006. 123(1): pp. 42–55.
49. McLin, V. A., S. A. Rankin, and A. M. Zorn, Repression of Wnt/beta-catenin signaling in the anterior endoderm is essential for liver and pancreas development. *Development*, 2007. 134(12): pp. 2207–17.
50. Wells, J. M. and D. A. Melton, Early mouse endoderm is patterned by soluble factors from adjacent germ layers. *Development*, 2000. 127(8): pp. 1563–72.
51. Wong, A. P. et al., Directed differentiation of human pluripotent stem cells into mature airway epithelia expressing functional CFTR protein. *Nat Biotechnol*, 2012. 30(9): pp. 876–82.
52. Okita, K. et al., A more efficient method to generate integration-free human iPS cells. *Nat Methods*, 2011. 8(5): pp. 409–12.
53. Cao, F. et al., Spatial and temporal kinetics of teratoma formation from murine embryonic stem cell transplantation. *Stem Cells Dev*, 2007. 16(6): pp. 883–91.
54. Lee, A. S. et al., Effects of cell number on teratoma formation by human embryonic stem cells. *Cell Cycle*, 2009. 8(16): pp. 2608–12.
55. Tang, C. et al., An antibody against SSEA-5 glycan on human pluripotent stem cells enables removal of teratoma-forming cells. *Nat Biotechnol*, 2011. 29(9): pp. 829–34.
56. Peterson, S. E. et al., Teratoma generation in the testis capsule. *J Vis Exp*, 2011(57): p. e3177.
57. Tam, P. and J. Rossant, Mouse embryonic chimeras: Tools for studying mammalian development. *Development*, 2003. 130(25): pp. 6155–63.
58. Kang, L. et al., iPS cells can support full-term development of tetraploid blastocyst-complemented embryos. *Cell Stem Cell*, 2009. 5(2): pp. 135–8.
59. Beier, A. F. et al., Effective surface-based cryopreservation of human embryonic stem cells by vitrification. *Cryobiology*, 2011. 63(3): pp. 175–85.
60. Imaizumi, K. et al., A simple and highly effective method for slow-freezing human pluripotent stem cells using dimethyl sulfoxide, hydroxyethyl starch and ethylene glycol. *PLoS One*, 2014. 9(2): p. e88696.
61. Boldt, J., Current results with slow freezing and vitrification of the human oocyte. *Reprod Biomed Online*, 2011. 23(3): pp. 314–22.
62. Katkov, I. I. et al., *DMSO*-free programmed cryopreservation of fully dissociated and adherent human induced pluripotent stem cells. *Stem Cells* Int, 2011. 2011: p. 981606.
63. Stacey, G. N. et al., Banking human induced pluripotent stem cells: Lessons learned from embryonic stem cells? *Cell Stem Cell*, 2013. 13(4): pp. 385–8.
64. Hirami, Y. et al., Generation of retinal cells from mouse and human induced pluripotent stem cells. *Neurosci Lett*, 2009. 458(3): pp. 126–31.
65. Osakada F., Y. Sasai, and M. Takahashi, Control of neural differentiation from pluripotent stem cells. *Inflamm Regen*, 2008. 28(3): pp. 166–173.
66. Cramer, A. O. and R. E. MacLaren, Translating induced pluripotent stem cells from bench to bedside: Application to retinal diseases. *Curr Gene Ther*, 2013. 13(2): pp. 139–51.

67. Sareen, D. et al., Inhibition of apoptosis blocks human motor neuron cell death in a stem cell model of spinal muscular atrophy. *PLoS One*, 2012. 7(6): p. e39113.
68. Brennand, K. J. et al., Modelling schizophrenia using human induced pluripotent stem cells. *Nature*, 2011. 473(7346): pp. 221–5.
69. Yoon, K. J. et al., Modeling a genetic risk for schizophrenia in iPSCs and mice reveals neural stem cell deficits associated with adherens junctions and polarity. *Cell Stem Cell*, 2014. 15(1): pp. 79–91.
70. Yagi, T. et al., Modeling familial Alzheimer's disease with induced pluripotent stem cells. *Hum Mol Genet*, 2011. 20(23): pp. 4530–9.
71. Kondo, T. et al., Modeling Alzheimer's disease with iPSCs reveals stress phenotypes associated with intracellular Abeta and differential drug responsiveness. *Cell Stem Cell*, 2013. 12(4): pp. 487–96.
72. Stepniewski, J. et al., Induced pluripotent stem cells as a model for diabetes investigation. *Sci Rep*, 2015. 5: p. 8597.
73. McCracken, K. W. et al., Modelling human development and disease in pluripotent stem-cell-derived gastric organoids. *Nature*, 2014. 516(7531): pp. 400–4.
74. Kim, C., Disease modeling and cell based therapy with iPSC: Future therapeutic option with fast and safe application. *Blood Res*, 2014. 49(1): pp. 7–14.
75. Dick, E. et al., Evaluating the utility of cardiomyocytes from human pluripotent stem cells for drug screening. *Biochem Soc Trans*, 2010. 38(4): pp. 1037–45.
76. Drawnel, F. M. et al., Disease modeling and phenotypic drug screening for diabetic cardiomyopathy using human induced pluripotent stem cells. *Cell Rep*, 2014. 9(3): pp. 810–21.
77. Xu, X. H. and Z. Zhong, Disease modeling and drug screening for neurological diseases using human induced pluripotent stem cells. *Acta Pharmacol Sin*, 2013. 34(6): pp. 755–64.
78. Choi, S. M. et al., Efficient drug screening and gene correction for treating liver disease using patient-specific stem cells. *Hepatology*, 2013. 57(6): pp. 2458–68.
79. Amariglio, N. et al., Donor-derived brain tumor following neural stem cell transplantation in an ataxia telangiectasia patient. *PLoS Med*, 2009. 6(2): p. e1000029.
80. Schwartz, S. D. et al., Embryonic stem cell trials for macular degeneration: A preliminary report. *Lancet*, 2012. 379(9817): pp. 713–20.

3

Genetic and Epigenetic Considerations in iPSC Technology

Yoshiaki Tanaka and In-Hyun Park

CONTENTS

3.1 Introduction

Genomic DNAs, RNAs, and histone proteins carry reversible covalent modifications to regulate various biological processes, such as transcription and replication. Terminally differentiated cells can be converted to pluripotent state by introducing defined transcription factors, Oct4, Sox2, Klf4, and c-Myc (1,2). In the course of reprogramming to the iPSC, somatic cells undergo multiple transcriptional and epigenetic changes. Although the efficiency of iPSC generation was reported to be initially extremely low (<0.1%), many efforts have succeeded in the improvement of iPSC reprogramming technique by introducing or depleting epigenetic modulators. Thus, this chapter will highlight the relationship of genetic and epigenetic modifications with somatic reprogramming and differentiation and will discuss future directions in iPSC technology (Figure 3.1).

3.2 DNA Methylation

In mammalian development, a zygote is differentiated into various types of cells and tissues. Each cell type regulates a distinct set of genes to generate the diversity of its phenotypes and functions. Covalent modification of genomic DNA is one of the major epigenetic changes and is highly orchestrated with normal embryonic development (3). The modification in carbon 5 position of cytosine is the most prevalent in mammalian genome and has multiple nucleotide variants, which display different functions (Figure 3.2). Here, we highlight the contribution of cytosine modification variants in PSCs.

3.2.1 5-Methylcytosine

From bacteria to eukaryotes, 5-methylcytosine (5mC) is widely distributed and has distinctive functions among organisms. In bacteria, their genomic DNA is protected by 5mC to be distinguished from invading viral DNA. The bacterial genome is immediately methylated after replication and avoids endonuclease cleavage by the host restriction enzyme system. In budding yeast and nematode, 5mC is rare or not detectable (4,5). In both mammals and plants, 5mC contributes to transcriptional regulation and DNA replication. In plant, 5mC is prevalent in the contexts of CpG, CHG, and CHH (H = A, T, or C). Although 5mC is generally limited to CpG in mammal, ESCs and iPSCs harbor significant amount of non-CpG methylation (6).

In human and mouse early embryonic developments, there are large waves of demethylation and remethylation (7–10). Upon fertilization, the

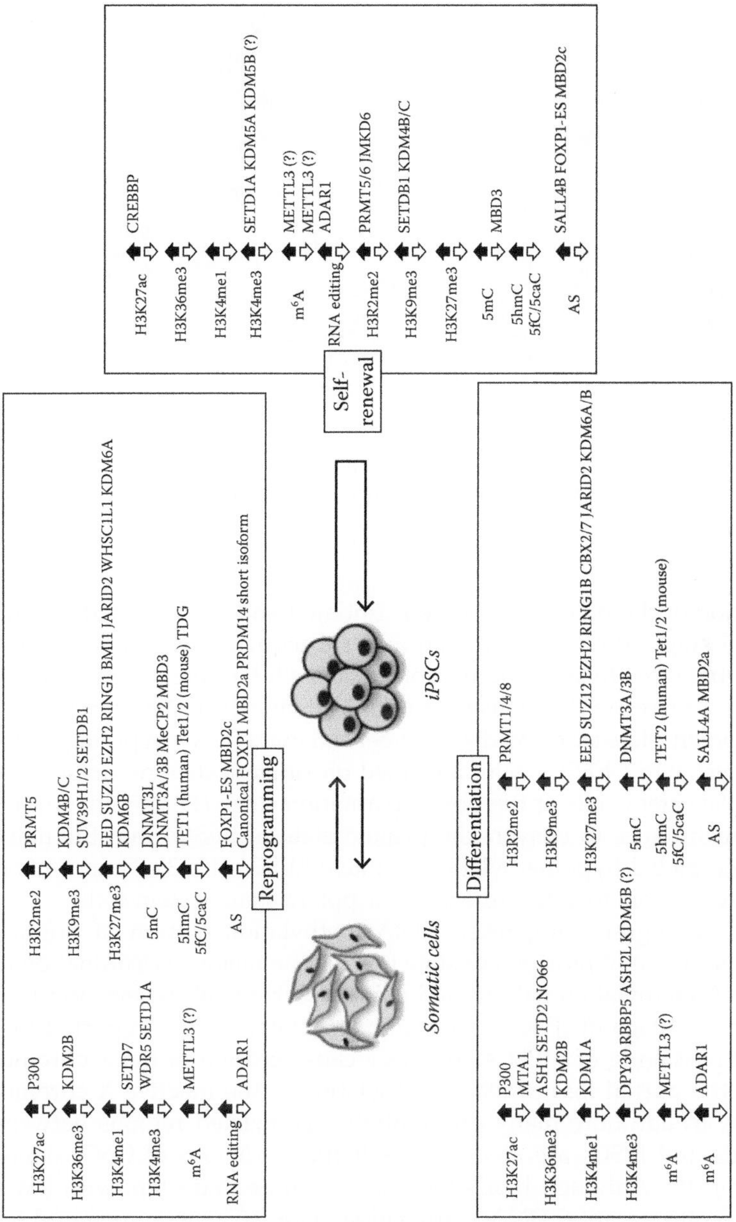

FIGURE 3.1
Relationships of genetic/epigenetic regulations with iPSC reprogramming and differentiation. Genetic and epigenetic modulators to affect reprogramming, self-renewal, and differentiation positively (black) and negatively (white) are shown.

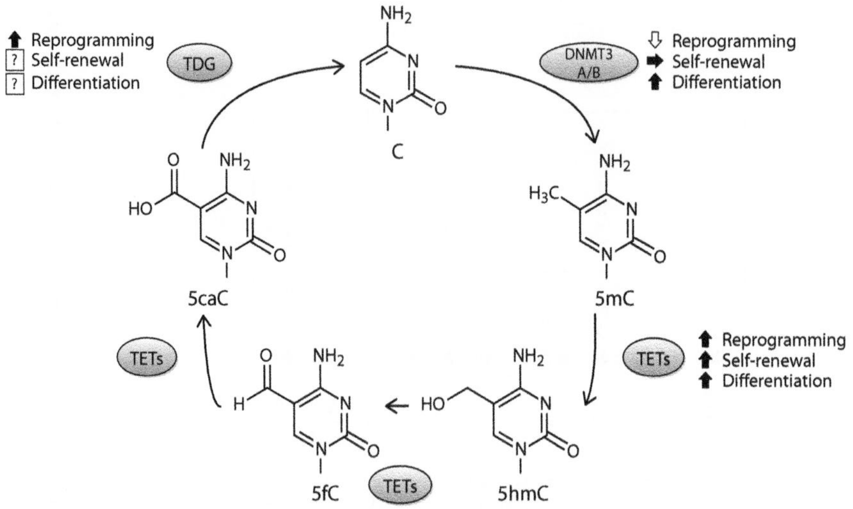

FIGURE 3.2
DNA modification derivatives and enzymes catalyzing the conversion of modification patterns. Effects of 5-carbon conversion in cytosine on reprogramming, self-renewal, and differentiation are shown.

demethylation of the parental genome is initiated and is completed around the two-cell stage. With implantation and subsequent gastrulation of the embryo, global remethylation occurs for loss of cellular potency and lineage specification. Interestingly, mouse embryonic stem cells (mESCs) represent naive pluripotent state associated with inner cell mass (ICM) in preimplantation blastocyst, while hESCs exhibit primed pluripotent state reminiscent of mouse epiblast stem cells at the postimplantation stage (11). Recent studies presented techniques to convert the primed state in hESCs into the naive state and the global hypomethylation in naive hESCs (12,13). Thus, 5mC is one of the key factors to determine the pluripotent state in stem cells.

The iPSC reprogramming resets DNA methylation pattern of original somatic cells to that of pluripotent state (14,15). The major discrimination of somatic DNA methylation pattern and gain of pluripotent pattern occurs at the late mature stage of iPSC reprogramming (16). It is also reported that *in vitro* cell passaging of hiPSCs and ESCs cause erosion of the X chromosome with the partial loss of DNA methylation at the inactive X chromosome (17,18). In addition, many differentially methylated regions between somatic cells and iPSCs are identified as flanking regions of CpG islands (CpGI shore) (19). Although iPSCs have very similar characters with ESCs, iPSCs still retain parental DNA methylation state (epigenetic memory) (6). The inadequate epigenetic changes were often observed near centromeric and telomeric regions and lead the differential expression of neighboring genes, including *TCERG1L*, *TMEM132D*, and *FAM19A5* (20,21).

In mammals, 5mC is catalyzed by two de novo DNA methyltransferases (DNMT3A and DNMT3B) using *S*-adenosyl methionine (SAM) as a substrate. In hESCs and mESCs, both DNMT3A and DNMT3B are highly expressed compared to differentiated embryoid bodies and adult tissues (22,23). During embryogenesis, Dnmt3b is specifically expressed in ICM, epiblast, and embryonic ectoderm until E9.5. In contrast, Dnmt3a is ubiquitously expressed after E10.5 (24). While Dnmt3a plays an important role in stem cell proliferation, Dnmt3b functions at the initiation of stem cell development by repressing Nanog and Oct4 expressions (25). Although the loss of de novo DNA methyltransferases is dispensable for iPSC reprogramming and maintaining the self-renewal in ESCs (26,27), the overexpression of Dnmt3a and Dnmt3b significantly decreases the efficiency of iPSC reprogramming (28). In addition, the inactivation of Dnmt3a and Dnmt3b in mESCs loses de novo DNA methylation and disrupts cell differentiation (29), indicating that 5mC is important for proper development.

The other methyltransferase DNMT1 induces DNA methylation at hemimethylated CpG after cell division. A SET and RING finger-associated domain protein, Np95, is essential for heterochromatin localization of Dnmt1, and the loss of Np95 or Dnmt1 shows comparable hypomethylation level in repetitive elements and imprinted genes in mESCs (30,31). Although ESCs are tolerant toward Dnmt1 inactivation (32), *Dnmt1* KO somatic cells induce p53-dependent apoptosis with global hypomethylation (33,34). Furthermore, *Np95* or *Dnmt1* KO embryos show developmental arrest with early gestational lethality. In long-term *in vitro* culture on MEF, Dnmt1-active hiPSCs lose pluripotency more easily than Dnmt1-silenced cells by inducing 5mC in *NANOG* and *OCT4* promoters (35). Since Nanog and Oct4 also directly regulate Dnmt1 expression, the balance of *Dnmt1* and pluripotent factor expression is important for the maintenance of pluripotency and the induction of differentiation (36). The function of maintaining DNA methylation by Dnmt1 is not perfect, and Dnmt3a and Dnmt3b are required to construct an appropriate global methylation pattern in ESCs (32).

DNMT3-like (DNMT3L) protein does not have a catalytic activity but can cooperate with DNMT3A and DNMT3B and is required for the establishment of maternal genomic imprinting and the suppression of retrotransposons (37–39). Chromatin localization of Dnmt3l is controlled by Dnmt3a (40). Dnmt3l is highly expressed in mESCs but not expressed in differentiated somatic cells (39,41). In preimplantation embryo, *Dnmt3l* promoter is unmethylated and targeted by Oct4 and Nanog (42). Zygotic Dnmt3l is dispensable for the establishment of methylation pattern in the embryo but can accelerate the initiation of embryonic de novo methylation (42). In postimplantation embryo, Dnmt3l is repressed with DNA methylation in its promoter by Dnmt3a, Dnmt3b, and Dnmt3l itself (43). Dnmt3b shows the most striking effect of the promoter methylation. Since Dnmt3l also represses Dnmt3a expression (42), these DNA methyltransferases are regulated by each other. During iPSC reprogramming, Dnmt3l is induced in the late phase with core pluripotent factors (16). DNMT3L-overexpressing HeLa cells do not induce NANOG but upregulate SOX2 expression and form ESC-like colony morphology (44), indicating that DNMT3L supports the acquisition of pluripotent state. Interestingly, a recent study demonstrated that Dnmt3l interacts with enhancer of zeste homologue 2 (Ezh2), which is a component of polycomb repressive complex 2 (PRC2), in competition with Dnmt3a and Dnmt3b and maintains low methylation level in the bivalent promoters, which contain both active (H3K4me3) and repressive histone modifications (H3K27me3) (41). Thus, Dnmt3l also contributes to the maintenance of hypomethylation as well as the stimulation of de novo methylation.

Methyl-CpG binding domain (MBD) is a functional domain that interacts with 5mC DNA and is identified in five proteins, MeCP2 and MBD1, MBD2, MBD3, and MBD4. MeCP2 is especially a critical factor for neuronal development and functions as a transcriptional repressor by recruiting NCoR/SMRT repressive complex (45). Loss of MeCP2 leads to neurodevelopmental disorder, the Rett syndrome (RTT), and several groups, including our laboratory, succeeded in generation of iPSCs from RTT patients (RTT-iPSC) (20,46). RTT-iPSCs show distinguishable transcriptome profiles from normal iPSCs, indicating that MeCP2 also affects transcriptional regulation in the stem cell stage. Furthermore, the deletion of MeCP2 increases the efficiency of iPSC generation (47). Overall, 5mC shows negative effect on iPSC reprogramming but is required to establish stem cells with proper differentiation capacities.

3.2.2 5-Hydroxymethylcytosine

The ten-eleven translocation (TET) protein family, TET1, TET2, and TET3, hydroxylates 5mC to 5-hydroxylmethylcytosine (5hmC) (48). A high level of 5hmC is detected in certain cell types, including ESCs and Purkinje neurons (49,50). 5hmC is enriched in transcriptional start sites (TSSs) (51), gene bodies, and enhancers of both high- and low-expressing genes (52–54). While genes with high or medium expression level show 5hmC enrichment in gene bodies, 5hmC is enriched in TSSs of low-expressing genes with PRC2 binding. These reports indicate that the potential role of 5hmC is both gene activation and repression.

The TET family shows distinct gene expression pattern between human and mouse PSCs. In human, only TET1 is highly expressed in ESCs and iPSCs, and TET2 and TET3 are relatively enriched in somatic tissues, such as the breast and the colon (55). Whereas TET1 is required for increasing the 5hmC level during hiPSC reprogramming (56), TET2 is essential for the lineage specification of hESCs (57). Although the functional role of TET3 in hESCs and hiPSCs is still unclear, the expression level is negatively correlated with OCT4 and NANOG during human EB differentiation, suggesting that TET3 may also be related to developmental processes in human (57). In mouse, Tet1 and Tet2 are highly expressed in mESCs, but Tet3 is enriched in the oocyte and the zygote (58,59). Both Tet1 and Tet2 can enhance mouse iPSC reprogramming with physical interaction to Nanog (60), but Tet3 has little effect on iPSC reprogramming (61). Tet1 and Tet2 also regulate the lineage differentiation by producing 5hmC in mESCs (62).

Several recent studies revealed that distinctive role of Tet1 and Tet2 in mESCs and reprogramming. Tet1, but not Tet2, is essential for the maintenance of ES self-renewal (63). Whereas Tet1 regulates 5hmC level in promoters, Tet2 deletion decreases 5hmC level in the gene body (64). At the early stage of iPSC reprogramming, Tet2 is required to accelerate Oct4 accessibility in the *Nanog* and *Esrrb* loci (65). In contrast, Tet1 is induced at the late stage of iPSC reprogramming (16).

In searching for candidates specifically recognizing 5hmC, several proteins have been identified (66). Uncharacterized proteins Wdr76 and Thy28 are detected as NPC-specific 5hmC readers. DNA glycosylase and helicase (e.g., Neli1 and Recql) also specifically bind to 5hmC. Interestingly, MeCP2 displays higher binding affinity with 5hmC than other methyl-CpG binding proteins, and RTT-causing mutation R133C loses preference to 5hmC (67). Furthermore, the abundance of 5hmC is negatively correlated with MeCP2 expression (50). MeCP2 targets L1 retrotransposons, and 5hmC is enriched in L1 promoters in mESCs (68). These reports suggest that MeCP2 is important for neuronal development and retrotransposon regulation via 5hmC.

3.2.3 Other DNA Modifications

5-Formylcytosine (5fC) and 5-caroboxylcytosine (5caC) are the other cytosine variants and are generated from 5hmC further catalyzed by TET family proteins (51). Both are considered as intermediates in the demethylation of cytosine. Although 5hmC and 5fC are detectable in various organs, 5caC is limited in ESCs and preimplantation embryos (51,69,70). Thymine-DNA glycosylase (TDG) specifically recognizes 5caC and excises carboxyl group, which results in non-modified cytosine (71). The deletion of TDG increases 5fC and 5caC levels and also prevents iPSC reprogramming (61,72). 5fC is enriched in poised enhancers, and the accumulation of 5fC in enhancers increases p300 binding in ESCs (73). In contrast, 5caC is transiently accumulated in the promoters of gene in neurodevelopmental genes, which are demethylated during neuronal differentiation (70). Together, both variants also regulate iPSC reprogramming, but 5caC is more associated with lineage specification.

3.3 Histone Modifications

Eukaryotic genomic DNA wraps around histone octamers and comprises higher-order structures of chromatins. N-terminal tail domain of histones

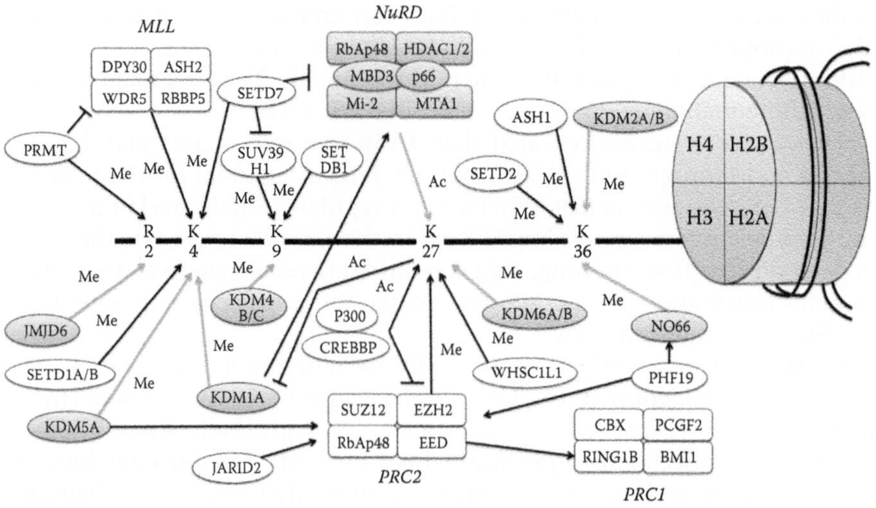

FIGURE 3.3
Histone modification derivatives and their catalytic enzymes. Black and gray represent methyl/acetyltransferase and demethyl/acetylase activities, respectively.

is subject to posttranslational modification, such as methylation and acetylation (Figure 3.3). Each modification displays distinct functions and distribution and plays an important role during cellular lineage specification. Chromatin immunoprecipitation with high-throughput approaches (ChIP-seq and ChIP-chip) has revealed that global changes of histone modifications are also involved during iPSC reprogramming (15). In the following, we will introduce several histone modifications and discuss their roles in ESCs and somatic reprogramming.

3.3.1 H3K4me3

Histone 3 lysine 4 trimethylation (H3K4me3) is associated with gene activation and enriched in active and poised promoter regions. Since the H3K4me3 level in the promoter is strongly correlated with gene expression, H3K4me3 plays an important role in gene activation (74). In somatic cells, H3K4me3 pattern is similar to that in ESCs, but several pluripotent genes do not harbor H3K4me3 in their promoters (75). During iPSC reprogramming, there are two big waves of global H3K4me3 changes (16). The first wave occurs at the early stage of reprogramming and coincides with both gain and loss of H3K4me3, while the second wave includes only gain of H3K4me3 at the late stage. For example, fibroblast growth factor 4, which is highly expressed in ESCs and essential for their lineage commitment (76), is induced by the first wave of H3K4me3 gain. Lin28, which is expressed in mature iPSCs and promotes iPSC reprogramming (77), is activated by the second H3K4me3 gain.

Several H3K4me3 writers and erasers have been reported to show critical functions in ESCs and iPSCs. Trithorax (Trx) group proteins are one family of H3K4me3 writers. MLL is one of Trx families in mammals and constructs a complex with Dpy30, Rbbp5, Ash2/2l, and Wdr5. Dpy30 and Rbbp5 are not essential for self-renewal but are crucial for differentiation into neuronal lineage (78). Ash2l is important to establish X chromosome inactivation (XCI) during ES differentiation (79). Furthermore, Wdr5 is targeted by Oct4 and Nanog and upregulated during iPSC reprogramming (80). Unlike the other components of the MLL complex, the deletion of Wdr5 results in loss of self-renewal and blocks iPSC reprogramming. Wdr5 also has many target genes shared with c-Myc and Oct4. Thus, Wdr5 has been proposed to be important to bridge c-Myc- and Oct4-based pluripotent networks.

Setd1a and Setd1b are the other members of the Trx group in mammals. Both enzymes are highly expressed in oocytes and gradually downregulated with embryonic development (81). The deletion of Setd1a fails gastrulation, and the establishment of ESC lines with loss of self-renewal. Setd1b-deleted embryo can grow until E11.5 but displays growth retardation. Regardless of the loss of cell proliferation and ESC-like morphology in Setd1a-deleted ESCs, Setd1b does not affect the cell growth and the morphology and cannot compensate for the cell proliferation defect of Setd1a-deleted ESC. Furthermore, the deletion of Setd1a, but not Setd1b, disturbs the derivation of iPSCs. Thus, Setd1a and Setd1b are involved in different molecular mechanism for the pluripotency.

Lysine (K)-specific demethylase 5 (KDM5) group is a family that removes tri- and dimethylation of H3K4. KDM5 proteins have Jumonji and ARID domains, which are required for H3K4 demethylation and DNA binding, respectively. Mammals contain four members, Kdm5a (also called Jarid1a or Rbp2), Kdm5b (Jarid1b or Plu1), Kdm5c (Jarid1c or Smcx), and Kdm5d (Jarid1d or Smcy). Kdm5a interacts with RPC2 by coordinating demethylation H3K4 and methylation of H3K27 and targets genes related to RNA metabolism and mitochondrial function in ESCs (82,83). In addition, Kdm5a binds to *Hox* genes, which is required for cell development (84). During ES differentiation, Kdm5a is displaced and H3K4me3 is increased in these loci, indicating that the potential function of Kdm5a is the maintenance of pluripotency by suppressing the induction of developmental genes.

The functional contribution of Kdm5b to ESCs is still controversial. Xie et al. demonstrated that the knockdown of *Kdm5b* triggered morphological changes and loss of pluripotency in ESC (85). This is supported by another group showing that the overexpression of Kdm5b reduced terminally differentiated cells and increased proliferating progenitors (86). However, Schmitz et al. demonstrated that Kdm5b-targeting short hairpin RNA (shRNA) treatment does not affect cell morphology, proliferation, and pluripotent marker expression in ESCs but impairs their neuronal differentiation capacity (87). Interestingly, another group showed that Kdm5b is important in focusing H3K4 methylation in promoters and enhancers in both ESCs and differentiated cells. The deletion of Kdm5b leads to the spreading of H3K4me3 to gene bodies and enhancer shores and results in defects of gene expression programs and decreasing of enhancer activity (88). Thus, Kdm5b may be involved in both pluripotency and cell differentiation by regulating the proper distribution of H3K4 methylation.

Kdm5c and Kdm5d are located in the X and Y chromosomes. In ESCs, Kdm5c represses promoter activity with local decreasing of H3K4me3 level but increases enhancer activity by increasing monomethylation of H3K4 (H3K4me1) (89). In addition, Kdm5c physically interact with c-Myc and regulate Oct4 gene expression in ESCs (90). Although the functional role of Kdm5d is not fully investigated, Kdm5d was shown to interact with Pcgf6 (also called Ring6a or MBLR) that is required for iPSC reprogramming and repression of mesodermal lineage-specific genes in ESCs (91,92).

3.3.2 H3K4me1

Similar to H3K4me3, H3K4me1 is also positively correlated with gene expression and enriched around TSSs (74). Since H3K4me1 is also enriched in distal regulatory sites, H3K4me1 is considered as one of the active enhancer markers.

Setd7 (Set7/9 or Kmt7) is known as H3K4me1 writer. Setd7-mediated H3K4me1 dissociates NuRD complex from H3 tail domain and blocks Suv39h1-mediated methylation of H3 lysine 9 (H3K9) (93). Setd7 is upregulated during differentiation and expressed in trophectoderm layer of embryo but not in ICM (94). Silencing of SETD7 assists in the somatic cell reprogramming in human fibroblasts.

Kdm1a (Lsd1) is a demethylase specifically targeting H3K4 and H3K9, and required for gastrulation during embryogenesis and silencing of several developmental genes (95,96). Kdm1a-deleted ESCs can maintain undifferentiated state but failed in EB differentiation. Kdm1a occupies active enhancers in ESCs with involvement to the NuRD complex, indicating that Kdm1a is essential to exit pluripotent state. The demethylase activity of Kdm1 is inhibited by acetylated histones (97). In ESCs, Lsd1 occupies Oct4-targeting active enhancers, but their H3K4me1 level is not changed, because of the presence of acetylated histones (98). During differentiation, the acetylation level is decreased and Kdm1a-mediated demethylation of H3K4me1 is substantially induced to silence ESC-specific enhancer. In addition, Kdm1a is colocalized with Oct4 and Nanog in bivalent promoter to control transcriptional activity of mesendodermal genes (95). Furthermore, 5hmC level is strongly correlated with active histone modifications with H3K4me1, and Kdm1a-lacked ESCs display significant reduction of 5mC (52,96), implying the interplay of histone modification with DNA methylation status. Together, Kdm1a plays important roles in suppressing pluripotent genes and balance of developmental genes by controlling H3K4me1 level in promoter and enhancer.

3.3.3 H3K27me3

Trimethylation of H3 lysine 27 (H3K27me3) is a repressive mark and localized in silent promoters and enhancers (74,99). Compared with somatic cells, H3K27me3 level is significantly higher in ESC with the construction of bivalent domain in developmental genes (100,101). Similar with H3K4me3, two large waves of H3K27me3 change occur during iPSC reprogramming (16). The first wave is larger than the second wave and constructs more than half of the bivalent domains, which are established in mature iPSCs. Since partially reprogrammed cells display insufficient gain of H3K27me3 patterns in promoters (15), the second wave may be important for maturation of iPSCs.

Two main families of PRC, PRC1 and PRC2, are related to H3K27me3-related gene silencing (102). PRC1 includes chromobox proteins (Cbx2, Cbx4, Cbx6, Cbx7, and Cbx8), ring finger proteins (Ring1 and Ring1b), B lymphoma Mo-MLV insertion region 1 (Bmi1), and polycomb group ring finger 2. Cbx proteins bind to H3K27me3, and ring proteins and Bmi1 are required for E3 ubiquitination of histone H2A lysine 119 (H2AK119ub), which retains RNA polymerase II in promoter (103). PRC2 is the most prevalent H3K27me3 writer, and its core components are Ezh2, embryonic ectoderm development (Eed), suppressor of zeste 12 homologue (Suz12), and retinoblastoma-associated proteins 48 (RbAp48 or Rbbp4) (104). Ezh2 has the histone methyltransferase activity, which is stimulated by Eed. Suz12 and RbAp48 are required for DNA and histone binding, respectively. Eed is also important in recruiting PRC1 to H3K27me3-enriched loci by introducing H2AK119ub (105). Eed- (106), Suz12- (107), Ezh2- (108), and Ring1b-deleted ESCs (109) increase the expression of developmental genes with global loss of H3K27me3 and display the aberrant differentiation potential but can maintain pluripotency with serial cell passaging. Interestingly, depleting the expression of components of PRC1 and PRC2 (Suz12, Ezh2, Eed, Ring1, and Bmi1) blocks iPSC reprogramming, indicating that H3K27me3 changes and H3K27me3-mediated regulation during iPSC reprogramming are essential steps for the acquisition of pluripotency and differentiation capacity (47). In contrast, different types of Cbx proteins show unique functions in ESCs (110). Although Cbx6 and Cbx7 are highly expressed in ESCs, only Cbx7 physically interacts with Ring1b and directly targets the developmental genes in ESCs. The deletion of Cbx7 does not affect pluripotency but produces smaller EBs with the downregulation of mesoderm- and endoderm-related genes. Thus, Cbx7 is required for the suppression of developmental genes in pluripotent stage and keeps the balance of commitment toward the germ cell layer. Whereas Cbx7 is downregulated during EB differentiation, the expression of Cbx2 and Cbx4 is increased. Cbx8 is not induced without retinoic acid treatment. In EBs, Ring1b strongly interacts with Cbx2 and Cbx 4, but not with the other Cbx proteins. Although Cbx2- and Cbx4-deleted ESCs also do not alter the pluripotency, the deletion of Cbx2, but not Cbx4, exhibits a smaller size of EB formation. Cbx2 regulates trophoblast, mesoderm, and endoderm markers, but Cbx4 only regulates mesodermal genes. Overall, despite the different molecular mechanism of gene regulation, Cbx proteins are essential for proper differentiation.

H3K27me3 also plays an important role in XCI in ESCs and differentiated cells (111). XCI is initiated by the activation of a non-coding RNA, X-inactive specific transcript (XIST), from either of two X chromosomes, and PRC2 is substantially introduced into XIST-expressing X chromosome. PRC1 is not necessary for the initiation and the maintenance of XCI (109). Although two X chromosomes are active in female mESC, hESCs exhibit either one or two active chromosomes (20,112–114). During reprogramming, the X chromosome is reactivated with the loss of H3K27me3 and XIST expression (115). After the completion of reprogramming, hiPSCs gain an inactive X chromosome. However, with serial cell passaging of hiPSCs, the inactive X chromosome starts to be eroded by losing H3K27me3 and XIST (18). The eroded X chromosome behaves as active X chromosome and derepresses X-linked genes. Importantly, the X chromosome erosion cannot be restored by secondary reprogramming and differentiation. In addition, ESCs with two active X chromosomes fail to exit from pluripotent state (116). The higher expression level of PRC2 components in XCI iPSCs than that in X-eroded iPSCs suggests the importance of PRC2-mediated H3K27me3 for the maintenance and the generation of high-quality hiPSCs as well as XCI (114).

Jarid2 (Jmj or Jumonji) recruits and stabilizes PRC2. Jarid2 is highly expressed and targets developmental genes together with PRC2 in ESCs (83). Similar with PRC2-deleted ESCs, *Jarid2* knockout ESCs can maintain self-renewal and pluripotency but display abnormal differentiation (117). In addition, the overexpression of Jarid2 with OSK (OCT4, SOX2, and KLF4) enhances the efficiency of iPSC reprogramming, supporting the functional roles of PRC2 in both differentiation and reprogramming (118).

Wolf–Hirschhorn syndrome candidate 1-like 1 (Whsc1l1) is a methyltransferase specific to H3K4 and H3K27 (119). Although the overexpression of Whsc1l1 with OSKM does not increase the efficiency of iPSC reprogramming, Whsc1l1 induction with OSK can improve the iPSC generation (120). This indicates that Whsc1l1 shares the functions with c-Myc in the reprogramming.

The KDM6 family is a group of proteins demethylating H3K27me3 and consists of three members, Kdm6a (Utx), Kdm6b (Jmjd3), and Kdm6al (Uty). Whereas Kdm6a can promote iPSC generation (121), Kdm6b negatively regulate the reprogramming (122). In contrast, both Kdm6a and Kdm6b are essential for proper differentiation to ectoderm and mesoderm (123–125). Although the functional role of Kdm6al is still unknown, Kdm6al cannot compensate the functional role of Kdm6a in reprogramming (121), suggesting that the erasure of H3K27me3 is required for both reprogramming and differentiation, but each demethylase targets a distinct set of genes.

3.3.4 H3K27ac

Acetylation of H3K27 (H3K27ac) is positively correlated with gene expression and is a counterpart of H3K27me3 (126). H3K27ac is also enriched in promoters and enhancers and switch from H3K27me3 to H3K27ac leads activation of poised regulatory sites (99). In ESCs, developmental enhancers harbor H3K27me3, and another enhancer marker, H3K4me1. During neuroectodermal differentiation, these poised developmental enhancers are activated by converting H3K27me3 into H3K27ac. Furthermore, H3K27ac is enriched in pluripotent factor binding sites and more abundant in ESC than somatic cells (127,128).

CREB-binding protein (CREBBP or CBP) and E1-binding protein p300 are proposed as histone acetyltransferases for H3K27ac (129). Both acetyltransferases antagonize PRC2, which results in the introduction of H3K27ac (130). CREBBP shows less effect on iPSC reprogramming but is required for long-term maintenance of ESC pluripotency (90,131). In contrast, the deletion of p300 does not affect ESC self-renewal but blocks iPSC generation and leads extraembryonic endoderm differentiation (90,132).

Histone deacetylase (HDAC) family includes 11 proteins in mammals (HDAC1-11). The NuRD complex is proposed to promote the deacetylation of H3K27ac by introducing HDACs. The core components of NuRD complex is composed of RbAp48, MBD3, HDAC1/2, Mi-2 (CHD4), p66, and MTA1 (133). Mi-2 contains chromodomain, which is required for ATP-dependent nucleosome remodeling. MTA1 may facilitate binding to DNA, and the role of p66 is still unknown. In ESCs, the knockdown of Mbd3 accelerates self-renewal in the absence of LIF, while Mta1 knockdown enhances differentiation (134). iPSC reprogramming is inhibited by Mbd3 (135) but is facilitated by the co-expression of Mbd3 and Nanog (136). In addition, NuRD-mediated H3K27 deacetylation is promoted by the recruitment of PRC2 (137). Overall, although the gain and the maintenance of H3K27ac are required for self-renewal, reprogramming and differentiation are each regulated by distinct enzymes.

3.3.5 H3K9me3

Trimethylation of H3 lysine 9 (H3K9me3) is another repressive histone mark and is distributed in the heterochromatin (74,75,138). The H3K9me3 domain size in ESCs is significantly larger than that in somatic cells and is the most discordant histone modification between ESCs and iPSCs (100). The differential H3K9me3 region is megabase scale and related to non-CG DNA methylation (6). In addition, H3K9me3 silences the expression of retrotransposons, which is associated with pluripotency and stem cell development (139,140). Since the aberrant H3K9me3 in iPSCs is refractory to OSKM (OCT4, SOX2, KLF4, and C-MYC)-mediated chromatin remodeling, additional factors to displace it is critical in producing truly ESC-like iPSCs (141).

OSKM induction cannot convert all somatic cells into fully reprogrammed iPSCs, and a part of them are stabilized in partially reprogrammed stage (pre-iPSCs). A bottleneck between pre-iPSCs and mature iPSCs is a barrier by H3K9me3, and the overexpression of demethylase for H3K9 is effective in converting pre-iPSCs into fully reprogrammed iPSCs (142). Bone morphogenetic protein (BMP) is present in serum and arrests cells in the prematured stage by increasing H3K9me3 level. The suppressor of variegation 3-9 homologues 1 and 2 (Suv39H1 and Suv39H2) are H3K9me3-specific histone methyltransferases. Although only the deletion of SUV39H1 effectively enhances iPSC reprogramming (47), both can rescue BMP-mediated inhibition of the iPSC maturation.

SET domain bifurcated is another group of H3K9me3-specific methyltransferases and includes two genes (*Setdb1* and *Setdb2* or *Kmt1e* and *Kmt1f*). Although both show less or small effect on iPSC reprogramming (47,120), Setdb1 is important for the repression of retrotransposons and developmental genes for ESC maintenance (143,144). In addition, BMP directly regulates Setdb1 expression, which results in the generation of pre-iPSCs (142). The functional role of Setdb2 in self-renewal and ESC development is largely unknown.

KDM4 family is a group of demethylases, which specifically targets H3K9me3. Among four types KDM4 (Kdm4a-d or Jmjd2a-d), both Kdm4b and Kdm4c are essential for ESC self-renewal but show distinct functions in ESCs (145). Kdm4b functions to construct a regulatory loop with Nanog, and Kdm4c interacts with PRC2 for transcriptional repression. *Kdm4b* and *Kdm4c* knockdown inhibits the conversion of pre-iPSC into mature iPSCs. The depletion of Kdm4c together with Kdm3a and Kdm3b, which is demethylase for H3K9me2/me1, increases the negative effect on the iPSC maturation. In contrast, only Kdm4b, but not Kdm3a and Kdm4c, can promote the iPSC maturation by overexpression experiment, indicating that Kdm4b is the most striking enzyme to overcome the barrier to full conversion.

3.3.6 H3K36me3

The trimethylation of H3 lysine 36 (H3K36me3) is distributed along gene bodies and is proposed to regulate transcriptional elongation (74). Although the total level of H3K36me3 is comparable between somatic cells and ESCs, H3K36me3 modulators are involved in iPSC reprogramming and differentiation.

Achaete–scute family bHLH transcription factor 1 (ASCL1 or ASH1) and SETD2 were identified as the methyltransferase of H3K36me3 (146). Both enzymes show less effect on iPSC reprogramming (90) but are required for neuronal and ectodermal differentiations of ESCs (147,148).

KDM2 family presents demethylase activity for H3K36me3 and includes two members, KDM2A (JHDM1A) and KDM2B (JHDM1B). Kdm2a is dispensable in ESCs (149). The deletion of Kdm2b de-represses lineage-specific genes and promotes early differentiation in ESC (150). Furthermore, Kdm2b is important at the early stage of iPSC reprogramming with cell cycle progression and suppression of cell senescence (151). No66 demethylates H3K4 and H3K36 (152). During ESC differentiation, No66 is recruited by Phf19 to suppress pluripotent genes with the loss of H3K36me3. Phf19 also recruits PRC2 to support transcriptional repression by inducing H3K27me3. Collectively, the gain of H3K36me3 more strongly affects differentiation, but loss of H3K36me3 is required for both reprogramming and differentiation.

3.3.7 H3R2me2

Compared to histone lysine methylation, many arginine methylations are still functionally unclear. The dimethylation of H3 arginine 2 includes two forms, asymmetric (H3R2me2a, type I) and symmetric (H3R2me2s, type II) (153). The protein arginine methyltransferase family includes eight members in mammals (PRMT1-8). It is known that PRMT1, PRMT3, PRMT6, and PRMT8 catalyze H3R2me2a, and PRMT5 and PRMT7 produce H3R2me2s. PRMT6-mediated H3R2me2a inhibits H3K4me3 by blocking the binding of WDR5, which is required for iPSC reprogramming (154). Whereas PRMT1, PRMT4, and PRMT8 regulate cell differentiation (155,156), Prmt5 and PRMT6 are related to the maintenance of self-renewal (157,158). In goat, PRMT5 enhances iPSC generation (159). PRMT2 may regulate metabolism-related genes (160).

The deletion of Jumonji domain containing 6 (JMJD6), which is a potential demethylase specific to H3R2, moderately exits from pluripotency (145). Altogether, not only lysine methylation, but also arginine methylation is important for the maintenance and the acquisition of pluripotency.

3.4 Alternative Splicing

Splicing is one of the mRNA processes done to exclude introns and connect exons from nascent pre-mRNA. Alternative splicing (AS) is a crucial regulation for producing multiple isoforms from a single gene by joining different sets of exons. Innovation of high-throughput sequencing and high-resolution microarray allows the genome-wide screening of unique splicing patterns in hESCs and mESCs (161–164). Recent studies identified several ESC-specific AS events and these transcriptional isoforms promote iPSC generation. Here, we highlight the functions and the importance of recently identified ESC-specific splicing variants (Table 3.1).

3.4.1 OCT4B1

OCT4 is one of the master regulators for pluripotency and have three splicing variants: OCT4A is the longest transcript and produces 45 kDa nucleus-localized transcription factor, OCT4B uses alternative first exon and encodes

TABLE 3.1
Examples of ESC-Specific Alternative Splicing

Gene	Splicing Pattern	Reference
OCT4	Alternative exons 1; intron 2 retention	(165)
NANOG	Alternative exons 1, 2, and 3; exons 1 and 2 skipping	(166)
SALL4	Exon 2 skipping	(167)
TCF7L1	New exon between exons 4 and 5	(161)
FOXP1	Alternative exons 18 (human) and 16 (mouse)	(168)
MBD2	Exons 3–7 skipping and alternative exon 3 inclusion	(169)
PRDM14	Exon 2 skipping	(170)

30 kDa protein, which is localized in the cytoplasm and related to cell stress response, and OCT4B1 uses the same first exon with OCT4B but retains the whole intron 2 sequence (165,171,172). Whereas OCT4A and OCT4B1 are highly expressed in hESCs, OCT4B is expressed in somatic cells as well as hESCs. OCT4B cannot bind OCT4A-dependent promoters and does not disturb OCT4A transcriptional activation (173). A recent study found that OCT4B1 is also expressed in gastric cancer and suppress its apoptosis (174). Although the functional role of OCT4B1 has not been fully investigated, the potential function of OCT4B1 in hESC seems to be related to cell survival.

3.4.2 Nanog a, b, and c

Nanog is one of the core pluripotent factors and controls pathways governing self-renewal, pluripotency, and cell fate determination (175). So far, five splicing variants of Nanog transcripts have been reported in ESCs (166,176). These variants show different patterns in splicing of exons 1–3. Three of them encode the same protein (Nanog a), and two other isoforms produce distinctive proteins (Nanog b and c). Nanog b and c interact with Nanog a, and all three Nanog variants bind with the pluripotent factors, such as Oct4 and Sall4. The ability of Nanog b for self-renewal maintenance in LIF-independent condition is weaker than that of Nanog a and c. Whereas Nanog a and b fully suppress trophectoderm marker expression, Nanog b and c cannot fully repress primitive endodermal differentiation markers. Together, the capacities of the two splicing variants Nanog b and c for self-renewal and pluripotency are lower than the main form Nanog a.

3.4.3 Tcf3(l)

Tcf3 (alias symbol is Tcf7l1) is a transcription repressor that inhibits ESC self-renewal by targeting Oct4 and Nanog. A downstream component of Wnt signaling pathway, β-catenin, blocks Tcf3 repressive activity and maintains ESC self-renewal (177,178). A recent study further showed that Tcf3 promotes iPSC reprogramming at the early stage but acts as a negative regulator at the late stage (179). In mESCs, the expression level of the longer isoform (Tcf3[l]) are higher than that of the shorter isoform (Tcf3[s]) (161). On EB differentiation, Tcf3(l) expression is downregulated. Both isoforms can repress Nanog gene expression and delay ESC differentiation without LIF condition. However, Oct4 expression is only regulated by Tcf3(l), but not by Tcf3(s). Furthermore, whereas Tcf3(s) suppresses most germ layer markers (e.g., Gata4 and Afp) and tissue-specific markers, Tcf3(l) targets cardiac and neuronal markers. Overall, AS in Tcf3 increases the diversity of the downstream targets.

3.4.4 SALL4A and SALL4B

Sall4 is a potential zinc finger transcription factor and modulates pluripotency by targeting *Oct4* enhancers (180). Two spliced isoforms of Sall4 (Sall4a and Sall4b) are expressed in ESCs and are downregulated during differentiation. While Sall4a is composed of five exons, Sall4b skips the exon 2 (167). Both isoforms can form homodimer or heterodimer with each other and individually interact with Nanog. A genome-wide location analysis of Sall4 isoforms revealed that Sall4a and Sall4b bind at different sites. While Sall4b binding and Sall4a/4b common sites show enrichment of H3K4me3 and H3K36me3, Sall4a's unique binding sites display enrichment of H3K27me3. Sall4b-deleted ESCs displays loss of ESC-like and AP-stained colonies and downregulation of pluripotency markers. Sall4a deletion retains ESC phenotypes and shows comparable Nanog and Oct4 expressions with wild type. However, Sall4a-deleted ESCs cannot suppress differentiation markers completely, indicating that Sall4b is required to maintain the pluripotent state and Sall4a helps to suppress differentiation markers.

3.4.5 FOXP1-ES

Forkhead box (FOX) is one of the DNA binding domains and often observed in transcription factor family regulating cell development (181). FOXP1 is one member of the FOXP subfamily and genetically acts as a transcriptional repressor. Gabut et al. performed high-resolution microarray in undifferentiated and differentiated H9 ESCs and found that unique exon 18 (exon 18b) is spliced in undifferentiated H9 hESCs (168). The inclusion of ESC-unique exon is also detectable in mESCs (exon 16b in mouse). The ESC-specific FOXP1 (FOXP1-ES) does not affect its secondary structure and dimerization function but displays distinct DNA-binding specificity from canonical FOXP1. Whereas the canonical FOXP1 binds consensus motif GTAAACAA, FOXP1-ES preferentially binds AATAAACA and CGATACAA. Han et al. recently identified that MBNL1 and MBNL2 suppress the inclusion of exon 18b and 16b of FOXP1 in human and mouse somatic cells (182). FOXP1-ES suppresses early developmental genes, such as WNT1 and GAS1, and promotes self-renewal and pluripotency by upregulating a part of pluripotent genes, OCT4, NANOG, NR5A2, GDF3, and TDGF1. Furthermore, the deletion of Foxp1-ES significantly decreases the efficiency iPSC reprogramming of MEF. Although the overexpression of Foxp1-ES with OSKM does not alter the reprogramming efficiency, the overexpression of canonical Foxp1 completely blocks iPSC generation. Altogether, ES-specific AS of FOXP1 is critical in iPSC formation.

3.4.6 MBD2c

MBD2 is a protein binding to methylated DNA and recruits histone deacetylase complex, such as Mi-2/NuRD, into gene promoters to repress their transcription (183). Canonical MBD2 (MBD2a) has a negative effect on iPSC reprogramming and is highly expressed in somatic cells and partially reprogrammed cells (47,184), suggesting that MBD2 is critical for iPSC maturation step. Lu et al. identified that a shorter isoform of MBD (MBD2c) is specifically expressed in H1 hESCs, while MBD2a is expressed in BJ fibroblast (169). MBD2c shares exons 1 and 2 with MBD2a but includes ESC-specific exon 3. The ESC-specific AS of MBD2 is regulated by a splicing factor SFRS2, which preferentially binds the exon–intron boundary of MBD2c, and is essential for embryonic development (185). The deletion of SFRS2 increases the expression of MBD2a and decreases that of MBD2c. OCT4 directly regulates SFRS2, and the deletion of OCT4 leads to MBD2c expression. In addition, MBD2a, but not MBD2c, is targeted by miR-302 miRNA family, and SFRP2 is targeted by miR-301 miRNA family. Interestingly, miR-301 and miR-302 are induced by OCT4. Whereas the overexpression of MBD2a promotes differentiation by repressing the expression of core pluripotent factors (OCT4, NANOG, and SOX2), the induction of MBD2c facilitates iPSC reprogramming. Furthermore, MBD2c loses the interaction with NuRD complexes. Altogether, they present a positive feedback loop where the core pluripotent factors control miRNAs and the splicing factor SFRS2 in order to induce MBD2 isoforms to support (MBD2a) and oppose (MBD2c) the expression of the core factors.

3.4.7 PRDM14 Short Isoform

PRDM14 is a transcriptional regulator controlling the expression of genes related to germ cell and PSC development (186). The overexpression of PRDM14 suppresses extraembryonic endoderm differentiation (187) and enhances iPSC reprogramming (90). PRDM14 suppresses de novo DNA methyltransferases and fibroblast growth factor receptor for the maintenance of ground state of pluripotency (188). Lu et al. recently found that spliceosome-associated factor (SON) is essential for the maintenance of hESC pluripotency by regulating proper splicing of pluripotent regulators (170). Normally, hESCs display high expression of canonical PRMD14 and low expression of exon 2-skipped PRDM14. The depletion of SON downregulates canonical PRDM14 and increases the expression of exon 2-skipped PRDM14. This shorter isoform fails the enhancement of iPSC reprogramming, suggesting that AS of PRDM14 in hESCs is essential in the maintenance of pluripotency.

3.4.8 Species Specificity of AS

Together, many studies demonstrated that ESC-specific AS is important for ESC maintenance and iPSC reprogramming. Regardless of similar sets of protein-coding genes, AS patterns among vertebrates are significantly different, indicating that AS events are essential process to gain organism-specific gene function (189). Although the ASs of SALL4, FOXP1, and MBD2 are commonly observed between human and mouse, OCT4B1 has not been identified in mouse and a splicing variant of Tcf3(l) is not detectable in hESCs (161,190). Interestingly, mESCs are closer to the ICM state in preimplantation embryo, and hESCs are similar to epiblast stem cells in postimplantation embryo. It is possible that the species-specific AS at early development was acquired during evolutionary processes.

3.5 RNA Modifications

RNA plays an important role in protein synthesis, posttranscriptional regulation, and structural scaffold of organelles. RNA molecules receive more than 100 types of posttranscriptional modifications in all four bases (A, C, G, and U) (Figure 3.4) (191). The RNA modifications are widely found from

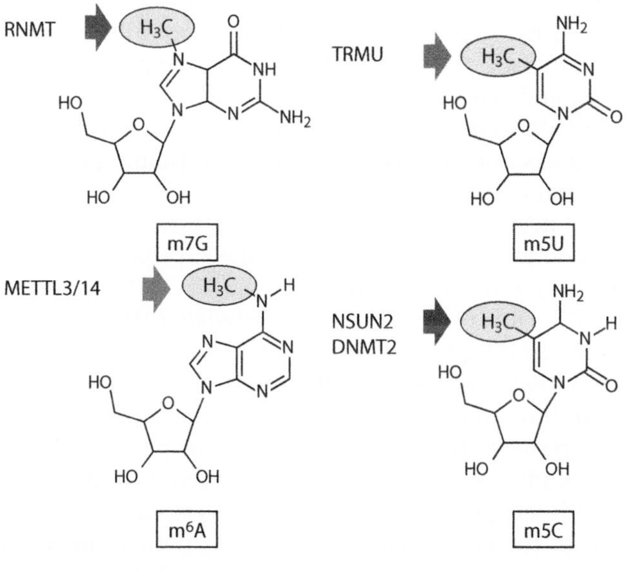

FIGURE 3.4
Examples of RNA modifications in all four bases. Putative methyltransferases to each modification are shown.

prokaryote to eukaryote and in various RNA subtypes (transfer RNA, ribosomal RNA, small nuclear RNA, and messenger RNA). Regardless of extensive studies in DNA and histone proteins in ESCs and iPSCs, the physiological relevance of most chemical modifications in RNA remains largely unknown.

3.5.1 N^6-Methyladenosine

N^6-Methyladenosine (m^6A) is the most abundant modifications in eukaryotic mRNA and occurs with a consensus sequence RGm^{6}ACU (192,193). METTL3, METTL14, WTAP, and KIAA1429 have been reported as components of m^6A RNA methyltransferase complex (194–196). Both METTL3 and METTL14 have SAM-binding domain and form a complex with the same stoichiometric ratio (195). In contrast, WTAP does not include SAM-binding domain but can bind with the METTL3–METTL14 complex to recruit it into nuclear speckles (197). KIAA1429 also does not have the methyltransferase activity, but its orthologous protein, virilizer, can interact with WTAP to regulate sex-dependent splicing in *Drosophila*. METTL3 and METTL14 preferentially bind GGAC motifs, whereas WTAP binds GACU (195).

The m^6A-methylated RNA immunoprecipitation by next-generation sequencing (MeRIP-seq) revealed transcriptome-wide distribution of m^6A sites (198–200). m^6A is found in more than 5000 mRNAs and non-coding RNAs and enriched near stop codon and 3′ untranslated regions and within long internal exons (>400 bp), but not in splice junctions. In addition, the m^6A sites are highly conserved between human and mouse. Around 70% of m^6A-containing genes in hESCs also display m^6A modification in mESCs. In both hESCs and mESCs, many master regulators for pluripotency including NANOG, SOX2, and NR5A2 harbor m^6A modification, but OCT4 does not (199). Developmental gene transcripts are also modified with m^6A, but not in housekeeping genes, such as genes encoding ribosomal proteins and mitochondrial membrane (194,196).

The half-life of m^6A-containing RNA is shorter than that of non-modified RNA. A YTH domain family protein, YTHDF2, preferentially binds m^6A-containing mRNA, and YTHDF2-mRNA complex is transported into processing body (201). HuR protein binds demethylated RNAs to block miRNA targeting (194), indicating that m^6A promotes RNA destabilization by using multiple degradation processes. Furthermore, the silencing of m^6A methyltransferases in human cancer cell line upregulates genes associated with splicing and introduces AS in genes involved in p53 signaling pathways. Overall, recent studies suggest that m^6A is involved in multiple molecular pathways to modulate posttranscriptional RNA regulation.

The deletion of the components of the m^6A methyltransferase complex significantly reduces m^6A level in many types of cell lines (194–196,199,200). However, the deletion of Metttl3 and Mettl14 shows controversial phenotypes in mESCs. Batista et al. depleted *Mettl3* gene in mESCs by CRISPR/Cas9 (clustered regularly interspaced short palindromic repeats/CRISPR-associated protein 9)-mediated knockout and shRNA and showed larger compact colonies and higher proliferation in Mettl3-deleted ESCs than wild type (199). In addition, they presented that *Mettl3* KO cells failed *in vitro* lineage differentiation and teratoma formation with a higher expression level of pluripotent genes and a lower level of lineage-specific genes. In contrast, Wang et al. reported that mESCs depleted of Mettl3 and Mettl14 by shRNA in display flatter morphology, lower self-renewal, decreasing level of pluripotent genes, and increasing of developmental regulators (194). Chen et al. also demonstrated that the overexpression of Mettl3 with OSKM significantly improves iPSC reprogramming efficiency (202). Although it is still unclear whether m^6A is important for the maintenance of pluripotency or the entrance into lineage development, m^6A is also a significant consideration in stem cell studies.

3.5.2 RNA Editing

RNA editing is a form of posttranscriptional processes to replace discrete nucleotide. The most frequent RNA editing in mammal is adenosine-to-inosine conversion (A-to-I editing) and occurs in double-stranded loop in RNA. The level of A-to-I editing is higher in hESCs but is decreased with neuronal differentiation (203). Adenosine deaminase acting on RNA protein 1 (ADAR1) is an enzyme to catalyze A-to-I editing and shows the highest gene expression in hESCs within *ADAR* gene family. The depletion of ADAR1 in hESCs upregulates developmental genes, suggesting the importance of A-to-I editing in maintaining pluripotency.

However, a recent study showed that the overexpression of ADAR1 significantly reduced the reprogramming efficiency and the depletion of ADAR1 slightly increases iPSC colonies (204). It has been demonstrated that ADAR family proteins suppress cell proliferation (205) and high cell proliferation is essential for iPSC reprogramming (206). Thus, it is hypothesized that the inhibitory function of cell cycles by ADAR1 blocks efficient iPSC generation. Although the deletion of ADAR1 does not largely affect iPSC reprogramming process, iPSC clones derived from ADAR1 knockdown cells cannot maintain their ESC-like morphology and failed proper EB differentiation. Furthermore, ADAR1-deleted iPSCs undergo oncogenic transformation. Taken together, A-to-I editing is essential for the long-term maintenance of safer ESCs and iPSCs.

Although A-to-I editing is a common phenomenon in eukaryotes, the amount of A-to-I editing events in human is larger than that in other species (207). Transcriptome studies revealed that A-to-I editing is dominant in the Alu element (139,208,209), which is a primate-specific retrotransposon. Therefore, it is supposed that A-to-I editing-mediated regulations may be newly acquired during evolution and may be involved in gaining functional diversity in hESCs and hiPSCs.

3.6 Considerations and Future Directions

Here, we have summarized recent findings that showed the functions of a diverse range of epigenetic modifiers in maintaining or inducing pluripotency. For most of genetic and epigenetic modifications, either gain or loss has been demonstrated being associated with iPSC reprogramming, ESC maintenance, and differentiation. Curiously, each member of the same protein family plays a distinct role, being induced at different stages and regulating distinct sets of genes. Thus, one of the most important considerations in reprogramming is to suppress or to activate subsets of enzymes at different stages to overcome the barriers in the course to iPSCs. For the purpose of ESCs/iPSCs in clinical utility, methods leaving no genome modification and thus minimizing tumorigenic potential of iPSCs will be critical when modulating the specific epigenetic factors. In the past few years, small molecules have been sought by several groups to improve iPSC generation. Examples include ascorbic acid-inducing Tet-mediated DNA demethylation and Kdm2-mediated demethylation of H3K36 (151,210). With more in-depth understanding of the function of epigenetic factors in PSCs, their derivatives and cells undergoing reprogramming together with searching for improving a quality of iPSCs will be essential in future utility of iPSC technology for clinical setting.

Acknowledgments

We thank all the Park laboratory members for helpful comments and discussion. In-Hyun Park was partly supported by the NIH (GM0099130-01, GM111667-01), the Connecticut Stem Cell Research Fund (12-SCB-YALE-11, 13-SCB-YALE-06), and the Korea Research Institute of Bioscience and Biotechnology/Korea Research Council of Fundamental Science and Technology (NAP-09-3) and by the Clinical and Translational Science Awards Grant (UL1 RR025750) from the National Center for Advancing Translational Science, a component of the NIH, and NIH Roadmap for Medical Research. Its contents are solely the responsibility of the authors and do not necessarily represent the official view of the NIH.

References

1. Takahashi K, Yamanaka S: Induction of pluripotent stem cells from mouse embryonic and adult fibroblast cultures by defined factors. *Cell* 2006, 126:663–676.
2. Park IH, Lerou PH, Zhao R, Huo H, Daley GQ: Generation of human-induced pluripotent stem cells. *Nat Protoc* 2008, 3:1180–1186.

3. Ehrlich M: Expression of various genes is controlled by DNA methylation during mammalian development. *J Cell Biochem* 2003, 88:899–10.
4. Colot V, Rossignol JL: Eukaryotic DNA methylation as an evolutionary device. *Bioessays* 1999, 21:402–411.
5. Tang Y, Gao XD, Wang Y, Yuan BF, Feng YQ: Widespread existence of cytosine methylation in yeast DNA measured by gas chromatography/mass spectrometry. *Anal Chem* 2012, 84:7249–7255.
6. Lister R, Pelizzola M, Kida YS, Hawkins RD, Nery JR, Hon G, Antosiewicz-Bourget J et al.: Hotspots of aberrant epigenomic reprogramming in human induced pluripotent stem cells. *Nature* 2011, 471:68–73.
7. Smith ZD, Chan MM, Mikkelsen TS, Gu H, Gnirke A, Regev A, Meissner A: A unique regulatory phase of DNA methylation in the early mammalian embryo. *Nature* 2012, 484:339–344.
8. Smith ZD, Chan MM, Humm KC, Karnik R, Mekhoubad S, Regev A, Eggan K, Meissner A: DNA methylation dynamics of the human preimplantation embryo. *Nature* 2014, 511:611–615.
9. Guo H, Zhu P, Yan L, Li R, Hu B, Lian Y, Yan J et al.: The DNA methylation landscape of human early embryos. *Nature* 2014, 511:606–610.
10. Borgel J, Guibert S, Li Y, Chiba H, Schübeler D, Sasaki H, Forné T, Weber M: Targets and dynamics of promoter DNA methylation during early mouse development. *Nat Genet* 2010, 42:1093–1100.
11. Nichols J, Smith A: Naive and primed pluripotent states. *Cell Stem Cell* 2009, 4:487–492.
12. Takashima Y, Guo G, Loos R, Nichols J, Ficz G, Krueger F, Oxley D et al.: Resetting Transcription Factor Control Circuitry toward Ground-State Pluripotency in Human. *Cell* 2014, 158:1254–1269.
13. Theunissen TW, Powell BE, Wang H, Mitalipova M, Faddah DA, Reddy J, Fan ZP et al.: Systematic Identification of Culture Conditions for Induction and Maintenance of Naive Human Pluripotency. *Cell Stem Cell* 2014.
14. Park IH, Zhao R, West JA, Yabuuchi A, Huo H, Ince TA, Lerou PH, Lensch MW, Daley GQ: Reprogramming of human somatic cells to pluripotency with defined factors. *Nature* 2008, 451:141–146.
15. Mikkelsen TS, Hanna J, Zhang X, Ku M, Wernig M, Schorderet P, Bernstein BE, Jaenisch R, Lander ES, Meissner A: Dissecting direct reprogramming through integrative genomic analysis. *Nature* 2008, 454:49–55.
16. Polo JM, Anderssen E, Walsh RM, Schwarz BA, Nefzger CM, Lim SM, Borkent M, Apostolou E, Alaei S, Cloutier J et al.: A molecular roadmap of reprogramming somatic cells into iPS cells. *Cell* 2012, 151:1617–1632.
17. Nazor KL, Altun G, Lynch C, Tran H, Harness JV, Slavin I, Garitaonandia I et al.: Recurrent variations in DNA methylation in human pluripotent stem cells and their differentiated derivatives. *Cell Stem Cell* 2012, 10:620–634.
18. Mekhoubad S, Bock C, de Boer AS, Kiskinis E, Meissner A, Eggan K: Erosion of dosage compensation impacts human iPSC disease modeling. *Cell Stem Cell* 2012, 10:595–609.
19. Doi A, Park IH, Wen B, Murakami P, Aryee MJ, Irizarry R, Herb B et al.: Differential methylation of tissue- and cancer-specific CpG island shores distinguishes human induced pluripotent stem cells, embryonic stem cells and fibroblasts. *Nat Genet* 2009, 41:1350–1353.

20. Tanaka Y, Kim KY, Zhong M, Pan X, Weissman SM, Park IH: Transcriptional regulation in pluripotent stem cells by methyl CpG-binding protein 2 (MeCP2). *Hum Mol Genet* 2013.
21. Ruiz S, Diep D, Gore A, Panopoulos AD, Montserrat N, Plongthongkum N, Kumar S et al.: Identification of a specific reprogramming-associated epigenetic signature in human induced pluripotent stem cells. *Proc Natl Acad Sci USA* 2012, 109:16196–16201.
22. Okano M, Xie S, Li E: Cloning and characterization of a family of novel mammalian DNA (cytosine-5) methyltransferases. *Nat Genet* 1998, 19:219–220.
23. Lister R, Pelizzola M, Dowen RH, Hawkins RD, Hon G, Tonti-Filippini J, Nery JR et al.: Human DNA methylomes at base resolution show widespread epigenomic differences. *Nature* 2009, 462:315–322.
24. Watanabe D, Suetake I, Tada T, Tajima S: Stage- and cell-specific expression of Dnmt3a and Dnmt3b during embryogenesis. *Mech Dev* 2002, 118:187–190.
25. Suetake I, Miyazaki J, Murakami C, Takeshima H, Tajima S: Distinct enzymatic properties of recombinant mouse DNA methyltransferases Dnmt3a and Dnmt3b. *J Biochem* 2003, 133:737–744.
26. Pawlak M, Jaenisch R: De novo DNA methylation by Dnmt3a and Dnmt3b is dispensable for nuclear reprogramming of somatic cells to a pluripotent state. *Genes Dev* 2011, 25:1035–1040.
27. Tsumura A, Hayakawa T, Kumaki Y, Takebayashi S, Sakaue M, Matsuoka C, Shimotohno K et al.: Maintenance of self-renewal ability of mouse embryonic stem cells in the absence of DNA methyltransferases Dnmt1, Dnmt3a and Dnmt3b. *Genes Cells* 2006, 11:805–814.
28. Guo X, Liu Q, Wang G, Zhu S, Gao L, Hong W, Chen Y et al.: microRNA-29b is a novel mediator of Sox2 function in the regulation of somatic cell reprogramming. *Cell Res* 2013, 23:142–156.
29. Okano M, Bell DW, Haber DA, Li E: DNA methyltransferases Dnmt3a and Dnmt3b are essential for de novo methylation and mammalian development. *Cell* 1999, 99:247–257.
30. Sharif J, Muto M, Takebayashi S, Suetake I, Iwamatsu A, Endo TA, Shinga J et al.: The SRA protein Np95 mediates epigenetic inheritance by recruiting Dnmt1 to methylated DNA. *Nature* 2007, 450:908–912.
31. Kim M, Trinh BN, Long TI, Oghamian S, Laird PW: Dnmt1 deficiency leads to enhanced microsatellite instability in mouse embryonic stem cells. *Nucleic Acids Res* 2004, 32:5742–5749.
32. Jackson M, Krassowska A, Gilbert N, Chevassut T, Forrester L, Ansell J, Ramsahoye B: Severe global DNA hypomethylation blocks differentiation and induces histone hyperacetylation in embryonic stem cells. *Mol Cell Biol* 2004, 24:8862–8871.
33. Jackson-Grusby L, Beard C, Possemato R, Tudor M, Fambrough D, Csankovszki G, Dausman J et al.: Loss of genomic methylation causes p53-dependent apoptosis and epigenetic deregulation. *Nat Genet* 2001, 27:31–39.
34. Georgia S, Kanji M, Bhushan A: DNMT1 represses p53 to maintain progenitor cell survival during pancreatic organogenesis. *Genes Dev* 2013, 27:372–377.
35. Chen Q, Qiu C, Huang Y, Jiang L, Huang Q, Guo L, Liu T: Human amniotic epithelial cell feeder layers maintain iPS cell pluripotency by inhibiting endogenous DNA methyltransferase 1. *Exp Ther Med* 2013, 6:1145–1154.

36. Tsai CC, Su PF, Huang YF, Yew TL, Hung SC: Oct4 and Nanog directly regulate Dnmt1 to maintain self-renewal and undifferentiated state in mesenchymal stem cells. *Mol Cell* 2012, 47:169–182.
37. Bourc'his D, Xu GL, Lin CS, Bollman B, Bestor TH: Dnmt3L and the establishment of maternal genomic imprints. *Science* 2001, 294:2536–2539.
38. Hata K, Okano M, Lei H, Li E: Dnmt3L cooperates with the Dnmt3 family of de novo DNA methyltransferases to establish maternal imprints in mice. *Development* 2002, 129:1983–1993.
39. Ooi SK, Wolf D, Hartung O, Agarwal S, Daley GQ, Goff SP, Bestor TH: Dynamic instability of genomic methylation patterns in pluripotent stem cells. *Epigenetics Chromatin* 2010, 3:17.
40. Nimura K, Ishida C, Koriyama H, Hata K, Yamanaka S, Li E, Ura K, Kaneda Y: Dnmt3a2 targets endogenous Dnmt3L to ES cell chromatin and induces regional DNA methylation. *Genes Cells* 2006, 11:1225–1237.
41. Neri F, Krepelova A, Incarnato D, Maldotti M, Parlato C, Galvagni F, Matarese F, Stunnenberg HG, Oliviero S: Dnmt3L antagonizes DNA methylation at bivalent promoters and favors DNA methylation at gene bodies in ESCs. *Cell* 2013, 155:121–134.
42. Guenatri M, Duffié R, Iranzo J, Fauque P, Bourc'his D: Plasticity in Dnmt3L-dependent and -independent modes of de novo methylation in the developing mouse embryo. *Development* 2013, 140:562–572.
43. Hu YG, Hirasawa R, Hu JL, Hata K, Li CL, Jin Y, Chen T et al.: Regulation of DNA methylation activity through Dnmt3L promoter methylation by Dnmt3 enzymes in embryonic development. *Hum Mol Genet* 2008, 17:2654–2664.
44. Gokul G, Ramakrishna G, Khosla S: Reprogramming of HeLa cells upon DNMT3L overexpression mimics carcinogenesis. *Epigenetics* 2009, 4:322–329.
45. Lyst MJ, Ekiert R, Ebert DH, Merusi C, Nowak J, Selfridge J, Guy J et al.: Rett syndrome mutations abolish the interaction of MeCP2 with the NCoR/SMRT co-repressor. *Nat Neurosci* 2013, 16:898–902.
46. Marchetto MC, Carromeu C, Acab A, Yu D, Yeo GW, Mu Y, Chen G, Gage FH, Muotri AR: A model for neural development and treatment of Rett syndrome using human induced pluripotent stem cells. *Cell* 2010, 143:527–539.
47. Onder TT, Kara N, Cherry A, Sinha AU, Zhu N, Bernt KM, Cahan P et al.: Chromatin-modifying enzymes as modulators of reprogramming. *Nature* 2012, 483:598–602.
48. Tahiliani M, Koh KP, Shen Y, Pastor WA, Bandukwala H, Brudno Y, Agarwal S, Iyer LM, Liu DR, Aravind L, Rao A: Conversion of 5–methylcytosine to 5–hydroxymethylcytosine in mammalian DNA by MLL partner TET1. *Science* 2009, 324:930–935.
49. Kriaucionis S, Heintz N: The nuclear DNA base 5–hydroxymethylcytosine is present in Purkinje neurons and the brain. *Science* 2009, 324:929–930.
50. Szulwach KE, Li X, Li Y, Song CX, Wu H, Dai Q, Irier H, Upadhyay AK et al.: 5–hmC-mediated epigenetic dynamics during postnatal neurodevelopment and aging. *Nat Neurosci* 2011, 14:1607–1616.
51. Ito S, Shen L, Dai Q, Wu SC, Collins LB, Swenberg JA, He C, Zhang Y: Tet proteins can convert 5–methylcytosine to 5–formylcytosine and 5–carboxylcytosine. *Science* 2011, 333:1300–1303.

52. Szulwach KE, Li X, Li Y, Song CX, Han JW, Kim S, Namburi S et al.: Integrating 5–hydroxymethylcytosine into the epigenomic landscape of human embryonic stem cells. *PLoS Genet* 2011, 7:e1002154.
53. Stroud H, Feng S, Morey Kinney S, Pradhan S, Jacobsen SE: 5–Hydroxymethyl cytosine is associated with enhancers and gene bodies in human embryonic stem cells. *Genome Biol* 2011, 12:R54.
54. Wu H, D'Alessio AC, Ito S, Wang Z, Cui K, Zhao K, Sun YE, Zhang Y: Genome-wide analysis of 5–hydroxymethylcytosine distribution reveals its dual function in transcriptional regulation in mouse embryonic stem cells. *Genes Dev* 2011, 25:679–684.
55. Nestor CE, Ottaviano R, Reddington J, Sproul D, Reinhardt D, Dunican D, Katz E, Dixon JM, Harrison DJ, Meehan RR: Tissue type is a major modifier of the 5–hydroxymethylcytosine content of human genes. *Genome Res* 2012, 22:467–477.
56. Wang T, Wu H, Li Y, Szulwach KE, Lin L, Li X, Chen IP et al.: Subtelomeric hotspots of aberrant 5–hydroxymethylcytosine-mediated epigenetic modifications during reprogramming to pluripotency. *Nat Cell Biol* 2013, 15:700–711.
57. Langlois T, da Costa Reis Monte-Mor B, Lenglet G, Droin N, Marty C, Le Couédic JP, Almire C et al.: TET2 deficiency inhibits mesoderm and hematopoietic differentiation in human embryonic stem cells. *Stem Cells* 2014, 32:2084–2097.
58. Gu TP, Guo F, Yang H, Wu HP, Xu GF, Liu W, Xie ZG, Shi L, He X, Jin SG et al.: The role of Tet3 DNA dioxygenase in epigenetic reprogramming by oocytes. *Nature* 2011, 477:606–610.
59. Wossidlo M, Nakamura T, Lepikhov K, Marques CJ, Zakhartchenko V, Boiani M, Arand J, Nakano T, Reik W, Walter J: 5–Hydroxymethylcytosine in the mammalian zygote is linked with epigenetic reprogramming. *Nat Commun* 2011, 2:241.
60. Costa Y, Ding J, Theunissen TW, Faiola F, Hore TA, Shliaha PV, Fidalgo M et al.: NANOG-dependent function of TET1 and TET2 in establishment of pluripotency. *Nature* 2013, 495:370–374.
61. Hu X, Zhang L, Mao SQ, Li Z, Chen J, Zhang RR, Wu HP et al.: Tet and TDG mediate DNA demethylation essential for mesenchymal-to-epithelial transition in somatic cell reprogramming. *Cell Stem Cell* 2014, 14:512–522.
62. Koh KP, Yabuuchi A, Rao S, Huang Y, Cunniff K, Nardone J, Laiho A et al.: Tet1 and Tet2 regulate 5–hydroxymethylcytosine production and cell lineage specification in mouse embryonic stem cells. *Cell Stem Cell* 2011, 8:200–213.
63. Ito S, D'Alessio AC, Taranova OV, Hong K, Sowers LC, Zhang Y: Role of Tet proteins in 5mC to 5hmC conversion, ES-cell self-renewal and inner cell mass specification. *Nature* 2010, 466:1129–1133.
64. Huang Y, Chavez L, Chang X, Wang X, Pastor WA, Kang J, Zepeda-Martínez JA, Pape UJ, Jacobsen SE, Peters B, Rao A: Distinct roles of the methylcytosine oxidases Tet1 and Tet2 in mouse embryonic stem cells. *Proc Natl Acad Sci USA* 2014, 111:1361–1366.
65. Doege CA, Inoue K, Yamashita T, Rhee DB, Travis S, Fujita R, Guarnieri P et al.: Early-stage epigenetic modification during somatic cell reprogramming by Parp1 and Tet2. *Nature* 2012, 488:652–655.
66. Spruijt CG, Gnerlich F, Smits AH, Pfaffeneder T, Jansen PW, Bauer C, Münzel M et al.: Dynamic readers for 5–(hydroxy)methylcytosine and its oxidized derivatives. *Cell* 2013, 152:1146–1159.

67. Mellén M, Ayata P, Dewell S, Kriaucionis S, Heintz N: MeCP2 binds to 5hmC enriched within active genes and accessible chromatin in the nervous system. *Cell* 2012, 151:1417–1430.
68. Muotri AR, Marchetto MC, Coufal NG, Oefner R, Yeo G, Nakashima K, Gage FH: L1 retrotransposition in neurons is modulated by MeCP2. *Nature* 2010, 468:443–446.
69. Inoue A, Shen L, Dai Q, He C, Zhang Y: Generation and replication-dependent dilution of 5fC and 5caC during mouse preimplantation development. *Cell Res* 2011, 21:1670–1676.
70. Wheldon LM, Abakir A, Ferjentsik Z, Dudnakova T, Strohbuecker S, Christie D, Dai N et al.: Transient accumulation of 5–carboxylcytosine indicates involvement of active demethylation in lineage specification of neural stem cells. *Cell Rep* 2014, 7:1353–1361.
71. He YF, Li BZ, Li Z, Liu P, Wang Y, Tang Q, Ding J et al.: Tet-mediated formation of 5-carboxylcytosine and its excision by TDG in mammalian DNA. *Science* 2011, 333:1303–1307.
72. Shen L, Wu H, Diep D, Yamaguchi S, D'Alessio AC, Fung HL, Zhang K, Zhang Y: Genome-wide analysis reveals TET- and TDG-dependent 5-methylcytosine oxidation dynamics. *Cell* 2013, 153:692–706.
73. Song CX, Szulwach KE, Dai Q, Fu Y, Mao SQ, Lin L, Street C et al.: Genome-wide profiling of 5–formylcytosine reveals its roles in epigenetic priming. *Cell* 2013, 153:678–691.
74. Barski A, Cuddapah S, Cui K, Roh TY, Schones DE, Wang Z, Wei G, Chepelev I, Zhao K: High-resolution profiling of histone methylations in the human genome. *Cell* 2007, 129:823–837.
75. Mikkelsen TS, Ku M, Jaffe DB, Issac B, Lieberman E, Giannoukos G, Alvarez P et al.: Genome-wide maps of chromatin state in pluripotent and lineage-committed cells. *Nature* 2007, 448:553–560.
76. Kunath T, Saba-El-Leil MK, Almousailleakh M, Wray J, Meloche S, Smith A: FGF stimulation of the Erk1/2 signalling cascade triggers transition of pluripotent embryonic stem cells from self-renewal to lineage commitment. *Development* 2007, 134:2895–2902.
77. Yu J, Vodyanik MA, Smuga-Otto K, Antosiewicz-Bourget J, Frane JL, Tian S, Nie J et al.: Induced pluripotent stem cell lines derived from human somatic cells. *Science* 2007, 318:1917–1920.
78. Jiang H, Shukla A, Wang X, Chen WY, Bernstein BE, Roeder RG: Role for Dpy-30 in ES cell-fate specification by regulation of H3K4 methylation within bivalent domains. *Cell* 2011, 144:513–525.
79. Pullirsch D, Härtel R, Kishimoto H, Leeb M, Steiner G, Wutz A: The Trithorax group protein Ash2l and Saf-A are recruited to the inactive X chromosome at the onset of stable X inactivation. *Development* 2010, 137:935–943.
80. Ang YS, Tsai SY, Lee DF, Monk J, Su J, Ratnakumar K, Ding J et al.: Wdr5 mediates self-renewal and reprogramming via the embryonic stem cell core transcriptional network. *Cell* 2011, 145:183–197.
81. Bledau AS, Schmidt K, Neumann K, Hill U, Ciotta G, Gupta A, Torres DC, Fu J, Kranz A, Stewart AF, Anastassiadis K: The H3K4 methyltransferase Setd1a is first required at the epiblast stage, whereas Setd1b becomes essential after gastrulation. *Development* 2014, 141:1022–1035.

82. Pasini D, Hansen KH, Christensen J, Agger K, Cloos PA, Helin K: Coordinated regulation of transcriptional repression by the RBP2 H3K4 demethylase and Polycomb-Repressive Complex 2. *Genes Dev* 2008, 22:1345–1355.
83. Peng JC, Valouev A, Swigut T, Zhang J, Zhao Y, Sidow A, Wysocka J: Jarid2/Jumonji coordinates control of PRC2 enzymatic activity and target gene occupancy in pluripotent cells. *Cell* 2009, 139:1290–1302.
84. Christensen J, Agger K, Cloos PA, Pasini D, Rose S, Sennels L, Rappsilber J, Hansen KH, Salcini AE, Helin K: RBP2 belongs to a family of demethylases, specific for tri-and dimethylated lysine 4 on histone 3. *Cell* 2007, 128: 1063–1076.
85. Xie L, Pelz C, Wang W, Bashar A, Varlamova O, Shadle S, Impey S: KDM5B regulates embryonic stem cell self-renewal and represses cryptic intragenic transcription. *EMBO J* 2011, 30:1473–1484.
86. Dey BK, Stalker L, Schnerch A, Bhatia M, Taylor-Papidimitriou J, Wynder C: The histone demethylase KDM5b/JARID1b plays a role in cell fate decisions by blocking terminal differentiation. *Mol Cell Biol* 2008, 28:5312–5327.
87. Schmitz SU, Albert M, Malatesta M, Morey L, Johansen JV, Bak M, Tommerup N, Abarrategui I, Helin K: Jarid1b targets genes regulating development and is involved in neural differentiation. *EMBO J* 2011, 30:4586–4600.
88. Kidder BL, Hu G, Zhao K: KDM5B focuses H3K4 methylation near promoters and enhancers during embryonic stem cell self-renewal and differentiation. *Genome Biol* 2014, 15:R32.
89. Outchkourov NS, Muiño JM, Kaufmann K, van Ijcken WF, Groot Koerkamp MJ, van Leenen D, de Graaf P, Holstege FC, Grosveld FG, Timmers HT: Balancing of histone H3K4 methylation states by the Kdm5c/SMCX histone demethylase modulates promoter and enhancer function. *Cell Rep* 2013, 3:1071–1079.
90. Chia NY, Chan YS, Feng B, Lu X, Orlov YL, Moreau D, Kumar P et al.: A genome-wide RNAi screen reveals determinants of human embryonic stem cell identity. *Nature* 2010, 468:316–320.
91. Zdzieblo D, Li X, Lin Q, Zenke M, Illich DJ, Becker M, Müller AM: Pcgf6, a polycomb group protein, regulates mesodermal lineage differentiation in murine ESCs and functions in iPS reprogramming. *Stem Cells* 2014, 32:3112–3125.
92. Lee MG, Norman J, Shilatifard A, Shiekhattar R: Physical and functional association of a trimethyl H3K4 demethylase and Ring6a/MBLR, a polycomb-like protein. *Cell* 2007, 128:877–887.
93. Nishioka K, Chuikov S, Sarma K, Erdjument-Bromage H, Allis CD, Tempst P, Reinberg D: Set9, a novel histone H3 methyltransferase that facilitates transcription by precluding histone tail modifications required for heterochromatin formation. *Genes Dev* 2002, 16:479–489.
94. Goyal A, Chavez SL, Reijo Pera RA: Generation of human induced pluripotent stem cells using epigenetic regulators reveals a germ cell-like identity in partially reprogrammed colonies. *PLoS One* 2013, 8:e82838.
95. Adamo A, Sesé B, Boue S, Castaño J, Paramonov I, Barrero MJ, Izpisua Belmonte JC: LSD1 regulates the balance between self-renewal and differentiation in human embryonic stem cells. *Nat Cell Biol* 2011, 13:652–659.
96. Wang J, Hevi S, Kurash JK, Lei H, Gay F, Bajko J, Su H et al.: The lysine demethylase LSD1 (KDM1) is required for maintenance of global DNA methylation. *Nat Genet* 2009, 41:125–129.

97. Lee MG, Wynder C, Bochar DA, Hakimi MA, Cooch N, Shiekhattar R: Functional interplay between histone demethylase and deacetylase enzymes. *Mol Cell Biol* 2006, 26:6395–6402.
98. Whyte WA, Bilodeau S, Orlando DA, Hoke HA, Frampton GM, Foster CT, Cowley SM, Young RA: Enhancer decommissioning by LSD1 during embryonic stem cell differentiation. *Nature* 2012, 482:221–225.
99. Rada-Iglesias A, Bajpai R, Swigut T, Brugmann SA, Flynn RA, Wysocka J: A unique chromatin signature uncovers early developmental enhancers in humans. *Nature* 2011, 470:279–283.
100. Hawkins RD, Hon GC, Lee LK, Ngo Q, Lister R, Pelizzola M, Edsall LE et al.: Distinct epigenomic landscapes of pluripotent and lineage-committed human cells. *Cell Stem Cell* 2010, 6:479–491.
101. Bernstein BE, Mikkelsen TS, Xie X, Kamal M, Huebert DJ, Cuff J, Fry B et al.: A bivalent chromatin structure marks key developmental genes in embryonic stem cells. *Cell* 2006, 125:315–326.
102. Simon JA, Kingston RE: Mechanisms of polycomb gene silencing: Knowns and unknowns. *Nat Rev Mol Cell Biol* 2009, 10:697–708.
103. Stock JK, Giadrossi S, Casanova M, Brookes E, Vidal M, Koseki H, Brockdorff N, Fisher AG, Pombo A: Ring1-mediated ubiquitination of H2A restrains poised RNA polymerase II at bivalent genes in mouse ES cells. *Nat Cell Biol* 2007, 9:1428–1435.
104. Cao R, Zhang Y: SUZ12 is required for both the histone methyltransferase activity and the silencing function of the EED-EZH2 complex. *Mol Cell* 2004, 15:57–67.
105. Cao Q, Wang X, Zhao M, Yang R, Malik R, Qiao Y, Poliakov A et al.: The central role of EED in the orchestration of polycomb group complexes. *Nat Commun* 2014, 5:3127.
106. Chamberlain SJ, Yee D, Magnuson T: Polycomb repressive complex 2 is dispensable for maintenance of embryonic stem cell pluripotency. *Stem Cells* 2008, 26:1496–1505.
107. Pasini D, Bracken AP, Hansen JB, Capillo M, Helin K: The polycomb group protein Suz12 is required for embryonic stem cell differentiation. *Mol Cell Biol* 2007, 27:3769–3779.
108. Shen X, Liu Y, Hsu YJ, Fujiwara Y, Kim J, Mao X, Yuan GC, Orkin SH: EZH1 mediates methylation on histone H3 lysine 27 and complements EZH2 in maintaining stem cell identity and executing pluripotency. *Mol Cell* 2008, 32:491–502.
109. Leeb M, Wutz A: Ring1B is crucial for the regulation of developmental control genes and PRC1 proteins but not X inactivation in embryonic cells. *J Cell Biol* 2007, 178:219–229.
110. Morey L, Pascual G, Cozzuto L, Roma G, Wutz A, Benitah SA, Di Croce L: Nonoverlapping functions of the Polycomb group Cbx family of proteins in embryonic stem cells. *Cell Stem Cell* 2012, 10:47–62.
111. Plath K, Fang J, Mlynarczyk-Evans SK, Cao R, Worringer KA, Wang H, de la Cruz CC, Otte AP, Panning B, Zhang Y: Role of histone H3 lysine 27 methylation in X inactivation. *Science* 2003, 300:131–135.
112. Tomoda K, Takahashi K, Leung K, Okada A, Narita M, Yamada NA, Eilertson KE et al.: Derivation conditions impact X-inactivation status in female human induced pluripotent stem cells. *Cell Stem Cell* 2012, 11:91–99.

113. Lengner CJ, Gimelbrant AA, Erwin JA, Cheng AW, Guenther MG, Welstead GG, Alagappan R et al.: Derivation of pre-X inactivation human embryonic stem cells under physiological oxygen concentrations. *Cell* 2010, 141:872–883.
114. Kim KY, Hysolli E, Park IH: Neuronal maturation defect in induced pluripotent stem cells from patients with Rett syndrome. *Proc Natl Acad Sci USA* 2011, 108:14169–14174.
115. Kim KY, Hysolli E, Tanaka Y, Wang B, Jung YW, Pan X, Weissman SM, Park IH: X chromosome of female cells shows dynamic changes in status during human somatic cell reprogramming. *Stem Cell Reports* 2014, 2:896–909.
116. Schulz EG, Meisig J, Nakamura T, Okamoto I, Sieber A, Picard C, Borensztein M, Saitou M, Blüthgen N, Heard E: The two active X chromosomes in female ESCs block exit from the pluripotent state by modulating the ESC signaling network. *Cell Stem Cell* 2014, 14:203–216.
117. Shen X, Kim W, Fujiwara Y, Simon MD, Liu Y, Mysliwiec MR, Yuan GC, Lee Y, Orkin SH: Jumonji modulates polycomb activity and self-renewal versus differentiation of stem cells. *Cell* 2009, 139:1303–1314.
118. Zhang Z, Jones A, Sun CW, Li C, Chang CW, Joo HY, Dai Q et al.: PRC2 complexes with JARID2, MTF2, and esPRC2p48 in ES cells to modulate ES cell pluripotency and somatic cell reprogramming. *Stem Cells* 2011, 29:229–240.
119. Kim SM, Kee HJ, Eom GH, Choe NW, Kim JY, Kim YS, Kim SK, Kook H, Seo SB: Characterization of a novel WHSC1-associated SET domain protein with H3K4 and H3K27 methyltransferase activity. *Biochem Biophys Res Commun* 2006, 345:318–323.
120. Nagamatsu G, Saito S, Kosaka T, Takubo K, Kinoshita T, Oya M, Horimoto K, Suda T: Optimal ratio of transcription factors for somatic cell reprogramming. *J Biol Chem* 2012, 287:36273–36282.
121. Mansour AA, Gafni O, Weinberger L, Zviran A, Ayyash M, Rais Y, Krupalnik V et al.: The H3K27 demethylase Utx regulates somatic and germ cell epigenetic reprogramming. *Nature* 2012, 488:409–413.
122. Zhao W, Li Q, Ayers S, Gu Y, Shi Z, Zhu Q, Chen Y, Wang HY, Wang RF: Jmjd3 inhibits reprogramming by upregulating expression of INK4a/Arf and targeting PHF20 for ubiquitination. *Cell* 2013, 152:1037–1050.
123. Morales Torres C, Laugesen A, Helin K: Utx is required for proper induction of ectoderm and mesoderm during differentiation of embryonic stem cells. *PLoS One* 2013, 8:e60020.
124. Ohtani K, Zhao C, Dobreva G, Manavski Y, Kluge B, Braun T, Rieger MA, Zeiher AM, Dimmeler S: Jmjd3 controls mesodermal and cardiovascular differentiation of embryonic stem cells. *Circ Res* 2013, 113:856–862.
125. Saha B, Home P, Ray S, Larson M, Paul A, Rajendran G, Behr B, Paul S: EED and KDM6B coordinate the first mammalian cell lineage commitment to ensure embryo implantation. *Mol Cell Biol* 2013, 33:2691–2705.
126. Wang Z, Zang C, Rosenfeld JA, Schones DE, Barski A, Cuddapah S, Cui K, Roh TY, Peng W, Zhang MQ, Zhao K: Combinatorial patterns of histone acetylations and methylations in the human genome. *Nat Genet* 2008, 40:897–903.
127. Mattout A, Biran A, Meshorer E: Global epigenetic changes during somatic cell reprogramming to iPS cells. *J Mol Cell Biol* 2011, 3:341–350.
128. Creyghton MP, Cheng AW, Welstead GG, Kooistra T, Carey BW, Steine EJ, Hanna J et al.: Histone H3K27ac separates active from poised enhancers and predicts developmental state. *Proc Natl Acad Sci USA* 2010, 107:21931–21936.

129. Tie F, Banerjee R, Stratton CA, Prasad-Sinha J, Stepanik V, Zlobin A, Diaz MO, Scacheri PC, Harte PJ: CBP-mediated acetylation of histone H3 lysine 27 antagonizes Drosophila Polycomb silencing. *Development* 2009, 136:3131–3141.
130. Pasini D, Malatesta M, Jung HR, Walfridsson J, Willer A, Olsson L, Skotte J, Wutz A, Porse B, Jensen ON, Helin K: Characterization of an antagonistic switch between histone H3 lysine 27 methylation and acetylation in the transcriptional regulation of Polycomb group target genes. *Nucleic Acids Res* 2010, 38:4958–4969.
131. Miyabayashi T, Teo JL, Yamamoto M, McMillan M, Nguyen C, Kahn M: Wnt/beta-catenin/CBP signaling maintains long-term murine embryonic stem cell pluripotency. *Proc Natl Acad Sci USA* 2007, 104:5668–5673.
132. Zhong X, Jin Y: Critical roles of coactivator p300 in mouse embryonic stem cell differentiation and Nanog expression. *J Biol Chem* 2009, 284:9168–9175.
133. Ahringer J: NuRD and SIN3 histone deacetylase complexes in development. *Trends Genet* 2000, 16:351–356.
134. Liang J, Wan M, Zhang Y, Gu P, Xin H, Jung SY, Qin J, Wong J, Cooney AJ, Liu D, Songyang Z: Nanog and Oct4 associate with unique transcriptional repression complexes in embryonic stem cells. *Nat Cell Biol* 2008, 10:731–739.
135. Luo M, Ling T, Xie W, Sun H, Zhou Y, Zhu Q, Shen M et al.: NuRD blocks reprogramming of mouse somatic cells into pluripotent stem cells. *Stem Cells* 2013, 31:1278–1286.
136. dos Santos RL, Tosti L, Radzisheuskaya A, Caballero IM, Kaji K, Hendrich B, Silva JC: MBD3/NuRD facilitates induction of pluripotency in a context-dependent manner. *Cell Stem Cell* 2014, 15:102–110.
137. Reynolds N, Salmon-Divon M, Dvinge H, Hynes-Allen A, Balasooriya G, Leaford D, Behrens A, Bertone P, Hendrich B: NuRD-mediated deacetylation of H3K27 facilitates recruitment of Polycomb Repressive Complex 2 to direct gene repression. *EMBO J* 2012, 31:593–605.
138. Peric-Hupkes D, Meuleman W, Pagie L, Bruggeman SW, Solovei I, Brugman W, Gräf S et al.: Molecular maps of the reorganization of genome-nuclear lamina interactions during differentiation. *Mol Cell* 2010, 38:603–613.
139. Tanaka Y, Chung L, Park IH: Impact of retrotransposons in pluripotent stem cells. *Mol Cells* 2012, 34:509–516.
140. Ohnuki M, Tanabe K, Sutou K, Teramoto I, Sawamura Y, Narita M, Nakamura M, Tokunaga Y, Watanabe A, Yamanaka S, Takahashi K: Dynamic regulation of human endogenous retroviruses mediates factor-induced reprogramming and differentiation potential. *Proc Natl Acad Sci USA* 2014, 111:12426–12431.
141. Soufi A, Donahue G, Zaret KS: Facilitators and impediments of the pluripotency reprogramming factors' initial engagement with the genome. *Cell* 2012, 151:994–1004.
142. Chen J, Liu H, Liu J, Qi J, Wei B, Yang J, Liang H et al.: H3K9 methylation is a barrier during somatic cell reprogramming into iPSCs. *Nat Genet* 2013, 45:34–42.
143. Karimi MM, Goyal P, Maksakova IA, Bilenky M, Leung D, Tang JX, Shinkai Y, Mager DL, Jones S, Hirst M, Lorincz MC: DNA methylation and SETDB1/H3K9me3 regulate predominantly distinct sets of genes, retroelements, and chimeric transcripts in mESCs. *Cell Stem Cell* 2011, 8:676–687.
144. Bilodeau S, Kagey MH, Frampton GM, Rahl PB, Young RA: SetDB1 contributes to repression of genes encoding developmental regulators and maintenance of ES cell state. *Genes Dev* 2009, 23:2484–2489.

145. Das PP, Shao Z, Beyaz S, Apostolou E, Pinello L, De Los Angeles A, O'Brien K et al.: Distinct and combinatorial functions of Jmjd2b/Kdm4b and Jmjd2c/Kdm4c in mouse embryonic stem cell identity. *Mol Cell* 2014, 53:32–48.
146. Wagner EJ, Carpenter PB: Understanding the language of Lys36 methylation at histone H3. *Nat Rev Mol Cell Biol* 2012, 13:115–126.
147. Zhang Y, Xie S, Zhou Y, Xie Y, Liu P, Sun M, Xiao H et al.: H3K36 histone methyltransferase Setd2 is required for murine embryonic stem cell differentiation toward endoderm. *Cell Rep* 2014, 8:1989–2002.
148. Hamada M, Yoshikawa H, Ueda Y, Kurokawa MS, Watanabe K, Sakakibara M, Tadokoro M, Akashi K, Aoki H, Suzuki N: Introduction of the MASH1 gene into mouse embryonic stem cells leads to differentiation of motoneuron precursors lacking Nogo receptor expression that can be applicable for transplantation to spinal cord injury. *Neurobiol Dis* 2006, 22:509–522.
149. Loh YH, Zhang W, Chen X, George J, Ng HH: Jmjd1a and Jmjd2c histone H3 Lys 9 demethylases regulate self-renewal in embryonic stem cells. *Genes Dev* 2007, 21:2545–2557.
150. He J, Shen L, Wan M, Taranova O, Wu H, Zhang Y: Kdm2b maintains murine embryonic stem cell status by recruiting PRC1 complex to CpG islands of developmental genes. *Nat Cell Biol* 2013, 15:373–384.
151. Wang T, Chen K, Zeng X, Yang J, Wu Y, Shi X, Qin B, Zeng L, Esteban MA, Pan G, Pei D: The histone demethylases Jhdm1a/1b enhance somatic cell reprogramming in a vitamin-C-dependent manner. *Cell Stem Cell* 2011, 9:575–587.
152. Brien GL, Gambero G, O'Connell DJ, Jerman E, Turner SA, Egan CM, Dunne EJ et al.: Polycomb PHF19 binds H3K36me3 and recruits PRC2 and demethylase NO66 to embryonic stem cell genes during differentiation. *Nat Struct Mol Biol* 2012, 19:1273–1281.
153. Bedford MT, Clarke SG: Protein arginine methylation in mammals: Who, what, and why. *Mol Cell* 2009, 33:1–13.
154. Guccione E, Bassi C, Casadio F, Martinato F, Cesaroni M, Schuchlautz H, Lüscher B, Amati B: Methylation of histone H3R2 by PRMT6 and H3K4 by an MLL complex are mutually exclusive. *Nature* 2007, 449:933–937.
155. Simandi Z, Czipa E, Horvath A, Koszeghy A, Bordas C, Póliska S, Juhász I et al.: PRMT1 and PRMT8 regulate retinoic acid-dependent neuronal differentiation with implications to neuropathology. *Stem Cells* 2015, 33:726–741.
156. Vu LP, Perna F, Wang L, Voza F, Figueroa ME, Tempst P, Erdjument-Bromage H et al.: PRMT4 blocks myeloid differentiation by assembling a methyl-RUNX1–dependent repressor complex. *Cell Rep* 2013, 5:1625–1638.
157. Lee YH, Ma H, Tan TZ, Ng SS, Soong R, Mori S, Fu XY, Zernicka-Goetz M, Wu Q: Protein arginine methyltransferase 6 regulates embryonic stem cell identity. *Stem Cells Dev* 2012, 21:2613–2622.
158. Tee WW, Pardo M, Theunissen TW, Yu L, Choudhary JS, Hajkova P, Surani MA: Prmt5 is essential for early mouse development and acts in the cytoplasm to maintain ES cell pluripotency. *Genes Dev* 2010, 24:2772–2777.
159. Chu Z, Niu B, Zhu H, He X, Bai C, Li G, Hua J: PRMT5 enhances generation of induced pluripotent stem cells from dairy goat embryonic fibroblasts via down-regulation of p53. *Cell Prolif* 2015, 48:29–38.

160. Iwasaki H, Kovacic JC, Olive M, Beers JK, Yoshimoto T, Crook MF, Tonelli LH, Nabel EG: Disruption of protein arginine N-methyltransferase 2 regulates leptin signaling and produces leanness in vivo through loss of STAT3 methylation. *Circ Res* 2010, 107:992–1001.
161. Salomonis N, Schlieve CR, Pereira L, Wahlquist C, Colas A, Zambon AC, Vranizan K et al.: Alternative splicing regulates mouse embryonic stem cell pluripotency and differentiation. *Proc Natl Acad Sci USA* 2010, 107:10514–10519.
162. Wu JQ, Habegger L, Noisa P, Szekely A, Qiu C, Hutchison S, Raha D et al.: Dynamic transcriptomes during neural differentiation of human embryonic stem cells revealed by short, long, and paired-end sequencing. *Proc Natl Acad Sci USA* 2010, 107:5254–5259.
163. Yeo GW, Xu X, Liang TY, Muotri AR, Carson CT, Coufal NG, Gage FH: Alternative splicing events identified in human embryonic stem cells and neural progenitors. *PLoS Comput Biol* 2007, 3:1951–1967.
164. Salomonis N, Nelson B, Vranizan K, Pico AR, Hanspers K, Kuchinsky A, Ta L, Mercola M, Conklin BR: Alternative splicing in the differentiation of human embryonic stem cells into cardiac precursors. *PLoS Comput Biol* 2009, 5:e1000553.
165. Atlasi Y, Mowla SJ, Ziaee SA, Gokhale PJ, Andrews PW: OCT4 spliced variants are differentially expressed in human pluripotent and nonpluripotent cells. *Stem Cells* 2008, 26:3068–3074.
166. Das S, Jena S, Levasseur DN: Alternative splicing produces Nanog protein variants with different capacities for self-renewal and pluripotency in embryonic stem cells. *J Biol Chem* 2011, 286:42690–42703.
167. Rao S, Zhen S, Roumiantsev S, McDonald LT, Yuan GC, Orkin SH: Differential roles of Sall4 isoforms in embryonic stem cell pluripotency. *Mol Cell Biol* 2010, 30:5364–5380.
168. Gabut M, Samavarchi-Tehrani P, Wang X, Slobodeniuc V, O'Hanlon D, Sung HK, Alvarez M et al.: An alternative splicing switch regulates embryonic stem cell pluripotency and reprogramming. *Cell* 2011, 147:132–146.
169. Lu Y, Loh YH, Li H, Cesana M, Ficarro SB, Parikh JR, Salomonis N et al.: Alternative splicing of MBD2 supports self-renewal in human pluripotent stem cells. *Cell Stem Cell* 2014, 15:92–101.
170. Lu X, Göke J, Sachs F, Jacques P, Liang H, Feng B, Bourque G, Bubulya PA, Ng HH: SON connects the splicing-regulatory network with pluripotency in human embryonic stem cells. *Nat Cell Biol* 2013, 15:1141–1152.
171. Cauffman G, Liebaers I, Van Steirteghem A, Van de Velde H: POU5F1 isoforms show different expression patterns in human embryonic stem cells and preimplantation embryos. *Stem Cells* 2006, 24:2685–2691.
172. Gao Y, Wei J, Han J, Wang X, Su G, Zhao Y, Chen B, Xiao Z, Cao J, Dai J: The novel function of OCT4B isoform-265 in genotoxic stress. *Stem Cells* 2012, 30:665–672.
173. Lee J, Kim HK, Rho JY, Han YM, Kim J: The human OCT-4 isoforms differ in their ability to confer self-renewal. *J Biol Chem* 2006, 281:33554–33565.
174. Asadi MH, Mowla SJ, Fathi F, Aleyasin A, Asadzadeh J, Atlasi Y: OCT4B1, a novel spliced variant of OCT4, is highly expressed in gastric cancer and acts as an antiapoptotic factor. *Int J Cancer* 2011, 128:2645–2652.
175. Loh YH, Wu Q, Chew JL, Vega VB, Zhang W, Chen X, Bourque G et al.: The Oct4 and Nanog transcription network regulates pluripotency in mouse embryonic stem cells. *Nat Genet* 2006, 38:431–440.

176. Das S, Jena S, Kim EM, Zavazava N, Levasseur DN: Transcriptional Regulation of Human NANOG by Alternate Promoters in Embryonic Stem Cells. *J Stem Cell Res Ther* 2012, Suppl 10:009.
177. Yi F, Pereira L, Hoffman JA, Shy BR, Yuen CM, Liu DR, Merrill BJ: Opposing effects of Tcf3 and Tcf1 control Wnt stimulation of embryonic stem cell self-renewal. *Nat Cell Biol* 2011, 13:762–770.
178. Lyashenko N, Winter M, Migliorini D, Biechele T, Moon RT, Hartmann C: Differential requirement for the dual functions of β-catenin in embryonic stem cell self-renewal and germ layer formation. *Nat Cell Biol* 2011, 13:753–761.
179. Ho R, Papp B, Hoffman JA, Merrill BJ, Plath K: Stage-specific regulation of reprogramming to induced pluripotent stem cells by Wnt signaling and T cell factor proteins. *Cell Rep* 2013, 3:2113–2126.
180. Zhang J, Tam WL, Tong GQ, Wu Q, Chan HY, Soh BS, Lou Y et al.: Sall4 modulates embryonic stem cell pluripotency and early embryonic development by the transcriptional regulation of Pou5f1. *Nat Cell Biol* 2006, 8:1114–1123.
181. Lam EW, Brosens JJ, Gomes AR, Koo CY: Forkhead box proteins: Tuning forks for transcriptional harmony. *Nat Rev Cancer* 2013, 13:482–495.
182. Han H, Irimia M, Ross PJ, Sung HK, Alipanahi B, David L, Golipour A et al.: MBNL proteins repress ES-cell-specific alternative splicing and reprogramming. *Nature* 2013, 498:241–245.
183. Le Guezennec X, Vermeulen M, Brinkman AB, Hoeijmakers WA, Cohen A, Lasonder E, Stunnenberg HG: MBD2/NuRD and MBD3/NuRD, two distinct complexes with different biochemical and functional properties. *Mol Cell Biol* 2006, 26:843–851.
184. Lee MR, Prasain N, Chae HD, Kim YJ, Mantel C, Yoder MC, Broxmeyer HE: Epigenetic regulation of NANOG by miR-302 cluster-MBD2 completes induced pluripotent stem cell reprogramming. *Stem Cells* 2013, 31:666–681.
185. Lin S, Coutinho-Mansfield G, Wang D, Pandit S, Fu XD: The splicing factor SC35 has an active role in transcriptional elongation. *Nat Struct Mol Biol* 2008, 15:819–826.
186. Yamaji M, Seki Y, Kurimoto K, Yabuta Y, Yuasa M, Shigeta M, Yamanaka K, Ohinata Y, Saitou M: Critical function of Prdm14 for the establishment of the germ cell lineage in mice. *Nat Genet* 2008, 40:1016–1022.
187. Ma Z, Swigut T, Valouev A, Rada-Iglesias A, Wysocka J: Sequence-specific regulator Prdm14 safeguards mouse ESCs from entering extraembryonic endoderm fates. *Nat Struct Mol Biol* 2011, 18:120–127.
188. Yamaji M, Ueda J, Hayashi K, Ohta H, Yabuta Y, Kurimoto K, Nakato R, Yamada Y, Shirahige K, Saitou M: PRDM14 ensures naive pluripotency through dual regulation of signaling and epigenetic pathways in mouse embryonic stem cells. *Cell Stem Cell* 2013, 12:368–382.
189. Barbosa-Morais NL, Irimia M, Pan Q, Xiong HY, Gueroussov S, Lee LJ, Slobodeniuc V et al.: The evolutionary landscape of alternative splicing in vertebrate species. *Science* 2012, 338:1587–1593.
190. Guo CL, Liu L, Jia YD, Zhao XY, Zhou Q, Wang L: A novel variant of Oct3/4 gene in mouse embryonic stem cells. *Stem Cell Res* 2012, 9:69–76.
191. Cantara WA, Crain PF, Rozenski J, McCloskey JA, Harris KA, Zhang X, Vendeix FA, Fabris D, Agris PF: The RNA Modification Database, RNAMDB: 2011 update. *Nucleic Acids Res* 2011, 39:D195–201.

192. Csepany T, Lin A, Baldick CJ, Beemon K: Sequence specificity of mRNA N6–adenosine methyltransferase. *J Biol Chem* 1990, 265:20117–20122.
193. Harper JE, Miceli SM, Roberts RJ, Manley JL: Sequence specificity of the human mRNA N6–adenosine methylase in vitro. *Nucleic Acids Res* 1990, 18: 5735–5741.
194. Wang Y, Li Y, Toth JI, Petroski MD, Zhang Z, Zhao JC: N6–methyladenosine modification destabilizes developmental regulators in embryonic stem cells. *Nat Cell Biol* 2014, 16:191–198.
195. Liu J, Yue Y, Han D, Wang X, Fu Y, Zhang L, Jia G et al.: A METTL3–METTL14 complex mediates mammalian nuclear RNA N6–adenosine methylation. *Nat Chem Biol* 2014, 10:93–95.
196. Schwartz S, Mumbach MR, Jovanovic M, Wang T, Maciag K, Bushkin GG, Mertins P et al.: Perturbation of m6A writers reveals two distinct classes of mRNA methylation at internal and 5′ sites. *Cell Rep* 2014, 8:284–296.
197. Ping XL, Sun BF, Wang L, Xiao W, Yang X, Wang WJ, Adhikari S et al.: Mammalian WTAP is a regulatory subunit of the RNA N6-methyladenosine methyltransferase. *Cell Res* 2014, 24:177–189.
198. Meyer KD, Saletore Y, Zumbo P, Elemento O, Mason CE, Jaffrey SR: Comprehensive analysis of mRNA methylation reveals enrichment in 3′ UTRs and near stop codons. *Cell* 2012, 149:1635–1646.
199. Batista PJ, Molinie B, Wang J, Qu K, Zhang J, Li L, Bouley DM et al.: m(6)A RNA modification controls cell fate transition in mammalian embryonic stem cells. *Cell Stem Cell* 2014, 15:707–719.
200. Dominissini D, Moshitch-Moshkovitz S, Schwartz S, Salmon-Divon M, Ungar L, Osenberg S, Cesarkas K et al.: Topology of the human and mouse m6A RNA methylomes revealed by m6A-seq. *Nature* 2012, 485:201–206.
201. Wang X, Lu Z, Gomez A, Hon GC, Yue Y, Han D, Fu Y et al.: N6–methyladenosine-dependent regulation of messenger RNA stability. *Nature* 2014, 505:117–120.
202. Chen T, Hao YJ, Zhang Y, Li MM, Wang M, Han W, Wu Y et al.: m(6)A RNA methylation is regulated by MicroRNAs and promotes reprogramming to pluripotency. *Cell Stem Cell* 2015.
203. Osenberg S, Paz Yaacov N, Safran M, Moshkovitz S, Shtrichman R, Sherf O, Jacob-Hirsch J, Keshet G, Amariglio N, Itskovitz-Eldor J, Rechavi G: Alu sequences in undifferentiated human embryonic stem cells display high levels of A-to-I RNA editing. *PLoS One* 2010, 5:e11173.
204. Germanguz I, Shtrichman R, Osenberg S, Ziskind A, Novak A, Domev H, Laevsky I, Jacob-Hirsch J, Feiler Y, Rechavi G, Itskovitz-Eldor J: ADAR1 is involved in the regulation of reprogramming human fibroblasts to induced pluripotent stem cells. *Stem Cells Dev* 2014, 23:443–456.
205. Hartner JC, Walkley CR, Lu J, Orkin SH: ADAR1 is essential for the maintenance of hematopoiesis and suppression of interferon signaling. *Nat Immunol* 2009, 10:109–115.
206. Ruiz S, Panopoulos AD, Herrerías A, Bissig KD, Lutz M, Berggren WT, Verma IM, Izpisua Belmonte JC: A high proliferation rate is required for cell reprogramming and maintenance of human embryonic stem cell identity. *Curr Biol* 2011, 21:45–52.
207. Eisenberg E, Nemzer S, Kinar Y, Sorek R, Rechavi G, Levanon EY: Is abundant A-to-I RNA editing primate-specific? *Trends Genet* 2005, 21:77–81.

208. Levanon EY, Eisenberg E, Yelin R, Nemzer S, Hallegger M, Shemesh R, Fligelman ZY et al.: Systematic identification of abundant A-to-I editing sites in the human transcriptome. *Nat Biotechnol* 2004, 22:1001–1005.
209. Peng Z, Cheng Y, Tan BC, Kang L, Tian Z, Zhu Y, Zhang W et al.: Comprehensive analysis of RNA-Seq data reveals extensive RNA editing in a human transcriptome. *Nat Biotechnol* 2012, 30:253–260.
210. Blaschke K, Ebata KT, Karimi MM, Zepeda-Martínez JA, Goyal P, Mahapatra S, Tam A et al.: Vitamin C induces Tet-dependent DNA demethylation and a blastocyst-like state in ES cells. *Nature* 2013, 500:222–226.

4

CRISPR-Based Genome Engineering in Human Stem Cells

Thelma Garcia and Deepak A. Lamba

CONTENTS

4.1 An Introduction on Different Genome Editing Technologies: ZFN, TALEN, and CRISPR

There are a number of different nuclease systems used for genome editing. The most common ones include zinc finger nucleases (ZFNs), transcription activator-like effector nucleases (TALENs), and RNA-guided engineered nucleases derived from the bacterial clustered regularly interspaced short palindromic repeat (CRISPR)–Cas9. However, these nucleases differ in several ways, including composition, targetable sites, specificities, and mutation signatures. One must choose the most appropriate tool based on the pros and the cons of each one.

ZFNs are a class of engineered DNA-binding proteins that facilitate targeted editing of the genome by creating double-strand breaks (DSBs) in the

DNA at user-specified locations (Carroll, 2011; Handel and Cathomen, 2011; Hansen et al., 2012; Hockemeyer et al., 2009; Swarthout et al., 2011). ZFN-mediated genome editing occurs in the nucleus when a ZFN pair targeting the user's gene of interest is delivered into a parental cell line, by either transfection, electroporation, or viral delivery. This method takes advantage of the cell's natural endogenous DNA repair machinery to precisely alter the genomes.

TALENs operate similarly as ZFNs (Ding et al., 2013a; Wright et al., 2014). They have two variable positions with a strong recognition for specific nucleotides. Genome editing can be accomplished by assembling the arrays of transcription activator–like (TALs) and fusing them to a Fokl nuclease. When two TALENs bind and meet, you get a DSB, which can inactivate a gene by creating a mutation or it can be used to insert a desired DNA to create a reporter line. Instead of recognizing DNA triplets like the ZFNs, TALENs recognize a single nucleotide. Hence, TALENs are more specific than ZFN but are larger and harder to deliver than ZFNs. ZNF and TALEN modifications have been engineered in zebrafish, fruit flies, nematodes, rats, livestock, and human stem cells (Biffi, 2015; Dreyer et al., 2015; Katsuyama et al., 2013; Kondo et al., 2014; Moghaddassi et al., 2014; Park et al., 2014; Shao et al., 2014; Tan et al., 2013; Tong et al., 2012; Wang et al., 2014, Wei et al., 2014; Zu et al., 2013).

CRISPR–Cas9 enables targeted genetic modifications in cultured cells and in whole animals and plants. This technology has the ability to perform site-specific DNA cleavage in the genome and repair (through endogenous mechanisms) which allows high-precision genome editing. They are RNA-based bacterial defense mechanisms designed to recognize and eliminate foreign DNA. A Cas endonuclease is directed to cleave a target sequence by a guide RNA (gRNA). A major advantage of the CRISPR technology is that the CRISPR gRNA is able to bind methylated DNA thereby allowing hypermethylated DNA sequences to be more efficiently targeted compared to TALENs, which cannot bind methylated cytosines (Hsu et al., 2013).

4.2 Different DNA Repair Mechanisms Involved in Genome Editing

These nucleases enable efficient and precise genetic modifications by inducing targeted DNA DSBs that stimulate the cellular DNA repair mechanisms, including error-prone nonhomologous end joining (NHEJ) and homology-directed repair (HDR).

4.2.1 NHEJ

NHEJ repairs DSBs in DNA. It is referred to as *nonhomologous* because it does not need a homologous template to guide the repair. It requires two DNA blunt ends (or overhangs) to join them together. Imprecise repair can occur when the overhangs are not compatible.

4.2.2 HDR

HDR is the most common double-strand repair mechanism in cells, and it is considered to be a more accurate mechanism than DSB repair because it requires a DNA template. Scientists have engineered plasmid-based systems that can target and cut specific sites to exploit this repair mechanism for genome editing.

4.3 Genome Editing in Human Pluripotent Stem Cells

The discovery of hESCs and of hiPSC reprogramming technology has created renewed excitement for the field of stem cell biology especially as a way to create better *in vitro* disease models and their subsequent use in focused drug discovery (Okita et al., 2011; Takahashi et al., 2007; Thomson et al., 1998; Yu et al., 2007). Patient-derived iPSC lines have been used to study disease development. However, a number of groups have reported variability in the appearance of the disease phenotype in the *in vitro* models even when different clones of the same donor line were used or even among clones derived from different tissues of the same donor (Bock et al., 2011; Kajiwara et al., 2012; Kyttala et al., 2016; Mills et al., 2013). There are other issues such as the right control cell line for comparison of phenotype. Labs have generally relied on iPSC lines generated from unaffected relative, but that may add another variable to the problem. One way to get around this problem can be the genomic engineering of disease mutations and having control isogenic lines to rule out any genetic background effects. All the discussed genome editing technologies have been applied to human stem cell lines (Byrne et al., 2014; Hockemeyer et al., 2011; Mali et al., 2013a; Ramalingam et al., 2013; Wang et al., 2012; Xue et al., 2016; Yang et al., 2013, 2014; Zhu et al., 2014). Of these, there is considerable excitement for CRISPR–Cas9-based genomic editing due the easy generation of guides and easy applicability in labs. The rest of the chapter will manly discuss the details and the applicability of this technology to human PSCs.

4.4 CRISPR Technology for Genome Editing

For the most part, CRISPR–Cas9 systems provide DNA-encoded, RNA-mediated, DNA- or RNA-targeting sequence-specific targeting. Cas9 is the signature protein for Type II CRISPR–Cas9 systems. A ribonucleoprotein complex consisting of a CRISPR RNA (crRNA) in combination with a trans-activating crRNA and a Cas9 nuclease targets complementary DNA flanked by a protospacer adjacent motif (PAM) to obtain desired gene editing. There are three different variants of Cas9 nuclease: double nick, single nick, and mutant nick.

4.4.1 Double Nick

The first is wild-type Cas9 or double nick. This one can site-specifically cleave double-stranded DNA, resulting in the activation of the DSB repair machinery using the NHEJ or the HDR pathways. As discussed previously, NHEJ may lead to insertions and/or deletions disrupting the targeted locus. By supplying a template, DSB may be repaired via the HDR pathway, which is more precise.

4.4.2 Single Nick

The second Cas9 type is called *Mutated Cas9* which makes a site-specific single-strand nick (Cong et al., 2013; Frock et al., 2015). Two sgRNA can then be utilized to incorporate a staggered DSB, which can then undergo HDR.

4.4.3 Mutant Nick

The third and last type of Cas9 is nuclease-deficient Cas9, which can be fused with various effector domains allowing specific localization. This can be used to allow for the transcriptional activation by fusing an activator peptide sequence to catalytically inactive cas9 (dCas9) (Gao et al., 2014; Lin et al., 2015; Mali et al., 2013b; Xu et al., 2014), transcriptional repression by the use of catalytically inactive Cas9 which interfered with transcription (Gilbert et al., 2014; Lebar et al., 2014), to visualize gene expression by fusion of dCas9 with a fluorescent tag (Chen et al., 2013; Ma et al., 2015; Ochiai et al., 2015).

4.5 CRISPR gRNA Designing

A number of useful online resources are available to help better design guide RNA (sgRNA) to direct endonuclease activity (Table 4.1). In general, CRISPR guide design does require stringent conditions as they are much more mismatch tolerant compared to TALENS (Fu et al., 2013). This can be tackled to some extent by reducing the size of sgRNA sequence to as low as 17 bps with minimal effects on cutting efficiency (Fu et al., 2014). Specificity can also be enhanced by the use of mutant nickase variants of Cas9. The Cas9-D10A

TABLE 4.1

In Silico Toolsets for CRISPR gRNA Designing

Online Resource	Reference
http://crispr.mit.edu	Hsu et al. (2013)
http://crispr.cos.uni-heidelberg.de	Stemmer et al. (2015)
https://crispr-bme-gatech-edu	Cradick et al. (2014)
http://zifit.partners.org/ZiFiT	Sander et al. (2010)
http://www.e-crisp.org/E-CRISP	Heigwer et al. (2014)
https://chopchop-rc-fas-harvard-edu	Montague et al. (2014)

mutant protein introduces a single-strand nick, instead of a DSB (Cong et al., 2013; Jinek et al., 2012). Also, the loci where CRISPR gRNAs bind are limited to those harboring PAM. Commonly used PAM motif NGG is found every 8–12 bp on average for the human genome (Cong et al., 2013). Recently, a number of non-conventional PAM sequences have been identified for variant Cas9 lines (Esvelt et al., 2013; Hou et al., 2013; Walsh and Hochedlinger, 2013).

4.6 Different Delivery Methods

A number of different methods have been attempted to deliver the gRNAs and the Cas9 to human PSCs including DNA, RNA, protein, and viral methods. The most conventional method involves using plasmids. Generally, due to size consideration, a two-plasmid delivery method is used with one containing Cas9 and the second, the gRNA sequence (Mali et al., 2013b), although a single plasmid containing both sequences has also been used (Ran et al., 2013). The plasmids are then delivered by electroporation. Using DNA delivery, high efficiency has been reported with CRISPRs generally as high as 50–79% (Ding et al., 2013b). However, recent reports suggest a much higher off-target mutation rate even when Cas9-D10A single or dual nickases were used (Merkle et al., 2015).

An alternative is to deliver the gRNA as a stabilized RNA sequence. The advantage here would be a much faster editing within hours of delivery and a much lower incidence of off-target effects, as they will not integrate into the genome. Chemically altered synthetic sgRNAs have been used for the genomic editing of human primary cells (Hendel et al., 2015). The group co-delivered the sgRNAs with Cas9 mRNA as well as Cas9 protein to induce genome editing with almost no toxicity. This study represents a good alternate for quick and streamlined genome editing. Although sgRNAs cannot be delivered as proteins, Cas9 protein delivery has been successfully used for quick and efficient genome editing (Kim et al., 2014; Liang et al., 2015; Zuris et al., 2015). Viral mediated delivery of CRISPR–Cas9 into human stem cells has been successful as well. This has been done using either adenoviruses (Cheng et al., 2014; Li et al., 2015a), adeno-associated virus (Karnan et al., 2015; Senis et al., 2014; Yang et al., 2016) or lentiviruses (Kabadi et al., 2014; Koike-Yusa et al., 2014; Shalem et al., 2014; Zhou et al., 2014).

4.7 Application of CRISPR Technology for Generation and Repair of Mutant Stem Cell Lines

A number of groups have looked at using the CRISPR–Cas9 technology for genomic engineering, and this section will review a few of those publications.

One of the first reports on the feasibility of CRISPR technology to generate disease models in hiPSC lines was described by the Ellerby group for studying Huntington's disease (An et al., 2014). The group used Cas9-mediated incorporation of polyQ repeats in the first exon of HTT gene. The group described recombination rates as high as 12% compared to less than 1% by other technologies. The first report on genetic correction of mutations in patient iPSC lines was reported by the Kan group (Xie et al., 2014). The group was looking at iPSC lines from patients with β-thalassemia, which is caused by mutations in the human hemoglobin beta (HBB) gene. iPSC lines were generated from the fibroblasts of a β-thalassemia patient doubly heterozygous for the –28 (A/G) mutation of the promoter and the 4-bp (TCTT) deletion at codons 41 and 42 of exon 2, both being common mutations in the Chinese population. They used three different gRNAs, which varied in efficiency in generating DSB from 5% to 23%, and the replacement gene was delivered using *piggyBac* transposons. They then differentiated these cell lines into the hematopoietic progenitors and erythroblasts to test the efficacy of gene correction. Upon analysis of HBB mRNA levels, they reported that the gene-corrected lines had 16-fold higher HBB levels compared to control lines. Similar results have been reported by others (Song et al., 2015). They also saw that gene-corrected lines had reduced reactive oxygen species production compared to control uncorrected iPSC lines. Another disease that has been corrected in iPSC-based *in vitro* models is severe combined immunodeficiency (SCID) (Chang et al., 2015). SCID is caused by mutations in the Janus family kinase member, JAK3. Correction of the JAK3 mutation by CRISPR–Cas9-enhanced gene targeting restored normal T cell development, including the production of mature T cell populations with a broad T cell receptor repertoire from the iPSCs following T cell-directed differentiation. The group also reported no off-target modification upon the whole genome sequencing. Others have modeled chronic granulomatous disease which is a rare genetic disease caused by the lack of an oxidative burst, normally performed by phagocytic cells (Flynn et al., 2015). The group used CRISPR-mediated homologous recombination to reintroduce a previously skipped exon in the cytochrome b-245 heavy chain protein. This resulted in the restoration of oxidative burst function in iPSC-derived phagocytes from the gene-corrected lines.

A study compared various corrections methods to repair gene defect in Duchenne muscular dystrophy (Li et al., 2015b). Duchenne muscular dystrophy is a severe form of muscle degenerative disease caused by a mutation in the dystrophin gene. The authors tried three correction methods including exon skipping, frameshifting, and exon knockin and reported that exon knockin works best. The group further confirmed that following differentiation, the generated skeletal muscles expressed full-length version of dystrophin protein. A recent publication reported modeling a number of kidney defects in three-dimensional (3D) culture systems following CRISPR-mediated gene knockouts (Freedman et al., 2015). The group reported that the CRISPR–Cas9 knockout of podocalyxin gene in the iPSCs caused junctional organization defects in podocyte-like cells in the 3D kidney cultures, while knocking out of the polycystic kidney disease genes PKD1 or PKD2 induced cyst formation from kidney tubules in these cultures. A new report recently looked at repairing a mutation associated with deafness (Chen et al., 2016). They generated iPSCs from members of a Chinese family carrying MYO15A c.4642G>A and c.8374G>A mutations. These iPSC lines upon differentiation generated hair cells with abnormal morphology. The authors then corrected the mutations in the patient iPSC lines using CRISPR, which resulted in the restoration of hair cell morphology and function in differentiated hair cells.

4.8 Conclusions

Gene editing using custom-designed nucleases in human PSCs is a powerful tool for disease modeling and repair and will likely have widespread implications for medical research. There have been incremental technological advances in the last few years in developing better tools both to increase efficiency and to minimize any off-target effects. These are the key concerns with this technology, and further advancement will allow for the generation of better models. This proxy system also has the advantage of testing gene therapy *in vitro* to identify efficacy and rule out unanticipated side effects prior to moving it to the clinic.

References

An, M. C., O'Brien, R. N., Zhang, N., Patra, B. N., De La Cruz, M., Ray, A., and Ellerby, L. M. (2014). Polyglutamine disease modeling: Epitope based screen for homologous recombination using CRISPR/Cas9 system. *PLoS Currents* 6, pii: ecurrents .hd.0242d2e7ad72225efa72f6964589369a.

Biffi, A. (2015). Clinical translation of TALENS: Treating SCID-X1 by gene editing in iPSCs. *Cell Stem Cell* 16, 348–349.

Bock, C., Kiskinis, E., Verstappen, G., Gu, H., Boulting, G., Smith, Z. D., Ziller, M. et al. (2011). Reference maps of human ES and iPS cell variation enable high-throughput characterization of pluripotent cell lines. *Cell* 144, 439–452.

Byrne, S. M., Mali, P., and Church, G. M. (2014). Genome editing in human stem cells. *Methods in Enzymology* 546, 119–138.

Carroll, D. (2011). Genome engineering with zinc-finger nucleases. *Genetics* 188, 773–782.

Chang, C. W., Lai, Y. S., Westin, E., Khodadadi-Jamayran, A., Pawlik, K. M., Lamb, L. S., Jr., Goldman, F. D., and Townes, T. M. (2015). Modeling human severe combined immunodeficiency and correction by CRISPR/Cas9-enhanced gene targeting. *Cell Reports* 12, 1668–1677.

Chen, B., Gilbert, L. A., Cimini, B. A., Schnitzbauer, J., Zhang, W., Li, G. W., Park, J. et al. (2013). Dynamic imaging of genomic loci in living human cells by an optimized CRISPR/Cas system. *Cell* 155, 1479–1491.

Chen, J. R., Tang, Z. H., Zheng, J., Shi, H. S., Ding, J., Qian, X. D., Zhang, C. et al. (2016). Effects of genetic correction on the differentiation of hair cell-like cells from iPSCs with MYO15A mutation. *Cell Death and Differentiation* 23(8), 1347–1357.

Cheng, R., Peng, J., Yan, Y., Cao, P., Wang, J., Qiu, C., Tang, L., Liu, D., Tang, L., Jin, J. et al. (2014). Efficient gene editing in adult mouse livers via adenoviral delivery of CRISPR/Cas9. *FEBS Letters* 588, 3954–3958.

Cong, L., Ran, F. A., Cox, D., Lin, S., Barretto, R., Habib, N., Hsu, P. D. et al. (2013). Multiplex genome engineering using CRISPR/Cas systems. *Science* 339, 819–823.

Cradick, T. J., Qiu, P., Lee, C. M., Fine, E. J., and Bao, G. (2014). COSMID: A web-based tool for identifying and validating CRISPR/Cas off-target sites. *Molecular Therapy Nucleic Acids* 3, e214.

Ding, Q., Lee, Y. K., Schaefer, E. A., Peters, D. T., Veres, A., Kim, K., Kuperwasser, N. et al. (2013a). A TALEN genome-editing system for generating human stem cell-based disease models. *Cell Stem Cell* 12, 238–251.

Ding, Q., Regan, S. N., Xia, Y., Oostrom, L. A., Cowan, C. A., and Musunuru, K. (2013b). Enhanced efficiency of human pluripotent stem cell genome editing through replacing TALENs with CRISPRs. *Cell Stem Cell* 12, 393–394.

Dreyer, A. K., Hoffmann, D., Lachmann, N., Ackermann, M., Steinemann, D., Timm, B., Siler, U. et al. (2015). TALEN-mediated functional correction of X-linked chronic granulomatous disease in patient-derived induced pluripotent stem cells. *Biomaterials* 69, 191–200.

Esvelt, K. M., Mali, P., Braff, J. L., Moosburner, M., Yaung, S. J., and Church, G. M. (2013). Orthogonal Cas9 proteins for RNA-guided gene regulation and editing. *Nature Methods* 10, 1116–1121.

Flynn, R., Grundmann, A., Renz, P., Hanseler, W., James, W. S., Cowley, S. A., and Moore, M. D. (2015). CRISPR-mediated genotypic and phenotypic correction of a chronic granulomatous disease mutation in human iPS cells. *Experimental Hematology* 43, 838—848,e833.

Freedman, B. S., Brooks, C. R., Lam, A. Q., Fu, H., Morizane, R., Agrawal, V., Saad, A. F. et al. (2015). Modelling kidney disease with CRISPR-mutant kidney organoids derived from human pluripotent epiblast spheroids. *Nature Communications* 6, 8715.

Frock, R. L., Hu, J., Meyers, R. M., Ho, Y. J., Kii, E., and Alt, F. W. (2015). Genome-wide detection of DNA double-stranded breaks induced by engineered nucleases. *Nature Biotechnology* 33, 179–186.

Fu, Y., Foden, J. A., Khayter, C., Maeder, M. L., Reyon, D., Joung, J. K., and Sander, J. D. (2013). High-frequency off-target mutagenesis induced by CRISPR-Cas nucleases in human cells. *Nature Biotechnology* 31, 822–826.

Fu, Y., Sander, J. D., Reyon, D., Cascio, V. M., and Joung, J. K. (2014). Improving CRISPR-Cas nuclease specificity using truncated guide RNAs. *Nature Biotechnology* 32, 279–284.

Gao, X., Tsang, J. C., Gaba, F., Wu, D., Lu, L., and Liu, P. (2014). Comparison of TALE designer transcription factors and the CRISPR/dCas9 in regulation of gene expression by targeting enhancers. *Nucleic Acids Research* 42, e155.

Gilbert, L. A., Horlbeck, M. A., Adamson, B., Villalta, J. E., Chen, Y., Whitehead, E. H., Guimaraes, C. et al. (2014). Genome-scale CRISPR-mediated control of gene repression and activation. *Cell* 159, 647–661.

Handel, E. M., and Cathomen, T. (2011). Zinc-finger nuclease based genome surgery: It's all about specificity. *Current Gene Therapy* 11, 28–37.

Hansen, K., Coussens, M. J., Sago, J., Subramanian, S., Gjoka, M., and Briner, D. (2012). Genome editing with CompoZr custom zinc finger nucleases (ZFNs). *Journal of Visualized Experiments* 14(64), e3304.

Heigwer, F., Kerr, G., and Boutros, M. (2014). E-CRISP: Fast CRISPR target site identification. *Nature Methods* 11, 122–123.

Hendel, A., Bak, R. O., Clark, J. T., Kennedy, A. B., Ryan, D. E., Roy, S., Steinfeld, I. et al. (2015). Chemically modified guide RNAs enhance CRISPR-Cas genome editing in human primary cells. *Nature Biotechnology* 33, 985–989.

Hockemeyer, D., Soldner, F., Beard, C., Gao, Q., Mitalipova, M., DeKelver, R. C., Katibah, G. E. et al. (2009). Efficient targeting of expressed and silent genes in human ESCs and iPSCs using zinc-finger nucleases. *Nature Biotechnology* 27, 851–857.

Hockemeyer, D., Wang, H., Kiani, S., Lai, C. S., Gao, Q., Cassady, J. P., Cost, G. J. et al. (2011). Genetic engineering of human pluripotent cells using TALE nucleases. *Nature Biotechnology* 29, 731–734.

Hou, Z., Zhang, Y., Propson, N. E., Howden, S. E., Chu, L. F., Sontheimer, E. J., and Thomson, J. A. (2013). Efficient genome engineering in human pluripotent stem cells using Cas9 from Neisseria meningitidis. *Proceedings of the National Academy of Sciences* 110, 15644–15649.

Hsu, P. D., Scott, D. A., Weinstein, J. A., Ran, F. A., Konermann, S., Agarwala, V., Li, Y. et al. (2013). DNA targeting specificity of RNA-guided Cas9 nucleases. *Nature Biotechnology* 31, 827–832.

Jinek, M., Chylinski, K., Fonfara, I., Hauer, M., Doudna, J. A., and Charpentier, E. (2012). A programmable dual-RNA-guided DNA endonuclease in adaptive bacterial immunity. *Science* 337, 816–821.

Kabadi, A. M., Ousterout, D. G., Hilton, I. B., and Gersbach, C. A. (2014). Multiplex CRISPR/Cas9-based genome engineering from a single lentiviral vector. *Nucleic Acids Research* 42, e147.

Kajiwara, M., Aoi, T., Okita, K., Takahashi, R., Inoue, H., Takayama, N., Endo, H. et al. (2012). Donor-dependent variations in hepatic differentiation from human-induced pluripotent stem cells. *Proceedings of the National Academy of Sciences* 109, 12538–12543.

Karnan, S., Ota, A., Konishi, Y., Wahiduzzaman, M., Hosokawa, Y., and Konishi, H. (2015). Improved methods of AAV-mediated gene targeting for human cell lines using ribosome-skipping 2A peptide. *Nucleic Acids Research.*

Katsuyama, T., Akmammedov, A., Seimiya, M., Hess, S. C., Sievers, C., and Paro, R. (2013). An efficient strategy for TALEN-mediated genome engineering in *Drosophila. Nucleic Acids Research* 41, e163.

Kim, S., Kim, D., Cho, S. W., Kim, J., and Kim, J. S. (2014). Highly efficient RNA-guided genome editing in human cells via delivery of purified Cas9 ribonucleoproteins. *Genome Research* 24, 1012–1019.

Koike-Yusa, H., Li, Y., Tan, E. P., Velasco-Herrera Mdel, C., and Yusa, K. (2014). Genome-wide recessive genetic screening in mammalian cells with a lentiviral CRISPR-guide RNA library. *Nature Biotechnology* 32, 267–273.

Kondo, T., Sakuma, T., Wada, H., Akimoto-Kato, A., Yamamoto, T., and Hayashi, S. (2014). TALEN-induced gene knock out in *Drosophila. Development, Growth & Differentiation* 56, 86–91.

Kyttala, A., Moraghebi, R., Valensisi, C., Kettunen, J., Andrus, C., Pasumarthy, K. K. et al. (2016). Genetic variability overrides the impact of parental cell type and determines iPSC differentiation potential. *Stem Cell Reports* 6, 200–212.

Lebar, T., Bezeljak, U., Golob, A., Jerala, M., Kadunc, L., Pirs, B., Strazar, M. et al. (2014). A bistable genetic switch based on designable DNA-binding domains. *Nature Communications* 5, 5007.

Li, C., Guan, X., Du, T., Jin, W., Wu, B., Liu, Y., Wang, P. et al. (2015a). Inhibition of HIV-1 infection of primary CD4+ T-cells by gene editing of CCR5 using adenovirus-delivered CRISPR/Cas9. *Journal of General Virology* 96, 2381–2393.

Li, H. L., Fujimoto, N., Sasakawa, N., Shirai, S., Ohkame, T., Sakuma, T., Tanaka, M. et al. (2015b). Precise correction of the dystrophin gene in duchenne muscular dystrophy patient induced pluripotent stem cells by TALEN and CRISPR-Cas9. *Stem Cell Reports* 4, 143–154.

Liang, X., Potter, J., Kumar, S., Zou, Y., Quintanilla, R., Sridharan, M., Carte, J. et al. (2015). Rapid and highly efficient mammalian cell engineering via Cas9 protein transfection. *Journal of Biotechnology* 208, 44–53.

Lin, S., Ewen-Campen, B., Ni, X., Housden, B. E., and Perrimon, N. (2015). In vivo transcriptional activation using CRISPR/Cas9 in *Drosophila. Genetics* 201, 433–442.

Ma, H., Naseri, A., Reyes-Gutierrez, P., Wolfe, S. A., Zhang, S., and Pederson, T. (2015). Multicolor CRISPR labeling of chromosomal loci in human cells. *Proceedings of the National Academy of Sciences* 112, 3002–3007.

Mali, P., Yang, L., Esvelt, K. M., Aach, J., Guell, M., DiCarlo, J. E., Norville, J. E., and Church, G. M. (2013a). RNA-guided human genome engineering via Cas9. *Science* 339, 823–826.

Mali, P., Aach, J., Stranges, P. B., Esvelt, K. M., Moosburner, M., Kosuri, S., Yang, L., and Church, G. M. (2013b). CAS9 transcriptional activators for target specificity screening and paired nickases for cooperative genome engineering. *Nature Biotechnology* 31, 833–838.

Merkle, F. T., Neuhausser, W. M., Santos, D., Valen, E., Gagnon, J. A., Maas, K., Sandoe, J., Schier, A. F., and Eggan, K. (2015). Efficient CRISPR-Cas9-mediated generation of knockin human pluripotent stem cells lacking undesired mutations at the targeted locus. *Cell Reports* 11, 875–883.

Mills, J. A., Wang, K., Paluru, P., Ying, L., Lu, L., Galvao, A. M., Xu, D. et al. (2013). Clonal genetic and hematopoietic heterogeneity among human-induced pluripotent stem cell lines. *Blood* 122, 2047–2051.

Moghaddassi, S., Eyestone, W., and Bishop, C. E. (2014). TALEN-mediated modification of the bovine genome for large-scale production of human serum albumin. *PLoS One* 9, e89631.

Montague, T. G., Cruz, J. M., Gagnon, J. A., Church, G. M., and Valen, E. (2014). CHOPCHOP: A CRISPR/Cas9 and TALEN web tool for genome editing. *Nucleic Acids Research* 42, W401–407.

Ochiai, H., Sugawara, T., and Yamamoto, T. (2015). Simultaneous live imaging of the transcription and nuclear position of specific genes. *Nucleic Acids Research* 43, e127.

Okita, K., Matsumura, Y., Sato, Y., Okada, A., Morizane, A., Okamoto, S., Hong, H. et al. (2011). A more efficient method to generate integration-free human iPS cells. *Nature Methods* 8, 409–412.

Park, C. Y., Kim, J., Kweon, J., Son, J. S., Lee, J. S., Yoo, J. E., Cho, S. R., Kim, J. H., Kim, J. S., and Kim, D. W. (2014). Targeted inversion and reversion of the blood coagulation factor 8 gene in human iPS cells using TALENs. *Proceedings of the National Academy of Sciences* 111, 9253–9258.

Ramalingam, S., London, V., Kandavelou, K., Cebotaru, L., Guggino, W., Civin, C., and Chandrasegaran, S. (2013). Generation and genetic engineering of human induced pluripotent stem cells using designed zinc finger nucleases. *Stem Cells and Development* 22, 595–610.

Ran, F. A., Hsu, P. D., Wright, J., Agarwala, V., Scott, D. A., and Zhang, F. (2013). Genome engineering using the CRISPR-Cas9 system. *Nature Protocols* 8, 2281–2308.

Sander, J. D., Maeder, M. L., Reyon, D., Voytas, D. F., Joung, J. K., and Dobbs, D. (2010). ZiFiT (Zinc Finger Targeter): An updated zinc finger engineering tool. *Nucleic Acids Research* 38, W462–468.

Senis, E., Fatouros, C., Grosse, S., Wiedtke, E., Niopek, D., Mueller, A. K., Borner, K., and Grimm, D. (2014). CRISPR/Cas9-mediated genome engineering: An adeno-associated viral (AAV) vector toolbox. *Biotechnology Journal* 9, 1402–1412.

Shalem, O., Sanjana, N. E., Hartenian, E., Shi, X., Scott, D. A., Mikkelsen, T. S., Heckl, D. et al. (2014). Genome-scale CRISPR-Cas9 knockout screening in human cells. *Science* 343, 84–87.

Shao, Y., Guan, Y., Wang, L., Qiu, Z., Liu, M., Chen, Y., Wu, L. et al. (2014). CRISPR/Cas-mediated genome editing in the rat via direct injection of one-cell embryos. *Nature Protocols* 9, 2493–2512.

Song, B., Fan, Y., He, W., Zhu, D., Niu, X., Wang, D., Ou, Z., Luo, M., and Sun, X. (2015). Improved hematopoietic differentiation efficiency of gene-corrected beta-thalassemia induced pluripotent stem cells by CRISPR/Cas9 system. *Stem Cells and Development* 24, 1053–1065.

Stemmer, M., Thumberger, T., Del Sol Keyer, M., Wittbrodt, J., and Mateo, J. L. (2015). CCTop: An intuitive, flexible and reliable CRISPR/Cas9 target prediction tool. *PLoS One* 10, e0124633.

Swarthout, J. T., Raisinghani, M., and Cui, X. (2011). Zinc finger nucleases: A new era for transgenic animals. *Annals of Neurosciences* 18, 25–28.

Takahashi, K., Tanabe, K., Ohnuki, M., Narita, M., Ichisaka, T., Tomoda, K., and Yamanaka, S. (2007). Induction of pluripotent stem cells from adult human fibroblasts by defined factors. *Cell* 131, 861–872.

Tan, W., Carlson, D. F., Lancto, C. A., Garbe, J. R., Webster, D. A., Hackett, P. B., and Fahrenkrug, S. C. (2013). Efficient nonmeiotic allele introgression in livestock using custom endonucleases. *Proceedings of the National Academy of Sciences* 110, 16526–16531.

Thomson, J. A., Itskovitz-Eldor, J., Shapiro, S. S., Waknitz, M. A., Swiergiel, J. J., Marshall, V. S., and Jones, J. M. (1998). Embryonic stem cell lines derived from human blastocysts. *Science* 282, 1145–1147.

Tong, C., Huang, G., Ashton, C., Wu, H., Yan, H., and Ying, Q. L. (2012). Rapid and cost-effective gene targeting in rat embryonic stem cells by TALENs. *Journal of Genetics and Genomics: Yi chuan xue bao* 39, 275–280.

Walsh, R. M., and Hochedlinger, K. (2013). A variant CRISPR-Cas9 system adds versatility to genome engineering. *Proceedings of the National Academy of Sciences* 110, 15514–15515.

Wang, X., Wang, Y., Huang, H., Chen, B., Chen, X., Hu, J., Chang, T., Lin, R. J., and Yee, J. K. (2014). Precise gene modification mediated by TALEN and single-stranded oligodeoxynucleotides in human cells. *PLoS One* 9, e93575.

Wang, Y., Zhang, W. Y., Hu, S., Lan, F., Lee, A. S., Huber, B., Lisowski, L. et al. (2012). Genome editing of human embryonic stem cells and induced pluripotent stem cells with zinc finger nucleases for cellular imaging. *Circulation Research* 111, 1494–1503.

Wei, Q., Shen, Y., Chen, X., Shifman, Y., and Ellis, R. E. (2014). Rapid creation of forward-genetics tools for *C. briggsae* using TALENs: Lessons for nonmodel organisms. *Molecular Biology and Evolution* 31, 468–473.

Wright, D. A., Li, T., Yang, B., and Spalding, M. H. (2014). TALEN-mediated genome editing: Prospects and perspectives. *The Biochemical Journal* 462, 15–24.

Xie, F., Ye, L., Chang, J. C., Beyer, A. I., Wang, J., Muench, M. O., and Kan, Y. W. (2014). Seamless gene correction of beta-thalassemia mutations in patient-specific iPSCs using CRISPR/Cas9 and piggyBac. *Genome Research* 24, 1526–1533.

Xu, T., Li, Y., Van Nostrand, J. D., He, Z., and Zhou, J. (2014). Cas9-based tools for targeted genome editing and transcriptional control. *Applied and Environmental Microbiology* 80, 1544–1552.

Xue, H., Wu, J., Li, S., Rao, M. S., and Liu, Y. (2016). Genetic modification in human pluripotent stem cells by homologous recombination and CRISPR/Cas9 system. *Methods in Molecular Biology* 1307, 173–190.

Yang, L., Guell, M., Byrne, S., Yang, J. L., De Los Angeles, A., Mali, P., Aach, J. et al. (2013). Optimization of scarless human stem cell genome editing. *Nucleic Acids Research* 41, 9049–9061.

Yang, L., Mali, P., Kim-Kiselak, C., and Church, G. (2014). CRISPR-Cas-mediated targeted genome editing in human cells. *Methods in Molecular Biology* 1114, 245–267.

Yang, Y., Wang, L., Bell, P., McMenamin, D., He, Z., White, J., Yu, H. et al. (2016). A dual AAV system enables the Cas9-mediated correction of a metabolic liver disease in newborn mice. *Nature Biotechnology* 34, 334–338.

Yu, J., Vodyanik, M. A., Smuga-Otto, K., Antosiewicz-Bourget, J., Frane, J. L., Tian, S., Nie, J. et al. (2007). Induced pluripotent stem cell lines derived from human somatic cells. *Science* 318, 1917–1920.

Zhou, Y., Zhu, S., Cai, C., Yuan, P., Li, C., Huang, Y., and Wei, W. (2014). High-throughput screening of a CRISPR/Cas9 library for functional genomics in human cells. *Nature* 509, 487–491.

Zhu, Z., Gonzalez, F., and Huangfu, D. (2014). The iCRISPR platform for rapid genome editing in human pluripotent stem cells. *Methods in Enzymology* 546, 215–250.

Zu, Y., Tong, X., Wang, Z., Liu, D., Pan, R., Li, Z., Hu, Y. et al. (2013). TALEN-mediated precise genome modification by homologous recombination in zebrafish. *Nature Methods* 10, 329–331.

Zuris, J. A., Thompson, D. B., Shu, Y., Guilinger, J. P., Bessen, J. L., Hu, J. H., Maeder, M. L., Joung, J. K., Chen, Z. Y., and Liu, D. R. (2015). Cationic lipid-mediated delivery of proteins enables efficient protein-based genome editing in vitro and in vivo. *Nature Biotechnology* 33, 73–80.

5

Stem Cells for Parkinson's Disease

Deepak A. Lamba

CONTENTS

5.1 Introduction

Parkinson's disease (PD) is one of the most common neurodegenerative disorders of aging first described in 1817 by James Parkinson. The disease is characterized by progressive degeneration and loss of dopaminergic neurons in the substantial nigra. This leads to a range of motor neuron symptoms, including muscle rigidity, resting tremors, and bradykinesia, along with secondary speech and movement defects, including a characteristic shuffling gait. According to the National Institute of Neurological Disorders and Stroke, 50,000 new cases are reported annually with an average age of onset being 60 years, and the numbers are expected to rise with the increasing aging population.

5.2 Etiology and Pathology

Not unlike a number of neurodegenerative disorders, PD is poorly understood. A majority of the cases are of unknown etiology, although approximately 7–8% of patients have well-characterized mutations. To date, approximately 16 gene loci have been linked to PD. One of the first genes found to cause familial PD was α-synuclein (SNCA). This has been associated with autosomal dominant PD. At least four mutations in SNCA have been described in literature by various groups (1–4). Two other mutations in genes result in autosomal dominant PD including those in the leucine-rich repeat kinase 2 (LRRK2) (5) and glucocerebrosidase (GBA) genes (6,7). The disease can also be inherited recessively due to homozygous mutations in PTEN-induced kinase 1 (PINK1) (8–10), Parkin (PARK2) (11,12), *ATP13A2*, and DJ1 (13–16). More recently, genome-wide association studies were performed on over 13,500 cases and almost 95,000 control samples (17). The study resulted in the identification of 24 loci with increased risk of Parkinson's, of which 6 were newly identified single nucleotide polymorphisms. In total, the study identified 28 independent risk variants for PD in 24 gene loci with risk profile analysis demonstrating substantial cumulative risk.

As stated earlier, PD results from the degeneration and the loss of dopaminergic neurons located in the substantia nigra region of the midbrain (18). These neurons typically project to the striatal region of the basal ganglia, which is involved in motor control. The loss of dopaminergic neurons is associated with the formation and the accumulation of Lewy bodies (19). These inclusions were first described by Friedrich H. Lewy in the nucleus basalis of Meynert in 1912 in PD patient autopsies. Two major components of these Lewy bodies include SCNA (coded by *SCNA*) and Tau protein (20,21). Until now, over 70 different components have been described to be present in these aggregates including some PD-related genes, including *DJ-1, LRRK1, PARKIN,* and *PINK1.* These are believed to negatively affect cell survival as they are often found in regions with neurodegeneration (22–25). The spread of Lewy bodies in the brain is strongly correlated with motor symptoms in these patients.

5.3 Generating Dopaminergic Neurons from Pluripotent Stem Cells

Numerous groups have reported the generation of tyrosine hydroxylase (TH)-positive dopaminergic neurons from human and mouse PSCs. In general, protocols have either focused on the (a) use of coculture methods using certain stromal cell line including PA6 and astrocytes or (b) systematic exposure to biomimicking factors that induce differentiation in a stepwise fashion following cues from developmental biology studies (Figure 5.1). A number of the protocols were initially developed for mESCs and later applied to hESCs and iPSCs.

The generation of dopaminergic neurons by coculturing mESCs with stromal cell lines including PA6 and MS5 was first reported by the Sasai lab (26).

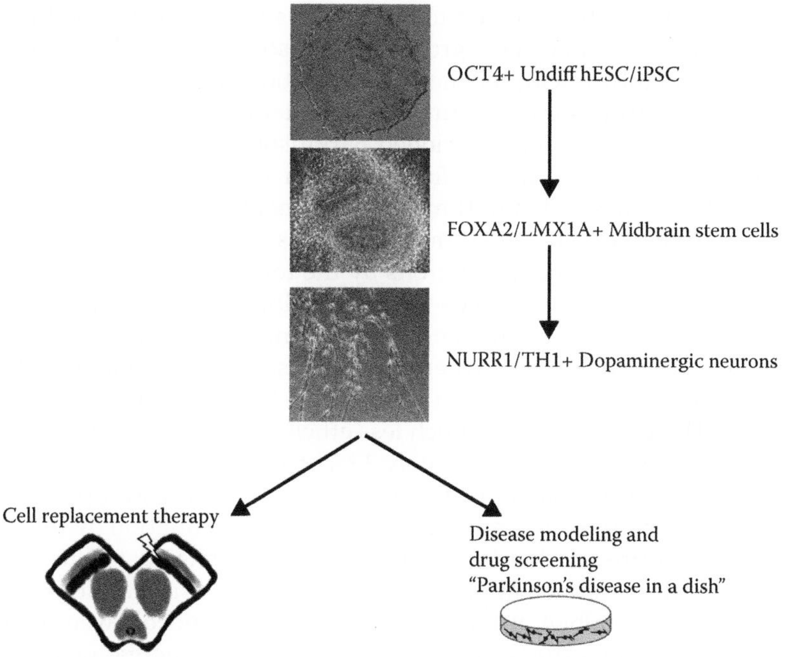

FIGURE 5.1
Schematic showing step-wise differentiation of pluripotent stem cells to dopaminergic neurons and potential uses of this technology.

This method was very efficient, with the group reporting more than 90% neuronal cells identified by neural cell adhesion molecule staining following one week of coculture. Approximately 30% of the Tuj1-positive neurons were TH-positive dopaminergic neurons after 12 days in coculture. The same group later published similar highly efficient dopaminergic neuron generation from nonhuman primate ESCs using the coculture method (27). This stromal coculture method has been modified for the generation of dopaminergic neurons from hESCs. hESCs could be differentiated into dopaminergic neurons following coculture with the mouse stromal cell line MS5 for four weeks followed by two to three weeks of Shh and FGF8 exposures in the absence of any stromal cells (28). This protocol resulted in the highly efficient generation of dopaminergic neurons (approximately 79%). Others have used the PA6 stromal line to show similar dopaminergic neuronal differentiation of hESCs in the presence of glial cell-derived neutrophic factor (GDNF) (29) as well as in the absence of any growth factors (30,31). These hESC-derived dopaminergic neurons also secreted dopamine into the media in response to stimulation with potassium chloride and exhibited typical neuronal electrophysiological profile.

These methods further evolved to the use of stromal conditioned media (32). Apart from stromal cells, groups have also generated dopaminergic neurons by coculture with telomerase-immortalized midbrain astrocytes (33). These cells were shown to functionally integrate following transplantation in 6-hydroxydopamine-lesioned parkinsonian rats. However, these cultures still contained a significant portion of undifferentiated hESC population, which resulted in tumor development.

As an alternative to using stromal cells, labs developed methodologies for generating dopaminergic neurons by the formation of embryoid bodies. This involved culturing cells *en masse* in low-attachment plates or as droplet cultures followed by plating, isolation, and expansion of neural precursors. These precursors where then exposed to a combination of several growth factors including Shh and FGF8 to bias their fate toward midbrain (34). The procedure was much less efficient compared to the stromal cocultures resulting in approximately 33% of TH+ dopaminergic neurons following exposure to Shh or FGF8, and ascorbic acid. The group showed a further enrichment of dopaminergic neuronal differentiation by the overexpression of transcription factor nuclear receptor-related 1 (Nurrl) (35). These methods were then applied to hESCs to induce similar dopaminergic differentiation albeit at low efficiency (36,37). Protocols were further refined to generate a higher number of hESC dopaminergic neurons by applying key secreted factors known to promote midbrain polarization of anterior neuroepithelium. These included the use of SHH and FGF8 at different culture periods to closely match midbrain development (38). These hESC-derived DA neurons had properties of mature and functional neurons upon electrophysiological analysis, and the cells released dopamine into the media upon stimulation. Alternate protocols using a variety of neuronal inducers including small molecules affecting the TGF-beta and Wnt pathways, as well as, supportive neurotrophic factors including BDNF and GDNF have been used to further enhance dopaminergic differentiation (39–43).

Others have worked on getting the protocols current Good Manufacturing Practice (cGMP) compliant. A good cGMP-compliant protocol would help fast track the cells to the clinic. The Xianmin Zeng lab has taken the lead on the refinement of the protocol for cGMP manufacture of these cells, and the lab published a four-step scalable process which is amenable to cGMP processing (44,45). The protocol allowed for the generation of cryopreservable neural stem cells and dopaminergic precursors and a final maturation step allowing for over 40% TH+ neurons. They further described a small molecule that could be added to the process to further eliminate any proliferating and undifferentiated cells without affecting the dopaminergic neurons (46). This would further promote the safety of the final product.

5.4 Cell Replacement Using Stem Cell-Derived Dopaminergic Neurons

Appropriate cell population for optimal integration and functional recovery is another critical area of research not just for PD but also for a whole host of degenerative disorders. It is critical to have a population of cells which are capable of surviving the dissociation and the transplantation procedure but at the same time do not contain any significant population of stem cells especially undifferentiated cells which could proliferate and take over the host cells and more important result in tumors. A study published by Pruszak et al. (47) demonstrated sorting methods of identifying the neuronal population by labeling neurons with synapsin1–GFP from undifferentiated embryonic cells using SSEA-1. Using this combinatorial positive and negative sorting strategy, they were able to purify differentiated dopaminergic neurons from a mixed culture thereby decreasing the risk of tumorigenesis following transplantation. In order to identify the most appropriate stage for maximal integration of ESC-derived dopaminergic precursors and neurons, Studer lab generated cells from ESCs derived from three different reporter lines matching three stages of dopaminergic neuron differentiation (early neural stem cell stage, middle dopaminergic precursors, and late mature dopaminergic neurons) following the induction of differentiation using Hes5-GFP, Nurrl-GFP, and Pitx3-YFP, respectively (48). Flourescence-activated cell sorting-purified cells from each line were then transplanted and analyzed for integration efficiency. Interestingly, they found that the mid-stage Nurrl+ cells had the best survival following transplantation in mice. They also tested for the recovery of motor deficits in a hemiparkinsonian mice model and confirmed that Nurrl-GFP cells could robustly restore motor function in these mice.

A few groups have recently published on efficient dopaminergic neuronal differentiation from human ESCs and their functional integration following transplantation (40,41,49). Using an approach based on a FOXA2-expressing neural stem cells, these publications showed highly efficient induction of dopaminergic neurons which, following engraftment into rodent models of PD, showed more robust survival and functional integration as far out as 18 weeks (Table 5.1). The group also compared and found equivalent integration capacities between hESC-derived dopaminergic neurons and those isolated from human fetal tissue. They also confirmed that the graft did not lead to teratoma formation and lost almost all of their proliferative potential by 18 weeks *in vivo*. Others also showed that the transplanted cells can project and connect across long distances required for human cortical integration and functional recovery which they confirmed using magnetic resonance and positron emission tomography imaging technologies (49).

TABLE 5.1

Publications Describing Various Patient-Derived iPSC Lines for PD Research

Gene	Publication	Mutation
SNCA	(50)	Triplication
	(51)	Triplication
	(52)	A53T, E46K
LRRK2	(53,54)	G2019S
	(55)	G2019S
	(56)	G2019S
	(57)	G2019S
	(58)	G2019S, R1441C
	(59)	L444P, N370S
PINK1	(60)	C1366T, C509G
	(61)	C509G
	(58)	Q456X
PARK2	(62)	Exon 3/5 deletion
	(63)	R42P, exon 3 deletion, exon 3/4 deletion, R275W
GBA1	(64)	N370S

5.5 Directly Reprogrammed Dopamine Neurons

Direct reprogramming or induced transdifferentiation refers to the process that allows cells to directly progress from one cell fate to another without an intermediate stem cell fate. The earliest landmark paper describing this was the direct conversion of fibroblasts cells to myoblasts using a single transcription factor MyoD1 (65). More recently, a similar transdifferentiation strategy to generate neurons from fibroblast has been described (66). The authors show that just three genes, ASCL1, BRN2, and MYT1L, are adequate to generate generic neurons from fibroblasts. The generation of these directly reprogrammed neurons from somatic cells might have significant implications for understanding neuronal development processes and disease development *in vitro* and in cell replacement therapies. Direct reprogramming gets around issues such as unwanted proliferation and tumor generation as well as time required to generate these cells from stem cells which can take weeks to months. Recent studies have reported the success of generating DA neurons by directly reprogramming the fibroblasts with different combinations of transcription factors Mash1 (Ascl1), Nurr1 (Nr4a2), Lmx1a, Ngn2, Sox2, and Pitx3 (67–69).

The first reported generation of the dopamine neurons directly reprogrammed from human and mouse fibroblasts was by the Broccoli group (67). They used a combination of three transcription factors, Ascl1, Nurr1, and Lmx1a, based on their known expression in midbrain dopaminergic neurons, to directly convert mouse and human fibroblasts to functional dopaminergic neurons. They have showed that the directly converted dopamine neurons had similar electrophysiological activity and expression pattern native to the dopaminergic neurons. Another group cross-compared various combinations of a group of eight different transcription factors including Ascl1, Mytl1, Brn2, Lmx1a, Lmx1b, Nurr1, Pitx3, and En1 in mouse fibroblasts to evaluate their capacity to induce midbrain dopaminergic neuron-like cells (69). They show that of all the genes tested Ascl1 and Pitx3 were necessary for inducing fibroblast to dopamine neurons. These cells showed typical voltage-dependant membrane currents. They then went on to transplant these induced dopaminergic neurons into a mouse model with PD and demonstrated that the transplanted cells survived in the host. These cells reenervated the lesioned striatum when analyzed eight weeks later. Upon testing for functional integration, they observed a partial rescue in the amphetamine-induced rotational behavior in the induced neuron transplant group but not in the control fibroblast-injected mice. One of the biggest disadvantages of this approach so far has been the extremely poor efficiency toward direct transdifferentiation. If this can be overcome, there is hope to both generate cells *in vitro* or even potentially directly reprogram surrounding glia and astrocytes to dopaminergic neurons *in vivo*.

5.6 Modeling PD Using iPSCs *In Vitro*

hiPSCs are an excellent system to study the molecular mechanisms underlying disease development *in vitro* as they are generated from patient somatic cells. The other advantage is the fact that these cells could also potentially be used then for drug discovery for personalized medicine. A number of reports have been published of generating iPSCs from PD patients, and some have provided valuable insight on the disorder.

As described earlier, one of the genes very strongly associated with the familial form of PD is SNCA. The SNCA protein is the principal component of the Lewy bodies which are the hallmark of the disorder. The exact role the protein plays in the disease context is unclear. It has however been suggested that it may interfere with the presynaptic release of dopamine. SNCA mutations may lead to alteration of the conformation of SNCA protein, resulting in misfolding, and aggregate formation, leading to neuronal dysfunction and death. A number of mutations in the gene have been identified including A53T mutation (3), A30P mutation (2), E46K mutations (4), and more recently H50Q mutation identified in exon 4 of the SNCA gene (1). In 2003, Singleton et al. (70) published a report of a patient family with triplication at the SCNA locus. These showed that this could result in excessive production of mutant protein and its aggregation. This family has since been used to generate a number of iPSC lines for disease modeling. The first of such reports demonstrated that following dopaminergic differentiation, these cells expressed much higher levels of the SNCA protein. The iPSC-derived dopaminergic neurons had twice the amount of SCNA protein as compared to controls (50). Interestingly, this phenotype was only observed in the differentiated lines and not in the iPSCs themselves. In another study from the same cohort, the disease lines increased the expression of effectors of reactive oxygen species (ROS)-associated NRF2 pathway including hemoxygenase 2 and monoamine oxidase (51). The group then went on to show that the patient-derived cells were highly susceptible to ROS stress especially in the absence of vitamin C.

Leucine-rich repeat kinase 2 (LRRK2) is the most common form of familial PD (5). The most common mutation is G2019S, found in the mixed lineage kinase-like domain, which results in a kinase gain-of-function phenotype (71). Dopamine neurons from iPSC lines generated from a patient carrying this mutation were analyzed (55). The group reported that the cells had an increased accumulation of SNCA protein. Additionally, there was an increase in the expression of ROS-associated genes, including *HSPB1*, and increased susceptibility to ROS-associated cell death after exposure to hydrogen peroxide. Another set of parallel studies by another group using iPSC lines with the above mutation also observed a similar SNCA protein accumulation (72). Additionally, they reported neuronal morphology phenotypes including reduced neurite length. Finally, Orenstein et al. (56) looked at the role of and identified LRRK2 as a target of chaperone-mediated autophagy (CMA). When they looked at G2019S-containing dopaminergic neurons, LRRK2 degradation was compromised and that this resulted in the blockage of the CMA degradation pathway. They also observed phenotypes described earlier including neurite shortening and accumulation of SNCA which colocalized with lysosome-associated membrane protein 2A (LAMP2A) at the lysosomal membranes. Cooper et al. (58) assessed mitochondrial dysfunction associated with LRRK2 mutations (G2019S and R1441C). Patient iPSC-derived neurons exhibited a reduced basal oxygen consumption rate compared to healthy controls. Upon exposure of these neurons to high concentrations of hydrogen peroxide, the cells produced less glutathione (GSH) compared to controls, suggesting that the LRRK2 mutant dopaminergic neurons have impaired ability to respond to ROS stress. Additionally, the lines also contained shorter mitochondria. *In vitro* gene therapy has also been demonstrated successfully (57). The authors used ZFN technology to have successfully corrected the G2019S point mutation in two patient-derived iPSC lines. The gene-corrected lines had a restored neurite length to levels similar to the controls. Additionally, the G2019S genetically corrected lines did not demonstrate an SNCA accumulation as observed in mutant lines.

Two mutations strongly associated with early onset PD include those in PINK1 and PARK2. PINK1 proteins play an important role in the recruitment of PARKIN protein, which is encoded by PARK2 at sites of mitochondrial damage. PARKIN is an E3 ubiquitin ligase which plays a critical role in mitophagy by ubiquitinating various mitochondrial proteins thereby promoting their degradation. iPSC lines have been generated from patients containing PINK1 missense mutation (60). Upon dopaminergic neuronal differentiation, they found that the cells expressed lower levels of PINK1 compared to healthy controls. The cells were then exposed to drugs promoting mitochondrial depolarization and damage. In the patient lines, interestingly, PARKIN failed to get recruited to sites of mitochondrial damage. The phenotype could then be rescued by viral-mediated overexpression of normal PINK1 protein in the cells. Similar studies were also carried out by the Klein lab (61). They compared Parkinson phenotypes in the patient fibroblasts and the iPSC-derived neurons. They interestingly discovered that patient lines had variable response such that they concluded that mitophagy differs between fibroblasts and neuronal cells and also between native and overexpressed PARKIN *in vitro* systems. Similarly, another study using PINK1 mutant lines from patients looked at mitochondrial dysfunction in original fibroblasts and in the iPSC-derived neurons (58). PINK1 iPSC-derived neurons exhibit lowered production of reduced GSH synthase upon exposure to either valinomycin or concanamycin, drugs which trigger mitochondrial depolarization, compared to control lines. Additionally, the treatment resulted in increased levels of ROS production in PINK1 iPSC-derived neurons but not in controls. The data confirmed the original ideas of increased susceptibility of dopaminergic neurons in Parkinson's patients to ROS. The group then sought to test compounds or drugs that may potentially rescue the phenotype. They found that exposure to rapamycin and GW5074, a LRRK2 inhibitor, prior to drug exposure, could reduce ROS production in the patient lines. Similarly, another antioxidant, Coenzyme Q, was neuroprotective in these cultures.

iPSC lines have also been derived from patients with loss-of-function exon deletions in PARK2 resulting in loss of PARKIN (62). Dopaminergic neurons differentiated from these iPSCs have impaired dopamine uptake compared to control lines. They found that this also strongly correlated with lower dopamine active transporter protein expression. Upon dopamine exposure in the culture media, the cells had elevated ROS levels resulting in protein carbonylation. This phenotype could then be reversed by viral-mediated overexpression by normal PARKIN. Recent work out of the Zeng lab looked at mitochondrial alterations in dopaminergic neurons in PARK2 patient-specific and in isogenic knockout lines (63). Upon direct comparison, PD patient-derived lines carrying various mutations in PARK2 generated far fewer dopaminergic neurons when compared with an age-matched control line. This reduction was accompanied by alterations in the mitochondrial volume fraction. The same phenotype was confirmed in isogenic PARK2 null lines generated using ZFN technology. Upon whole genome expression analysis, they observed an upregulation of SNCA in these cultures along with the upregulation of a host of autophagy-related genes as well as BCL2 family member, HARAKIRI.

Mutations in GBA1, a gene normally associated with lysosomal disorder Gaucher's disease, has also been shown to result in PD. Two heterozygous mutations, N370S and L444P, have been strongly associated with the early onset familial form of the disease (7,73,74). Dopaminergic neurons have been derived from N370S mutation bearing patient iPSCs and healthy controls (64). The group observed a significant increase in SNCA levels in the neurons derived from GBA1-mutant iPSC lines. Additionally, the neurons also had reduced GBA activity compared to healthy controls. The group replicated the findings by shRNA-mediated knockdown of GBA-1 in primary cortical neurons in culture. They also observed an increased trapping of GBA in the endoplasmic reticulum, which led to further SNCA accumulation. Interestingly, A53T SNCA transgenic mice also exhibit reduced GBA activity (75) suggesting the SNCA accumulation leads to GBA dysfunction in PD and may be important for disease progression.

5.7 Conclusions

PD is an extremely debilitating condition. Although there are a few drugs in the market offering some symptomatic relief, a more permanent solution is being highly sought. Cell replacement therapy is an ideal solution for these patients to provide a long-term relief. There are a number of cell sources that could potentially benefit these patients including iPSCs, ESCs, and induced neurons, which have a distinct advantage over the previously tried human fetal tissue replacement strategies in terms of both cell supply issues and ethical issues associated with using human fetal tissue. Although there are no ongoing clinical trials yet, the preclinical data from multiple groups have so far been extremely encouraging. It however remains to be seen if patients will have similar benefits or have some of the debilitating side effects such as dyskinesias which have been reported in human fetal transplant trials.

The iPSC field has also revolutionized personalized medicine. The ability to now be able to create a disease in a dish will allow for more focused drug-screening approaches. This should ideally lead to new better-acting drugs and to fewer failures in late phase clinical trials benefitting both the patients and the pharmaceutical industry.

References

1. Appel-Cresswell, S. et al., Alpha-synuclein p.H50Q, a novel pathogenic mutation for Parkinson's disease. *Mov Disord*, 2013. 28(6): pp. 811–3.
2. Kruger, R. et al., Ala30Pro mutation in the gene encoding alpha-synuclein in Parkinson's disease. *Nat Genet*, 1998. 18(2): pp. 106–8.

3. Polymeropoulos, M. H. et al., Mutation in the alpha-synuclein gene identified in families with Parkinson's disease. *Science*, 1997. 276(5321): pp. 2045–7.
4. Zarranz, J. J. et al., The new mutation, E46K, of alpha-synuclein causes Parkinson and Lewy body dementia. *Ann Neurol*, 2004. 55(2): pp. 164–73.
5. Zimprich, A. et al., The PARK8 locus in autosomal dominant parkinsonism: Confirmation of linkage and further delineation of the disease-containing interval. *Am J Hum Genet*, 2004. 74(1): pp. 11–9.
6. Aharon-Peretz, J., H. Rosenbaum, and R. Gershoni-Baruch, Mutations in the glucocerebrosidase gene and Parkinson's disease in Ashkenazi Jews. *N Engl J Med*, 2004. 351(19): pp. 1972–7.
7. Neumann, J. et al., Glucocerebrosidase mutations in clinical and pathologically proven Parkinson's disease. *Brain*, 2009. 132(Pt 7): pp. 1783–94.
8. Valente, E. M. et al., Hereditary early-onset Parkinson's disease caused by mutations in PINK1. *Science*, 2004. 304(5674): pp. 1158–60.
9. Rogaeva, E. et al., Analysis of the PINK1 gene in a large cohort of cases with Parkinson disease. *Arch Neurol*, 2004. 61(12): pp. 1898–904.
10. Groen, J. L. et al., Genetic association study of PINK1 coding polymorphisms in Parkinson's disease. *Neurosci Lett*, 2004. 372(3): pp. 226–9.
11. Kitada, T. et al., Mutations in the parkin gene cause autosomal recessive juvenile parkinsonism. *Nature*, 1998. 392(6676): pp. 605–8.
12. Hattori, N. et al., Molecular genetic analysis of a novel Parkin gene in Japanese families with autosomal recessive juvenile parkinsonism: Evidence for variable homozygous deletions in the Parkin gene in affected individuals. *Ann Neurol*, 1998. 44(6): pp. 935–41.
13. Miller, D. W. et al., L166P mutant DJ-1, causative for recessive Parkinson's disease, is degraded through the ubiquitinproteasome system. *J Biol Chem*, 2003. 278(38): pp. 36588–95.
14. Macedo, M. G. et al., The DJ-1L166P mutant protein associated with early onset Parkinson's disease is unstable and forms higher-order protein complexes. *Hum Mol Genet*, 2003. 12(21): pp. 2807–16.
15. Hague, S. et al., Early-onset Parkinson's disease caused by a compound heterozygous DJ-1 mutation. *Ann Neurol*, 2003. 54(2): pp. 271–4.
16. Bonifati, V. et al., Mutations in the DJ-1 gene associated with autosomal recessive early-onset parkinsonism. *Science*, 2003. 299(5604): pp. 256–9.
17. Nalls, M. A. et al., Large-scale meta-analysis of genome-wide association data identifies six new risk loci for Parkinson's disease. *Nat Genet*, 2014. 46(9): pp. 989–93.
18. Damier, P. et al., The substantia nigra of the human brain. II. Patterns of loss of dopamine-containing neurons in Parkinson's disease. *Brain*, 1999. 122 (Pt 8): pp. 1437–48.
19. Wakabayashi, K. et al., The Lewy body in Parkinson's disease: Molecules implicated in the formation and degradation of alpha-synuclein aggregates. *Neuropathology*, 2007. 27(5): pp. 494–506.
20. Galloway, P. G., C. Bergeron, and G. Perry, The presence of tau distinguishes Lewy bodies of diffuse Lewy body disease from those of idiopathic Parkinson disease. *Neurosci Lett*, 1989. 100(1–3): pp. 6–10.
21. Ishizawa, T. et al., Colocalization of tau and alpha-synuclein epitopes in Lewy bodies. *J Neuropathol Exp Neurol*, 2003. 62(4): pp. 389–97.

22. Redeker, V. et al., Identification of protein interfaces between alpha-synuclein, the principal component of Lewy bodies in Parkinson disease, and the molecular chaperones human Hsc70 and the yeast Ssa1pp. *J Biol Chem*, 2012. 287(39): pp. 32630–9.
23. Johansen, K. K. et al., Biomarkers: Parkinson disease with dementia and dementia with Lewy bodies. *Parkinsonism Relat Disord*, 2010. 16(5): pp. 307–15.
24. Klein, J. C. et al., Neurotransmitter changes in dementia with Lewy bodies and Parkinson disease dementia in vivo. *Neurology*, 2010. 74(11): pp. 885–92.
25. Trojanowski, J. Q. and V. M. Lee, Aggregation of neurofilament and alpha-synuclein proteins in Lewy bodies: Implications for the pathogenesis of Parkinson disease and Lewy body dementia. *Arch Neurol*, 1998. 55(2): pp. 151–2.
26. Kawasaki, H. et al., Induction of midbrain dopaminergic neurons from ES cells by stromal cell-derived inducing activity. *Neuron*, 2000. 28(1): pp. 31–40.
27. Takagi, Y. et al., Dopaminergic neurons generated from monkey embryonic stem cells function in a Parkinson primate model. *J Clin Invest*, 2005. 115(1): pp. 102–9.
28. Perrier, A. L. et al., Derivation of midbrain dopamine neurons from human embryonic stem cells. *Proc Natl Acad Sci USA*, 2004. 101(34): pp. 12543–8.
29. Buytaert-Hoefen, K. A., E. Alvarez, and C. R. Freed, Generation of tyrosine hydroxylase positive neurons from human embryonic stem cells after coculture with cellular substrates and exposure to GDNF. *Stem Cells*, 2004. 22(5): pp. 669–74.
30. Zeng, X. et al., Dopaminergic differentiation of human embryonic stem cells. *Stem Cells*, 2004. 22(6): pp. 925–40.
31. Carpenter, M. et al., Derivation and characterization of neuronal precursors and dopaminergic neurons from human embryonic stem cells in vitro. *Methods Mol Biol*, 2006. 331: pp. 153–67.
32. Swistowska, A. M. et al., Stage-specific role for shh in dopaminergic differentiation of human embryonic stem cells induced by stromal cells. *Stem Cells Dev*, 2010. 19(1): pp. 71–82.
33. Roy, N. S. et al., Functional engraftment of human ES cell-derived dopaminergic neurons enriched by coculture with telomerase-immortalized midbrain astrocytes. *Nat Med*, 2006. 12(11): pp. 1259–68.
34. Lee, S. H. et al., Efficient generation of midbrain and hindbrain neurons from mouse embryonic stem cells. *Nat Biotechnol*, 2000. 18(6): pp. 675–9.
35. Kim, J. H. et al., Dopamine neurons derived from embryonic stem cells function in an animal model of Parkinson's disease. *Nature*, 2002. 418(6893): pp. 50–6.
36. Reubinoff, B. E. et al., Neural progenitors from human embryonic stem cells. *Nat Biotechnol*, 2001. 19(12): pp. 1134–40.
37. Zhang, S. C. et al., In vitro differentiation of transplantable neural precursors from human embryonic stem cells. *Nat Biotechnol*, 2001. 19(12): pp. 1129–33.
38. Yan, Y. et al., Directed differentiation of dopaminergic neuronal subtypes from human embryonic stem cells. *Stem Cells*, 2005. 23(6): pp. 781–90.
39. Chambers, S. M. et al., Highly efficient neural conversion of human ES and iPS cells by dual inhibition of SMAD signaling. *Nat Biotechnol*, 2009. 27(3): pp. 275–80.
40. Kriks, S. et al., Dopamine neurons derived from human ES cells efficiently engraft in animal models of Parkinson's disease. *Nature*, 2011. 480(7378): pp. 547–51.

41. Kirkeby, A. et al., Generation of regionally specified neural progenitors and functional neurons from human embryonic stem cells under defined conditions. *Cell Rep*, 2012. 1(6): pp. 703–14.
42. Rhee, Y. H. et al., Protein-based human iPS cells efficiently generate functional dopamine neurons and can treat a rat model of Parkinson disease. *J Clin Invest*, 2011. 121(6): pp. 2326–35.
43. Sonntag, K. C. et al., Enhanced yield of neuroepithelial precursors and midbrain-like dopaminergic neurons from human embryonic stem cells using the bone morphogenic protein antagonist noggin. *Stem Cells*, 2007. 25(2): pp. 411–8.
44. Swistowski, A. et al., Xeno-free defined conditions for culture of human embryonic stem cells, neural stem cells and dopaminergic neurons derived from them. *PLoS One*, 2009. 4(7): p. e6233.
45. Swistowski, A. and X. Zeng, Scalable production of transplantable dopaminergic neurons from hESCs and iPSCs in xeno-free defined conditions. *Curr Protoc Stem Cell Biol*, 2012. *Chapter* 2: p. Unit2D 12.
46. Han, Y. et al., Identification by automated screening of a small molecule that selectively eliminates neural stem cells derived from hESCs but not dopamine neurons. *PLoS One*, 2009. 4(9): p. e7155.
47. Pruszak, J. et al., Markers and methods for cell sorting of human embryonic stem cell-derived neural cell populations. *Stem Cells*, 2007. 25(9): pp. 2257–68.
48. Ganat, Y. M. et al., Identification of embryonic stem cell-derived midbrain dopaminergic neurons for engraftment. *J Clin Invest*, 2012. 122(8): pp. 2928–39.
49. Grealish, S. et al., Human ESC-derived dopamine neurons show similar preclinical efficacy and potency to fetal neurons when grafted in a rat model of Parkinson's disease. *Cell Stem Cell*, 2014. 15(5): pp. 653–65.
50. Devine, M. J. et al., Parkinson's disease induced pluripotent stem cells with triplication of the alpha-synuclein locus. *Nat Commun*, 2011. 2: p. 440.
51. Byers, B. et al., SNCA triplication Parkinson's patient's iPSC-derived DA neurons accumulate alpha-synuclein and are susceptible to oxidative stress. *PLoS One*, 2011. 6(11): p. e26159.
52. Soldner, F. et al., Generation of isogenic pluripotent stem cells differing exclusively at two early onset Parkinson point mutations. *Cell*, 2011. Jul 22; 146(2): pp. 318–31.
53. Sánchez-Danés, A. et al., Efficient generation of A9 midbrain dopaminergic neurons by lentiviral delivery of LMX1A in human embryonic stem cells and induced pluripotent stem cells. *Hum Gene Ther*, 2012. Jan; 23(1): pp. 56–69.
54. Sánchez-Danés, A. et al., Disease-specific phenotypes in dopamine neurons from human iPS-based models of genetic and sporadic Parkinson's disease. *EMBO Mol Med*, 2012. May; 4(5): pp. 380–95.
55. Nguyen, H. N. et al., LRRK2 mutant iPSC-derived DA neurons demonstrate increased susceptibility to oxidative stress. *Cell Stem Cell*, 2011. 8(3): pp. 267–80.
56. Orenstein, S. J. et al., Interplay of LRRK2 with chaperone-mediated autophagy. *Nat Neurosci*, 2013. 16(4): pp. 394–406.
57. Reinhardt, P. et al., Genetic correction of a LRRK2 mutation in human iPSCs links parkinsonian neurodegeneration to ERK-dependent changes in gene expression. *Cell Stem Cell*, 2013. 12(3): pp. 354–67.
58. Cooper, O. et al., Pharmacological rescue of mitochondrial deficits in iPSC-derived neural cells from patients with familial Parkinson's disease. *Sci Transl Med*, 2012. 4(141): pp. 141ra90.

59. Panicker, L. M. et al., Induced pluripotent stem cell model recapitulates pathologic hallmarks of Gaucher disease. *Proc Natl Acad Sci USA*, 2012. Oct 30; 109(44): pp. 18054–9.
60. Seibler, P. et al., Mitochondrial Parkin recruitment is impaired in neurons derived from mutant PINK1 induced pluripotent stem cells. *J Neurosci*, 2011. 31(16): pp. 5970–6.
61. Rakovic, A. et al., Phosphatase and tensin homolog (PTEN)-induced putative kinase 1 (PINK1)-dependent ubiquitination of endogenous Parkin attenuates mitophagy: Study in human primary fibroblasts and induced pluripotent stem cell-derived neurons. *J Biol Chem*, 2013. 288(4): pp. 2223–37.
62. Jiang, H. et al., Parkin controls dopamine utilization in human midbrain dopaminergic neurons derived from induced pluripotent stem cells. *Nat Commun*, 2012. 3: p. 668.
63. Shaltouki, A. et al., Mitochondrial alterations by PARKIN in dopaminergic neurons using PARK2 patient-specific and PARK2 knockout isogenic iPSC lines. *Stem Cell Reports*, 2015. 4(5): pp. 847–59.
64. Mazzulli, J. R. et al., Gaucher disease glucocerebrosidase and alpha-synuclein form a bidirectional pathogenic loop in synucleinopathies. *Cell*, 2011. 146(1): pp. 37–52.
65. Davis, R. L., H. Weintraub, and A. B. Lassar, Expression of a single transfected cDNA converts fibroblasts to myoblasts. *Cell*, 1987. 51(6): pp. 987–1000.
66. Vierbuchen, T. et al., Direct conversion of fibroblasts to functional neurons by defined factors. *Nature*, 2010. 463(7284): pp. 1035–41.
67. Caiazzo, M. et al., Direct generation of functional dopaminergic neurons from mouse and human fibroblasts. *Nature*, 2011. 476(7359): pp. 224–7.
68. Liu, X. et al., Direct reprogramming of human fibroblasts into dopaminergic neuron-like cells. *Cell Res*, 2012. 22(2): pp. 321–32.
69. Kim, J. et al., Functional integration of dopaminergic neurons directly converted from mouse fibroblasts. *Cell Stem Cell*, 2011. 9(5): pp. 413–9.
70. Singleton, A. B. et al., alpha-Synuclein locus triplication causes Parkinson's disease. *Science*, 2003. 302(5646): pp. 841.
71. West, A. B. et al., Parkinson's disease-associated mutations in leucine-rich repeat kinase 2 augment kinase activity. *Proc Natl Acad Sci USA*, 2005. 102(46): pp. 16842–7.
72. Sanchez-Danes, A. et al., Disease-specific phenotypes in dopamine neurons from human iPS-based models of genetic and sporadic Parkinson's disease. *EMBO Mol Med*, 2012. 4(5): pp. 380–95.
73. DePaolo, J. et al., The association between mutations in the lysosomal protein glucocerebrosidase and parkinsonism. *Mov Disord*, 2009. 24(11): pp. 1571–8.
74. Sidransky, E., G. M. Pastores, and M. Mori, Dosing enzyme replacement therapy for Gaucher disease: Older, but are we wiser? *Genet Med*, 2009. 11(2): pp. 90–1.
75. Sardi, S. P. et al., Augmenting CNS glucocerebrosidase activity as a therapeutic strategy for parkinsonism and other Gaucher-related synucleinopathies. *Proc Natl Acad Sci USA*, 2013. 110(9): pp. 3537–42.

6

Huntington's Disease and Stem Cells

Karen Ring, Robert O'Brien, Ningzhe Zhang, and Lisa M. Ellerby

CONTENTS

6.1 Introduction on Huntington's Disease

Huntington's disease (HD) is a progressive neurodegenerative disorder that affects between 5 and 10 out of 100,000 individuals throughout the world (1). HD is a dominantly inherited, late-onset disorder with symptoms manifesting between the ages of 30 and 50 years (2). Patients afflicted with HD experience loss in muscle coordination, chorea, personality changes, impaired cognitive abilities, psychiatric problems, and consequently premature death (3).

HD is caused by an abnormal expansion of CAG repeats, which encode an extended polyglutamine tract, within the first exon of the Huntingtin gene (*HTT*) located on the short arm of human chromosome 4. Individuals with 35 CAG repeats or fewer do not present with HD symptoms, while individuals with 40 repeats or more acquire HD at some point in their lives (4). Disease onset and severity are correlated with an increase in repeat number, and genetic anticipation can occur with each successive generation manifesting the disease earlier than the previous generation due to spontaneous increase in CAG repeat number (Figure 6.1).

HD neuropathology has been well characterized in postmortem HD brains. Typical pathology presents as atrophy of the caudate and the putamen, which together make up the striatum and are located in the basal ganglia of the brain (Figure 6.1). The basal ganglia control motor movement and cognitive processes. It receives inputs from the cortex layers II–VI through cortical pyramidal projection neurons. The cerebral cortex is also degenerated in HD

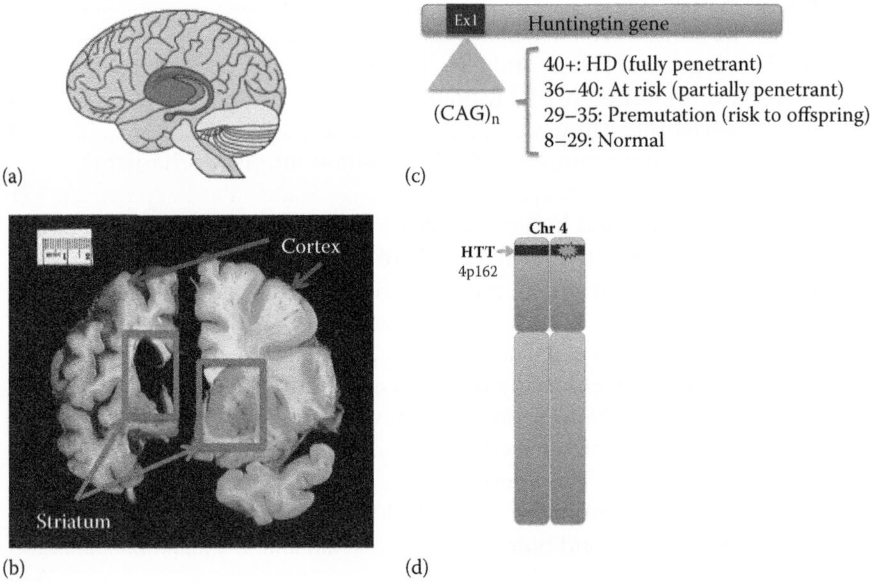

FIGURE 6.1
(See color insert.) (a) The striatum is located below the cortex in the forebrain in a structure called the *basal ganglia*. Shown in pink are the caudate nucleus and putamen, which are part of the striatum. In orange is the thalamus, which separates the cortex from the midbrain. The basal ganglia play an important role in controlling movement and behavior, and patients with HD experience disturbances in both those actions. (b) Postmortem brain tissue from (*left*) a Huntington's disease patient and (*right*) a normal individual. The HD brain has substantial atrophy in the striatum due to the loss of striatal MSNs and also has atrophy in the cortex due to loss of cortical projection neurons that innervate in the striatum. (c) HD is caused by a mutation in the Huntingtin (HTT) gene. The mutation is a CAG trinucleotide repeat expansion located in exon 1 of HTT. Normal individuals have 29 repeats or less while those with 40 repeats or more will get HD. (d) The HTT gene is located on the short arm of human chromosome 4.

brains. Medium spiny neurons (MSNs) in the striatum and pyramidal neurons in the cortex are the main cell types that are lost in HD. In early stages of the disease, loss of cortical pyramidal neuron projections to the striatum due to axonal degeneration is observed, suggesting a disruption of corticostriatal connectivity in HD (5). Other areas of the brain that exhibit HD-induced pathology to a lesser extent include the hippocampus, the substantia nigra, the thalamus, the cerebellum, and the telencephalic white matter (6). One consideration recently highlighted by Waldvogel et al. (7) is the fact that the HD pathology has a variable pattern, and perhaps this corresponds to the variability of symptoms in HD. Not all patients have the same symptoms of HD. Besides atrophy, other neuropathologies observed in HD brains include gliosis, intranuclear inclusions in neurons, and neuropil aggregates in neuronal processes (8,9).

Neuroimaging techniques have been utilized to monitor changes in the brain in both asymptomatic and symptomatic HD patients (10). The most commonly utilized techniques are positron emission tomography (PET) and magnetic resonance imaging (MRI). The atrophy in the striatum and the cortex and the enlargement of the anterior horns of the lateral ventricles can be severe and easily visualized by MRI (Figure 6.2). MRI detects a progressive loss of volume in the striatum even 15–20 years before the onset of the disease symptoms in HD patients. MRI also detects cortical thinning in HD patients along with changes in white matter and whole brain. Most importantly, gray matter and striatal atrophy are predictors of clinical diagnosis of HD. PET studies in HD patients reveal disruption of the postsynaptic dopaminergic

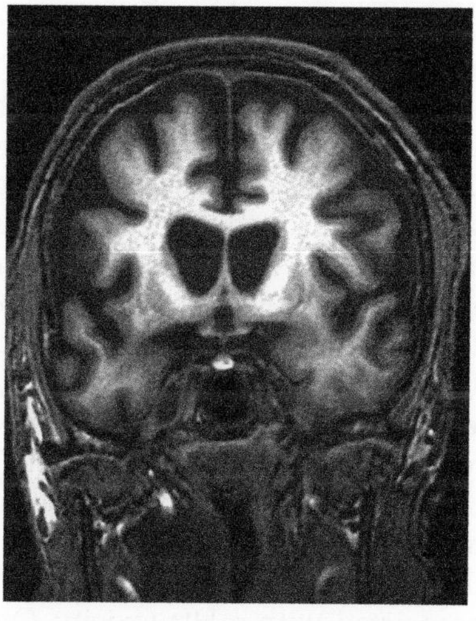

FIGURE 6.2
MRI of HD patient brain reveals significant brain atrophy in the striatum and the cerebral cortex and enlargement of the frontal horns of the lateral ventricles. This image represents a patient at the end stages of HD. (Courtesy of Frank Galland used under the GNU Free Documentation License, Version 1.3.)

system. Dopamine D1 receptor density is reduced in HD patients when compared to controls using radioligand SCH23390 (10).

HTT encodes a large protein called *huntingtin* (Htt). The protein has 3144 amino acids, and the 3D structure of the full-length protein has not been solved. However, it is predicted to contain huntingtin, EF3, PP2A, and TOR1 (HEAT) repeats that form domains. Molecular modeling predicts that the protein may be involved in mechanical force through movement of the HEAT repeat domains (11). The posttranslational modifications of the protein are extensive and include proteolysis, phosphorylation, sumoylation, acetylation, and lipidation (12–14). Many of the posttranslational modifications have been linked to Htt function or toxicity of the polyQ-expanded form of the protein. Htt is involved in numerous cellular processes including transcriptional regulation, vesicular transport, protein trafficking, synaptic transmission, and inhibition of apoptosis (15). Htt is essential during mammalian development, and mice lacking the protein do not survive (16). Htt is ubiquitously expressed throughout the body but is most abundant in the brain (17). Accordingly, Htt plays important roles in neural cells. Htt directly regulates levels of brain-derived neurotrophic factor (BDNF) and interacts with many other important proteins and transcription factors in the brain such as NeuroD and REST/NRSF complex in order to maintain a neuron-specific transcriptional profile during development and differentiation (5,18–23). However, its exact functions are unknown, making it difficult to determine whether dominant negative properties of mutant Htt (mHtt) lead to loss of function of normal Htt or whether the gain of function in mHtt through interactions with protein-binding partners is responsible for HD pathogenesis.

mHtt has an altered protein structure compared to normal Htt caused by protein misfolding due to the polyglutamine expansion (24). mHtt's altered structure affects its normal function and leads to abnormal protein interactions; both of which cause toxicity and death in specific neuronal and glial cells in the brain, including GABAergic MSNs located in the striatum of the basal ganglia, pyramidal neurons located in the cortex, and astrocytes (9,25–28). The mechanism by which mHtt causes neurodegeneration and HD is not clearly understood. However, known phenotypes resulting from mHtt expression involve reduction in BDNF levels, excitotoxicity, impaired proteolysis and autophagy, mitochondrial dysfunction, and transcriptional dysregulation (29). Interestingly, there appears to be events upstream of the reduction of BDNF levels that involve alterations in the TrkB signaling in the striatum (30). In this case, postsynaptic signaling mechanisms were altered due to inappropriate engagement by TrkB receptors.

While the exact mechanisms of HD pathogenesis are unknown, a number of players have been associated with mHtt toxicity. One example is mHtt's role in transcriptional dysregulation by its interaction with the CREB-binding protein (CBP), which acts as a coactivator of numerous gene promoters in the survival pathway and is necessary for cell function (31,32). The structure of the CBP protein contains 18 glutamines, which interact directly with the glutamine tract in mHtt. Thus, mHtt sequesters CBP and inhibits it from conducting its normal function, which negatively impacts gene transcription and results in neurotoxicity. Another pathway mediating mHtt toxicity is the Rhes protein, which is a small GTPase located specifically in the striatum (33). Rhes directly binds to mHtt and facilitates its sumoylation by acting as an E3 ligase. The sumoylated form of mHtt leads to disaggregation of mHTT and elevates the levels of soluble intracellular mHtt, which leads to neurotoxicity and neurodegeneration. The role of Rhes in HD pathogenesis has been supported with *in vivo* studies that discovered that HD transgenic mice crossed with Rhes knockout mice had delayed onset of HD symptoms (34).

Another mechanism of mHtt toxicity is the toxic fragments generated from cleavage of the N-terminal region of mHtt, which accumulate and form intraneuronal inclusions within neurons or neuropil aggregates in neuronal processes. These inclusions and aggregates are thought to impair normal cellular processes such as gene expression and protein folding, degradation, and clearance (9). However, some believe that mHtt aggregates play a neuroprotective role in HD and that soluble monomeric mHtt is the toxic agent in HD (35–37). Because mHtt interacts with many proteins including chaperone proteins, mHtt can cause misfolding of non-Htt interacting proteins. Thus, sequestering mHTT into inclusion bodies or aggregates could be a protective cellular mechanism for maintaining the function of the proteostasis network. One study revealed that MSNs containing mHtt inclusion bodies had better survival rates than neurons without unaggregated, soluble mHtt (38). Another study by Lu et al. (35) found that soluble monomeric mHtt was toxic in neurons derived from HD iPSCs and that aggregated and oligomeric forms of mHtt were not toxic to neurons. They discovered that knocking down the expression of the monomeric form of mHtt by as little as 10% was sufficient to rescue neurotoxicity. Imbalance of tau isoforms and nuclear rods with filamentous ultrastructure is involved in HD pathogenesis as motor deficits are attenuated in mouse HD models when tau is knocked out (39).

Currently, there are no treatments to halt HD progression, and existing pharmacological approaches such as antipsychotics, antidepressants, anticonvulsants, and acetylcholinesterase inhibitors only partially ameliorate psychological, physical, and cognitive impairments seen in HD (40,41). The only FDA-approved drug to treat symptoms of chorea in HD patients is Tetrabenazine, which was approved in 2008. Tetrabenazine inhibits human vesicular monoamine transporter 2 (42). Scientists have developed numerous *in vitro* and *in vivo* models of HD in hopes of identifying disease mechanisms and drug therapies to treat HD. Attempts to model HD initially focused on transgenic animal models, immortalized human cell lines, and postmortem tissue from HD patients. While these models address some questions regarding HD pathogenesis, they do not fully represent HD pathology in humans. Further, there are over 100 small molecules that have been evaluated in HD mouse models; many show therapeutic benefits, yet few have translated into human HD therapies. This may suggest that there are limitations to existing HD models.

A promising complementary approach is the use of hESCs and iPSCs derived from HD patient cells to model HD (43,44). HD iPSCs and ESCs engineered to harbor the huntingtin mutation have the potential to model HD more accurately, as they are untransformed human cells that are capable of differentiating into multiple types of neural tissue susceptible in HD. A limitation of these cells was the diversity of genetic backgrounds when comparing control cells with cells harboring the disease gene. Several approaches have been used to reduce these differences, including deriving control cell lines from healthy sibling controls (45). The search for better control cell lines has led scientists to use homologous recombination and genome editing techniques to generate isogenic iPSCs with the mutation in the huntingtin allele repaired (46). These cells provide a platform for systemic genomic profiling and drug screens and are a promising tool for cellular replacement therapy in HD patients (47,48).

6.2 Cell Types Affected in HD

As previously mentioned, the areas of the brain most affected by HD are the striatum and the cortex.

Striatal MSNs are the predominant cell type that dies in HD; however, a number of other cell types are also affected. Recent studies suggest that cortical neurons, astrocytes, and microglia are all affected in HD (Figure 6.3). A summary of each of these cell types, how they are affected by HD, and their pathology are as follows.

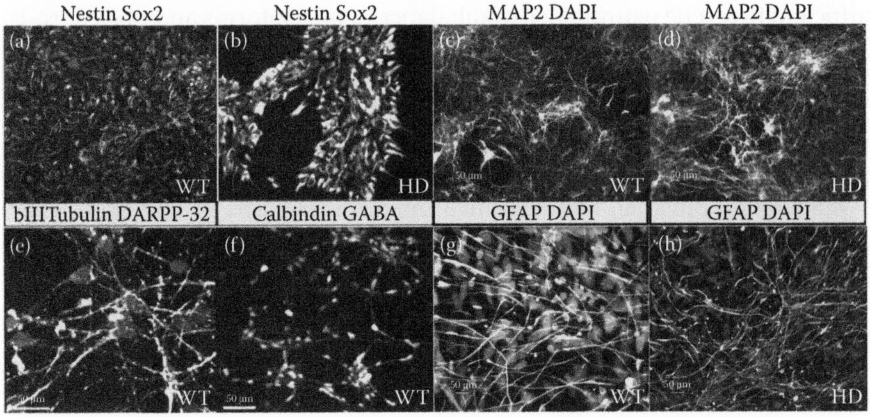

FIGURE 6.3
(See color insert.) Immunocytochemical staining of neural cell types derived from HD and genetically corrected wild type (WT) iPSCs. iPSCs can differentiate into (a, b) NSCs, (c–f) neurons, and (g, h) astrocytes. Striatal neurons expressing DARPP-32 and GABA can also be derived from (e, f) iPSCs.

6.2.1 Striatal Medium Spiny Neurons

Within the striatum, 90–95% of neurons are MSNs, which are inhibitory GABAergic neurons. The other 5–10% of neurons in the striatum are striatal interneurons, which have densely arborized dendrites that make synaptic connections to MSNs. These neurons have a medium-sized cell body, dendrites with spiny protrusions, and long axons (49–51). There are two classes of MSNs. The first class is in the direct striatonigral pathway, expresses dopamine D1 receptors and peptide enkaphalin, and project to dopaminergic neurons of the substantia nigra. The second class of MSNs is in the indirect or striatopallidal pathway, expresses dopamine D2 receptors and substance P, and project to the globus pallidus. Interestingly, D2 expressing MSNs are more susceptible to cell death than D1 MSNs (52–54). However, the exact causes for MSN death are unknown. It is thought though that elevated levels of the excitatory neurotransmitter glutamate and activation of excitatory NMDA receptors on the surface of MSNs are toxic events that induce cell death specifically in MSNs. Interestingly, striatal interneurons remain largely unaffected by HD toxicity.

6.2.2 Cortical Neurons

The cortex sends information to the striatum through cortical projection neurons called *pyramidal neurons*. During HD disease progression, cortical neurons are also susceptible and can degenerate by as much as 30% in the cortical layers III, V, and VI (55,56). Layers V and V1 contain pyramidal projection neurons that extend axons to the striatum. Cortical interneurons are largely untouched by HD similarly to interneurons located in the striatum (56,57). Although further refined studies suggest that HD cases with major mood disorder may have loss of cortical interneurons, in cases of major motor dysfunction, there is no loss (58).

6.2.3 Astrocytes

Astrocytes play an important role in maintaining brain function. They regulate extracellular ion and neurotransmitter concentrations in the brain, modulate synaptic function, and remove excess glutamate from the extracellular space. Astrocytes proliferate in response to injury and neurodegeneration in the brain. They switch to a reactive state in which they upregulate the expression of astrocyte genes such as GFAP and GLAST and are unable to maintain their regular neuroprotective functions in the brain. Neuropathological analysis of late stage HD patient brains revealed widespread astrogliosis in both the striatum and the cortex (9). Further analysis revealed that astrocytes accumulate mHtt, which decreases the expression of glutamate transporters on the surface of astrocytes (59) and consequently inhibits glutamate uptake by astrocytes and exacerbates neuronal excitotoxicity (59). When normal astrocytes are cocultured with MSNs, they are able to protect neurons from mHtt-mediated neurotoxicity. However when HD astrocytes are cocultured with MSNs, they are not able to protect neurons from glutamate-induced excitotoxicity, and MSNs die (59). A separate study discovered a defect in extracellular potassium homeostasis in astrocytes due to a downregulation of the potassium channel Kir4.1 (60). They found that in the transgenic R6/2 mouse model of HD, mHtt lowered levels of the Kir4.1 channel, specifically in astrocytes located in the striatum. Interestingly, they found that astrocyte depolarization was directly linked to the onset of HD motor symptoms. This directly elevated extracellular potassium levels and increased MSN excitability. At the cellular level, *in vitro* human models revealed vacuolization in mHtt and extended repeat astrocytes derived from HD patient iPSCs (27). It is unclear whether this phenotype is pathogenic.

6.2.4 Microglia

Many studies have identified an altered immune response in HD brains even before typical HD symptoms appear. mHTT can induce an inflammatory response, which results in microglial activation. Microglia act as the resident immune cells of the brain. They exist in a quiescent state in normal healthy brains but are activated upon inflammation and neurodegeneration. Microglia activation sets off a cascade of immune responses including overproduction of cytokines, chemokines, and ROS, mitochondrial dysfunction, glutamate-induced excitotoxicity, caspase activation, and ultimately neuronal cell death. Microglia originate from hematopoietic stem cells in the bone marrow, and during development, they cross into the brain through the blood brain barrier where they reside and proliferate upon immune stimuli. mHtt has been implicated in microglial dysfunction in HD mouse models. For instance, the expression of mHTT in primary microglia isolated from postnatal HD mice impaired microglial migration in response to chemotaxic stimuli (61,62). Studies of mHTT expression in microglia reveal impairments in migration in response to laser-induced injury. Migration defects are likely due to defects in actin remodeling caused by mHtt (61). Microglial activation is also used as a presymptomatic biomarker for HD pathogenesis. Correlations between microglial activation and striatal MSN dysfunction and between activated microglial numbers, and HD disease progression were identified (63–65).

6.2.5 Peripheral Tissue and Cell Types

Given that the Htt protein is ubiquitously expressed throughout the body, it is not surprising that the mHtt impacts the function and the health of tissues outside the central nervous system (66). The skeletal muscle of HD patients and mouse models has pathology, and there is degeneration of the myofibers (67). Circulating markers of muscle injury (sTnI, FABP3, and Myl3) are increased in mouse serum (68). There are also significant energy deficit phenotypes in HD. The brown adipose tissue is reduced in HD and consistent with the altered thermoregulation found in the disease (69). In a recent study of peripheral organs, it appears that the heart and muscle tissues are associated with altered autophagy, whereas dysfunction of ubiquitin–proteasome system was observed in the liver and the lung (70). Along with changes in organ function, the peripheral innate immune system is altered in HD and numerous inflammatory cytokines are detected (71).

6.3 Overview on Different HD Models

Animal and cell-based models of HD have been used since before the identification of the HD mutation in human patients to characterize aspects of the disease as well as to test potential therapeutic interventions in the disease.

6.3.1 Vertebrate Models

Prior to the discovery of the HD mutation, nongenetic approaches were used to recreate aspects of the disease seen in human patients. One such approach was the use of quinolinic acid (QA)-mediated excitotoxicity to produce striatal lesions in rodent models of HD (72). QA is a derivative of kynurerine and is produced by microglia in the brain. QA acts as an agonist of N-methyl-D-aspartate (NMDA) receptors and is proinflammatory, gliotoxic, and neurotoxic. When injected into the striatum of rats, QA produces lesions and motor phenotypes that resemble degeneration seen in human patients in HD. However, it should be noted that the molecular changes of QA lesions are very distinct form HD genetic models, making this a crude model at best (73,74).

Since the discovery of the HD mutation in 1993 (75), multiple genetic models of HD have been made using the fragment of HTT or the full-length protein. Among the various models of HD, the most commonly used are genetic mouse models of HD. The first and best-characterized mouse model of HD was generated by Gillian Bates. This model expresses exon 1 of the human Htt under the control of a small region of the human HTT promoter (R6/1, R6/2) (76). Shortly after the R2/1 model was generated, another fragment model was generated under the control of prion promoter and expresses N171 of the human HTT protein. Full-length Htt protein models include BACHD (100 CAG) and YACHD (128 CAG) transgenic mice, which express human Htt under the control of the human Htt promoter. In addition, there are numerous knockin mouse models of HD with varying CAG repeat lengths and cell specific promoters to help understand the cell-autonomous and cell-nonautonomous aspects of HD. The R6/2 model has a robust and reproducible phenotype and has a shortened lifespan. The R6/2 mice show brain atrophy, alterations in neurotransmitter receptors, and aggregation formation, as well as other similar features of HD including motor behavioral deficit and shorter lifespan. This model has been utilized for the evaluation of numerous therapeutic compounds to treat HD. The disadvantage of this model is that the protein expressed is not full-length human Htt. The zQ175 knockin mouse model, which expresses the full-length Htt protein with polyQ expansion at endogenous levels, has behavioral and neuroanatomic abnormalities consistent with HD (77). They have a decrease in brain weight and striatal and cortical volume. Stereological counts demonstrate a 15% decrease in the number of striatal cells and thinning of the cortical layer, with no change in cell numbers in the cerebellum. This model is particularly appropriate for evaluating proteolysis of Htt as wild-type and mHtt are expressed at endogenous levels. The zQ175 model has reproducible behavioral and cognitive deficits on rotarod and open-field testing. The YAC128 and BACHD models have similar features and comparison of the various models has been reported (78). The wide diversity of the mouse genetic models of HD that have been developed reflects the fact that no single model recapitulates all aspects of HD seen in human patients (for review see Lee et al. [79]).

Recently, there have been efforts to make large-animal models of HD in nonhuman primates (*Macaca mulatta*), sheep (*Ovis aries*), and miniature pig (*Sus scrofa*), in an effort to better recapitulate the aspects of the disease (80). Carter et al. have derived iPSC from HD monkey models and show reversal of cellular phenotypes in neural cells (81). Other genetic models of HD and huntingtin loss of function mutations have been generated in nonmammalian systems such as *Drosophila* and *Caenorhabditis elegans*. As with the mice, these models do not fully recapitulate human HD phenotypes, but they have proven useful to model some aspects of the disease and to perform genetic screens for suppressors of HD phenotypes (82).

6.3.2 Cell-Autonomous and Cell-Nonautonomous Mechanisms in HD

HTT is ubiquitously expressed across somatic tissues, and the overexpression of full-length and fragments of mHTT is sufficient to increase apoptosis in multiple cell lines derived from diverse tissue types, suggesting that mHTT is, to some degree, toxic to the cell expressing it. One hypothesis is that these cell types affected in HD are particularly sensitive to mHTT-based toxicity, while another postulates that dysfunctional signaling, synaptic connections, or interactions with immune cells sensitizes these affected cells to degeneration and death. There is evidence to support both hypotheses, and they are not mutually exclusive. Although MSNs are the most vulnerable neuronal subtype in HD, the cortex and other regions are also subject to dysfunction and neurodegeneration. It has become increasingly clear that both cell-autonomous and cell-nonautonomous mechanisms contribute to the HD phenotype, and vulnerable cells include MSNs, cortical neurons, microglia, and astrocytes (reviewed by Ehrlich [83] and Bradford et al. [84]).

The expression of mHTT in cortical or striatal neurons, or astrocytes, leads to abnormalities in mice, but the most severe phenotype is seen in mice in which all cell types express the mutation. Most notably, Brown et al. (85), Thomas et al. (86), and Kim et al. (87) demonstrated that the expression in MSNs alone leads to motor and transcriptional abnormalities similar to pan-cellular models, but enhanced sensitivity of MSNs to excitotoxicity and elements of the electrophysiological abnormalities dependent on corticostriatal interactions are, as would be anticipated, not replicated. Conversely, the expression in cortical neurons alone leads to a much less severe phenotype (88,89). Most recently, Wang et al. (90) reported that "to ameliorate disease in a mouse model of HD ... reduction of mHTT expression in cortical or striatal neurons partially ameliorates corticostriatal synaptic deficits, further restoration of striatal synaptic function can be achieved by reduction of mHTT expression in both neuronal cell types" consistent with a distinct role for cell–cell interaction between cortical and striatal neurons in the disease. Therefore, it is likely that not only cell-autonomous mechanisms that lead to HD phenotypes, but also alterations in cell–cell interactions (specifically synaptic connections) are involved in the disease progression.

6.4 Using Embryonic and Induced Pluripotent Stem Cells to Model HD

6.4.1 Embryonic Stem Cell Models of HD

While much was learned from immortalized HD cell lines (including striatal cell lines derived from wild-type and mutant Hdh(Q111) knockin mice), hESCs were an early human *in vitro* cellular models of HD (91). HD ESCs have been derived from preimplantation genetic diagnosis screened human embryos that have abnormal numbers of CAG repeats (92,93). These cell lines can differentiate into any type of brain cell, making them ideal starting material for modeling HD.

As an alternative approach, scientists have taken existing ESC lines such as the commonly used H9 cell line and genetically manipulate these cells to harbor a m*HTT* allele. A study by Lu and Palacino (35) developed a transgenic HD neuronal model derived from isogenic ESCs expressing varying lengths of polyQ repeats. The authors modified the H9 hESC line by stable transfection with cDNA containing HTT exon 1 fragments with varying CAG repeat number (Q23, normal; Q73 and Q145, HD). They discovered that mHTT formed insoluble aggregates in neurons and that polyQ length correlated strongly with monomeric soluble mHTT levels and severity of neurodegeneration, which they quantified by neuronal cell death. By comparing mHTT cell lines to their wild-type isogenic counterparts, the authors determined that mHTT toxicity could be rescued by knocking down the expression of soluble monomeric mHTT.

While ESC models of HD provide valuable *in vitro* models of the disease, ethical and practical issues limit this technology for future research and therapy development. Screening embryos and deriving ESC lines takes more time than newer methods such as generating iPSC lines from patient skin samples. Furthermore, deriving ESC lines requires the destruction of embryos, which has been politically controversial in the United States. Lastly, potential tissue generated from ESC lines for cell replacement therapy will require allogeneic transplantation rather than autologous. Thus, patients would risk tissue rejection or face a lifetime of immunosuppressants in order to maintain transplants. Thus, the future of cell replacement approaches to treat HD may lie in the generation of patient-specific stem-cell derived transplants.

6.4.2 Induced Pluripotent Stem Cell Technology and Disease Modeling

Derivation of iPSCs from human somatic tissue is an alternative stem cell technique that avoids the difficulty and the controversy of generating ESCs from discarded embryos. Dr. Shinya Yamanaka developed this technique in 2006 (44). He discovered that four transcription factors, Oct-4, Sox2, KLF4, and c-Myc, when ectopically expressed in mouse and human fibroblasts, could reprogram these cells back to a PSC state (Figure 6.4). iPSCs are practically

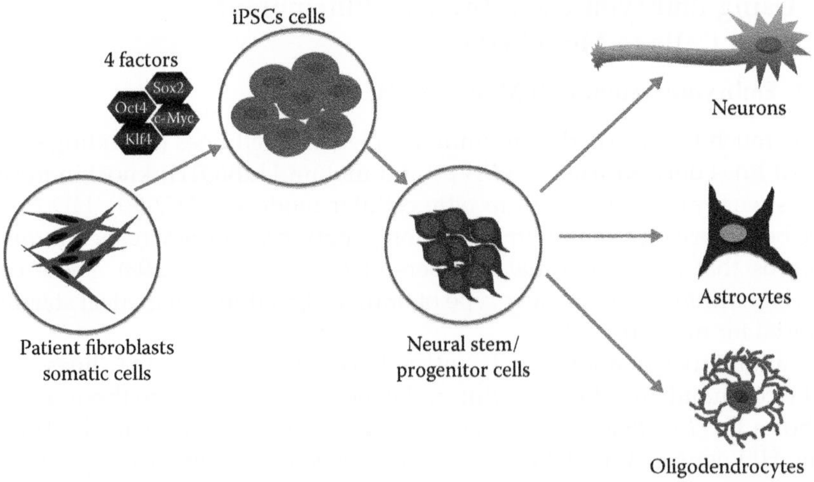

FIGURE 6.4
(See color insert.) Human skin cells can be reprogrammed back to a PSC state by ectopically expressing the transcription factors Oct4, Sox2, KLF4, and c-Myc. These cells are called iPSCs and have identical characteristics and differentiation ability compared to hESCs. iPSCs can differentiate into any cell type in the body. In the case of HD, iPSCs can be differentiated into NSCs or neural progenitor cells, which can then be differentiated into mature brain cell types including neurons, astrocytes, and oligodendrocytes.

indistinguishable from PSCs and have the ability to differentiate into any cell type in the body. iPSCs provide an invaluable source for disease modeling and drug screening and for future autologous transplantation of cells and tissue. Multiple reprogramming techniques have been published since Yamanaka's initial paper focusing on improving delivery methods of the factors and safer ways to generate iPSCs by omitting potentially oncogenic factors like myc. Early experiments used retroviral and lentiviral methods to generate iPSCs; however, newer technologies use recombinant proteins, miRNAs, and mRNAs to generate integration free iPSCs (94).

Numerous disease-specific iPSCs have been generated from patient fibroblasts. Certain countries with low genetic variability in their populations, such as Japan, are making iPSC banks. Under the guidance of Shinya Yamanaka, the Riken Center in Japan aims to develop an iPSC bank of 75 human leukocyte antigen-matched donor lines for cell transplantation therapies. These 75 lines would cover 80% of the Japanese population, which has a more homogenic genetic pool compared to other countries (95). An iPSC bank of this magnitude is a potentially valuable resource for both basic research and transplantation applications. Potential disease-related studies with these cells include identification of disease-associated single nucleotide polymorphisms (SNPs), genome-wide association study (GWAS) analysis of the transcriptome, and drug screening. Potential transplantation applications relate to the high regulatory bars required to validate a cell line as appropriate for transplantation. Stem cells for transplantation must be maintained and cultured using GMP methods and characterized for aneuploidy, viral infection, and other potential sources of tumorogenicity. By reducing the number of cell lines needed for transplantation to fewer than 100, the cost and the difficulty of validating cells for transplantation is reduced.

For neurodegenerative diseases, Park et al. (96) generated numerous lines from various patients with several diseases. These studies not only have advanced the field, but also have identified novel disease mechanisms and potential drug targets. Furthermore, with the revolution in genome editing, scientists have genetically corrected disease-causing mutations in patient iPSCs and developed isogenic wild-type iPSCs with the same genetic background to better compare differences at the genomic and proteomic levels between disease and healthy states (46).

6.4.3 iPSC Models of HD

A number of HD iPSC models exist and are well characterized. Recent studies have shown that HD iPSC-derived cells exhibit pathological phenotypes that model phenotypes observed in HD (27,84,97–101). While HD and wild-type iPSCs show little phenotypic difference when pluripotent, gene expression profiling reveals differences in cadherin and TGF-β signaling pathways in HD cells in comparison to genetically corrected isogenic iPSCs. Specifically, the expression of cadherin signaling genes is lower in HD iPSCs, while the expression of TGF-β signaling genes is elevated in HD iPSCs relative to corrected iPSCs. When HD and corrected iPSC-derived neural stem cells (NSCs) are stressed by the removal of growth factors from neural media, HD NSCs display elevated caspase-3/7 activity, a readout for apoptosis, compared to corrected control lines. The genetic correction of mHTT restored normal levels of caspase-3/7 activity, elevated BDNF levels, prevented apoptosis, and rescued mitochondrial dysfunction (Figure 6.5) (98). In a separate study, HD iPSCs and neurons displayed elevated lysosomal activity indicating a disruption in cellular maintenance and protein degradation (99).

Finally, a study conducted by the HD iPSC consortium identified key functional differences in striatal MSNs generated from HD and control patient iPSCs (97). They generated 14 iPSC lines derived from HD patients and determined whether *in vitro* models of HD derived from these iPSCs were sufficient to model HD phenotypes. The authors generated NSCs, neurons, and astrocytes from various HD iPSCs and wild-type age-matched controls. Using GWASs, they discovered 1601 genes that were differentially expressed in HD against wild-type NSCs. They observed differences in cell adhesion properties, cytoskeletal markers, and mitochondrial respiration. Differentiated cell types revealed neurons with differences in electrical activity including action potential firing, increased HD neural cell death, and sensitivity to BDNF withdrawal. HD MSNs display altered electrophysiological properties including differences in their ability to fire spontaneous and evoked action potentials and to regulate intracellular calcium signaling. When these neurons were further matured to DARPP-32 (a marker of striatal neurons) positive cells, they exhibited increased cell death and susceptibility to stress such as growth factor removal and exposure to glutamate (97).

Two recent studies also used the same 72 repeat HD iPSC line to generate human HD models in a dish. A study by Chae et al. (100) conducted a proteomic analysis of undifferentiated H9 ESCs, normal iPSCs, and HD-72 iPSCs and identified 26 proteins with altered expression. These proteins are involved in cellular processes such as oxidative stress, programmed cell death, and cellular biogenesis. Within this protein group, oxidative stress proteins such as SOD1, GST, and Gpx1, were downregulated in HD iPSCs suggesting that HD iPSCs are more susceptible to oxidative stress. Cytoskeleton-associated proteins were also downregulated in HD iPSCs including Cfl1-, Stmn-1, Facn-1, and Sept-2. These proteins are important for neuronal differentiation thus suggesting potential defects in neuronal differentiation. A separate study by Jeon et al. (101) used HD-72 iPSCs for transplantation experiments into rat models of HD. This study will be further discussed in a separate section.

One consideration in disease modeling using stem cells relates to the quality of the cells used to generate the cell types affected in the disease. We mentioned that HD has both cell-autonomous and cell-nonautonomous mechanisms that contribute to the HD phenotype, and vulnerable cells include MSNs, cortical neurons, microglia, and astrocytes. Additionally,

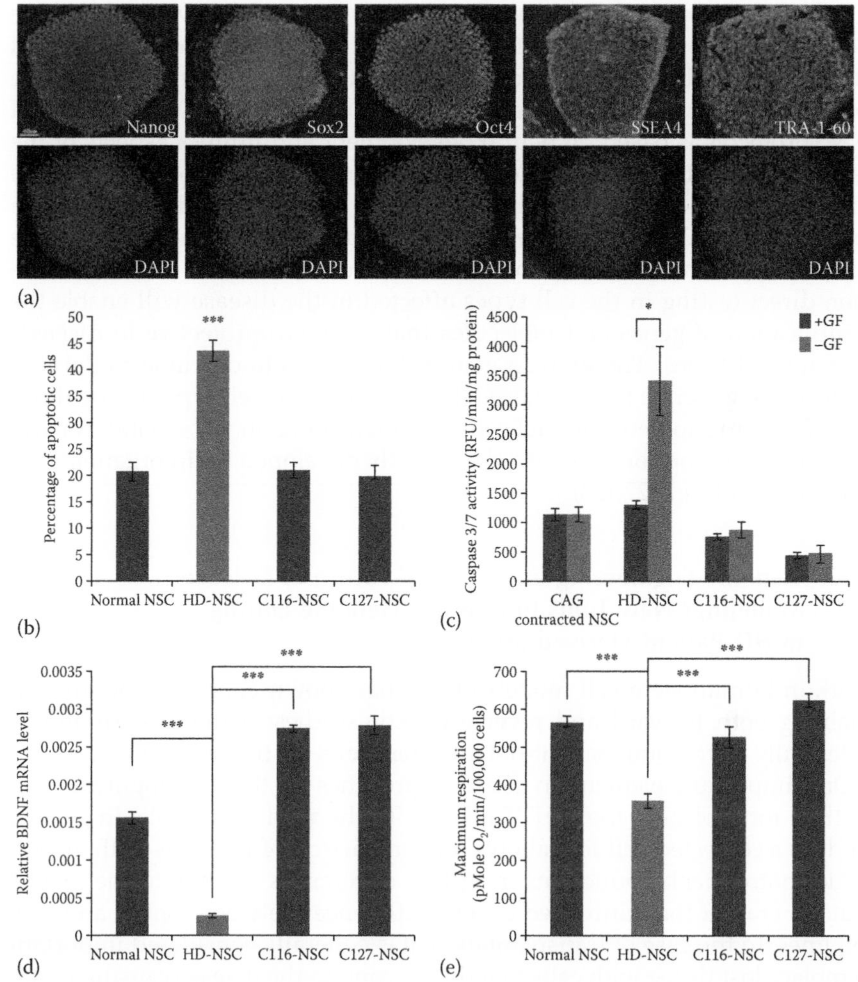

FIGURE 6.5

(See color insert.) Using iPSCs to model HD. (a) HD iPSCs with 72 CAG repeats were genetically corrected using homologous recombination and characterized. Corrected iPSCs expressed hallmark pluripotency markers including Nanog, Sox2, Oct4, SSEA4, and TRA-1-60. (b–e) NSCs derived from HD iPSCs exhibit HD phenotypes including increased cellular apoptosis, elevated Caspase 3/7 activity, reduced BDNF expression, and reduced mitochondrial respiration compared to wild-type and genetically corrected (C116 or C127) NSCs. (Reprinted from *Cell Stem Cell*, 11, An, M. C. et al., Genetic correction of Huntington's disease phenotypes in induced pluripotent stem cells, 253–63, Copyright 2012, with permission from Elsevier.)

certain MSNs and cortical neurons are more vulnerable than others. Therefore, to model HD, it may take careful and continued efforts to produce pure and mixed populations of MSNs and cortical neurons that are appropriate to model the disease. Unlike Parkinson's disease where dopaminergic neurons are produced in high yields (102–106), the development of protocols to generate the specialized types of cells for HD requires more work. There are now four to five published methods to generate striatal MSNs, but the details of how comparable their characteristics are relative to *in vivo* neurons in humans still warrant further work (107–112). Each of these protocols is based on neural development and mimics aspects of the development of the striatum. Cortical and astrocytic protocols are available but are time-consuming, and the subtypes of cells generated have not been fully defined (27,113–117). Screens for modifiers of HD phenotypes have been undertaken in animal models and immortalized cells (118–121); however, direct testing in the cell types affected in the disease will enable the identification of genes and molecules that are neuroprotective in disease-relevant cell types. Therefore, a goal in the field is to continue to develop protocols to generate individual cell types and mixed cell types in coculture and 3D culture to better model cell–cell interactions in the striatal–cortical network. Other options are also the recently developed brain organoids for disease modeling (122,123).

6.4.4 Genome Engineering of iPSCs and Future of Disease Modeling? New Tools to Improve Genome Editing in HD Patient-Derived iPSCs

Work in human stem cell models of HD has shown enormous promise in enabling both forward and reverse genetic studies of the disease and in potentially developing patient-derived stem cells for transplantation.

One important element in these approaches is the development of a method for replacing the mHTT allele with the wild-type allele. In genetic studies, a corrected cell line allows for comparison of HD cells with an isogenic control background, ensuring that differences are due to the mHTT mutation rather than unrelated genetic differences between control and case cell lines. In the case of transplantation, it is generally considered important to replace lost tissue with cells that do not contain the disease causing mutation and thus are not subject to mHTT-mediated toxicity. Genetic correction of mHTT iPSCs has been achieved using traditional homologous recombination approaches, which utilize donor plasmids containing large regions of homology with the HTT exon 1 locus coupled with selection in order to identify putative recombined iPSC colonies (Figure 6.6) (46). This method relies on random events that result in damage to the genome to activate DNA repair machinery and homologous recombination. Using this approach, HR is very inefficient, with ~1 in 100 antibiotic resistant colonies identified after selection showing true homologous recombination; the remaining colonies are usually present due to random integration of the resistance cassette into the genome. It is estimated that of 1×10^6 cells electroporated, 2 are found to be recombinants (efficiency is thus $1{:}10^6$) (46).

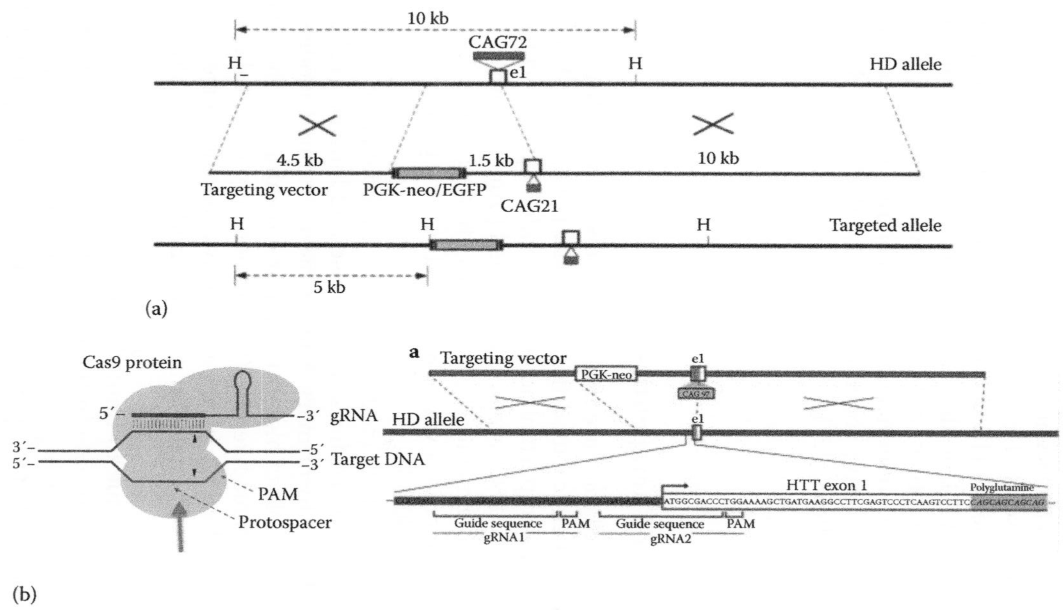

FIGURE 6.6
(See color insert.) Generation of genetically corrected iPSC lines from HD iPSCs. (a) Using the traditional method of homologous recombination, HD IPSCs with 72 CAG repeats were corrected to have only 21 repeats. (b) Using the CRISPR/Cas9 genome targeting technique, the HD mutation can be targeted and corrected, or longer repeat sequences can be inserted. Specific gRNA sequences are selected to target exon 1 of the HTT gene, and donor sequences containing either the corrected allele or a longer mutant allele can be used. This method has a much higher efficiency than homologous recombination. (From An, M. C. et al. *PLoS Curr.* 6, 2014; reprinted from *Cell Stem Cell*, 11, An, M. C. et al., Genetic correction of Huntington's disease phenotypes in induced pluripotent stem cells, 253–63, Copyright 2012, with permission from Elsevier.)

Increasing the frequency of homologous recombination has been achieved by targeting DNA damage to specific regions of the genome (Figure 6.6). The specific tools to do this vary: artificially designed zinc finger proteins or transcription activator-like effectors (TALE) transcription factors designed to interact with a specific sequence of DNA and fused to the FOK1 nuclease or the Cas9 nuclease targeted to the genome by gRNAs that base pair with their complementary sequence in the genome have all been shown to dramatically increase the absolute frequency of homologous recombination (124–126). Absolute efficiencies of HR in nuclease-targeted iPSCs have been reported at between 1% and 5% of all cells transfected (124). The practical effect of this is that the generation of cell lines with new genetic corrections can be achieved using smaller donors and less screening. As a result, new iPSC and ESC HD lines are being generated such as iPSCs with different CAG repeat-length mHTT alleles (127).

6.5 iPSCs to Screen for Therapeutics

Cell culture models of the disease are important tools for high throughput screening for genetic and chemical modifiers of disease phenotypes (128,129). As such, iPSCs represent an important advance in modeling HD in cell culture. Cells derived from HD iPSCs are untransformed primary human cells and can be differentiated to generate the cell types affected in the disease. One powerful use for these cells is via forward and reverse genetic screens for modifiers of cell death (130–132).

Prior to the advent of HD iPSC models, cell-based modeling of HD has been limited to immortalized cell lines from mouse and humans (120,133), to animal models of HD, or to primary human tissue. Animal models of HD are not generally amenable to high throughput screening. On the other hand, models that are amenable like flies (134) do not fully recapitulate the disease (135). Immortalized cell lines are not always representative of the cells affected in the disease. Primary human postmortem tissue, although important for understanding the pathology of the disease, is not an effective tool for drug screening due to limited availability and unpredictable nature of the severity of the disease in the tissue that is acquired. In most cases, it is impossible to get viable primary cells such as neurons or astrocytes from living patients, so iPSCs represent an optimal source to model the disease in human cells. iPSCs are disease- and patient-specific and are a scalable model that can model multiple cell types (niche or system) and can be genetically modified to make isogenic corrected iPSCs with the same genetic background. Neural cells derived from iPSCs can be used for drug screening platforms either looking for morphological changes, cellular changes, or reduction of phenotypes such as cell death. HD phenotypes include changes in cell morphology, soma size, neurite outgrowth and length, electrophysiological function of neurons, cell death, cell adhesion, metabolism, mitochondrial bioenergetics, expression of growth factors (BDNF), etc. In HD, one can also look for changes in mRNA and protein expression, as well as biochemical changes including the aggregation of mHTT and the generation of toxic fragments resulting from cleavage of mHTT.

Screening for HD therapeutics can be done in a number of different cell types. Striatal MSNs are the likely candidate for drug development as these are the neurons most strongly affected in HD. HD MSNs display differences in electrophysiological function and survival that could potentially be used as readouts for drug screens. Current protocols to derive MSNs from ESCs and iPSCs are time consuming and inefficient. In order to perform high throughput screening, new methods to generate large numbers of MSNs are needed. An alternative is NSCs derived from HD iPSCs. HD NSCs exhibit HD phenotypes such as elevated caspase activity, reduced mitochondrial biogenesis, and reduction in BDNF levels. NSCs can easily be generated in large numbers and used in high throughput format to identify compounds that correct HD-associated phenotypes. In the future, as differentiation protocols improve striatal progenitor cells as well as other cell types affected in HD (cortical neurons, astrocytes, microglia) are developed, these cells can be used to screen for new compounds to treat HD.

6.6 iPSCs for Transplantation in HD Models

iPSCs represent a powerful technology to potentially generate tissue to replace cells lost in HD; however, there are safety concerns associated with tumor formation and overproliferation of NSCs upon transplantation. Thus, some scientists have argued that it is safer to transplant neural precursor cells (NPCs) derived from ESCs rather than iPSCs. However, recent studies using iPSCs for transplantation have shown promise as a potential therapeutic without tumorigenic side effects.

In a study, Mu et al. transplanted iPSCs into a rat model of HD (101). These rats were treated with QA. QA as mentioned previously is a drug that specifically targets and kills off MSNs in the striatum and is used in animal models to mimic phenotypes of HD. Transplanted iPSCs improved learning and memory as well as motor function in QA-treated rats. Other studies have used similar approaches (136,137). In another study, Jeon et al. injected NSCs derived from ESCs and iPSCs into a rat model of HD (138). They generated NPCs from HD iPSCs and H9 ESCs or F5 control iPSCs and transplanted these cells into the striatum of rats treated with QA. Interestingly, the authors discovered that transplantation of H9, F5, or HD-NPCs improved behavior in QA-treated rats. Histological analysis of QA-treated rat brains 12 weeks posttransplantation revealed that HD NPCs differentiated into GABAergic neurons; some of which expressed striatal neuron markers such as DARPP-32 and also synaptophysin, indicating potential for synaptic connections of human cells. Importantly, no signs of tumors or excessive proliferation of NPCs was observed in any of the 40 transplanted rats. However, when the authors looked for hallmarks of HD neuropathology in their human transplants, they did not observe intranuclear aggregates at 12 weeks posttransplantation using the EM48 antibody, which recognizes mHtt aggregates. They hypothesized that 12 weeks was not long enough for aggregates to form, so they transplanted HD-NPCs in the lateral ventricles of neonatal mice (P2) and observed EM48 expression at 33 and 40 weeks post transplantation, although they did not determine what cell types contained mHtt aggregates. The authors also did not observe formation of intranuclear inclusions in neurons generated *in vitro* from HD iPSCs and hypothesized that it could be due to the neurons not having aged enough to show this phenotype. To address this issue, they treated HD iPSCs with a proteasome inhibitor, MG132, and observed EM48-positive aggregates at higher doses of MG132 in HD iPSCcs compared to control iPSCs. They concluded from these results that HD-72 iPSCs are highly susceptible to proteasome inhibition and can exhibit typical HD pathology including formation of mHtt aggregates at later stages of the disease. This same group also transplanted HD-72 iPSC-derived NPCs into the striatum of YAC128 mice and observed improvement in motor function as assessed by rotarod (83). However, they did not see formation of mHtt aggregates by 12 weeks posttransplantation in grafted GABA-positive neurons thus indicating that HD pathology was not apparent by 12 weeks *in vivo*.

The overall goal for iPSC transplantation therapy for HD would be to repopulate the areas affected by HD, specifically the striatum and the cortex, with healthy striatal and cortical neurons, thus reestablishing the neural networks that are degenerated in HD. Simply transplanting HD NSCs or neurons into a patient might temporarily improve HD symptoms or disease progression; however, the underlying issue of the HD mutation remains within these cells. The ideal strategy would be to genetically correct the HD mutation in patient-derived iPSCs using published strategies such as Cas9-mediated genome editing and transplant corrected iPSC-derived NSCs into patient brains. These studies are already being conducted in animal models and will show much promise in the future.

In the present, iPSC therapy is reaching clinical trials. Japan has made progressive moves to change their regulations on the use of stem cells in clinical trials (139). They currently have a pilot study using human iPSCs to treat a retinal disease called *macular degeneration*. It is likely that similar studies might be conducted in the future to treat patients with neurodegenerative disorders like HD and PD.

6.7 Modeling HD: Beyond Stem Cells

What is the future of HD disease modeling? We have come a long way since the mutation causing the disease was initially discovered in 1993 (75). From animal models to *in vitro* stem cell models, scientists have generated a myriad of tools to study HD from multiple angles. While ESC and iPSC models of HD have drastically improved our understanding of HD pathogenesis, future studies will need to continue to develop more accurate and relevant models. For instance, the efficiency of generating important cell types affected in HD is still low. Developing more robust protocols for making MSNs, astrocytes, and microglia would be game changing.

Other improvements will likely involve generating coculture models of HD in a dish as well as 3D models. As previously mentioned, multiple cell types within the brain are affected at various stages of HD progression in patients. Besides the well-known targeted striatal neurons, cortical neurons, astrocytes, microglia, and even oligodendrocytes are affected in HD to some extent. Thus, generating coculture models that more accurately represent the brain's anatomy would help tease apart cell-autonomous against cell-nonautonomous effects caused by HD and identify new phenotypes in cell types that have only recently come into the spotlight such as glia. While coculture models have worked with great success in other neurodegenerative models like amyotrophic lateral sclerosis (140), it is

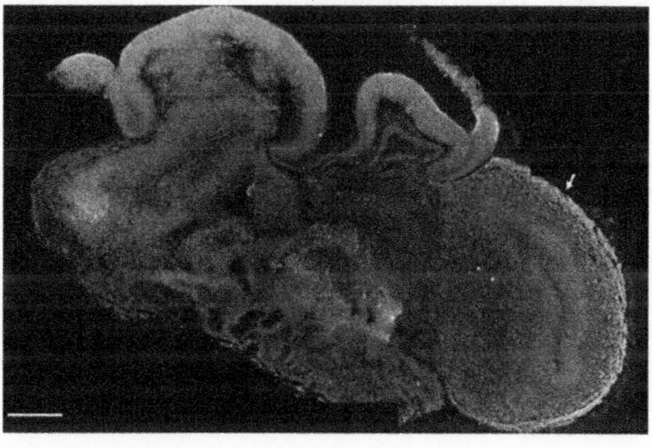

FIGURE 6.7
(See color insert.) Human cerebral organoids can be generated in culture. These organoids are a 3D representation of cortical development in the human brain and contain most of the commonly found brain cells including neurons and astrocytes. (Reprinted by permission from Macmillan Publishers Ltd. *Nature*, Lancaster, M. A. et al., 501(7467): pp. 373–9, copyright 2013.)

unclear whether such models will have such drastic and obvious phenotypes in HD. Thus, the idea of generating a 3D culture system including all relevant cell types (not just two) would be ideal. The ability to generate a 3D HD human cell model is already possible with the discovery of human cerebral organoids (Figure 6.7) (122,123). Cerebral organoids have been successfully used to model both brain development and human disease such as microencephaly. Two main protocols currently exist to develop these "mini-brains"; however, both predominantly focused on cortical development and generation of cerebral cortex structures and progenitor zones. It will be interesting to see if, in the future, these protocols can be manipulated to generate other structures of the brain such as the striatum in the case of HD.

Lastly, HD is not just a disease of aging. It is likely that HTT affects important processes during early stages of human development (141,142). It will be worth using both animal and stem cell models of HD to further investigate molecular mechanisms that are altered or disturbed during brain development. This could potentially open up new therapeutic targets and opportunities to prevent or delay HD onset.

References

1. Driver-Dunckley, E. and J. N. Caviness, Huntington's disease. In *Schapira AHV Neurology and Clinical Neuroscience.* Maryland Heights, MO: Mosby Elsevier. 2007: pp. 879–885.
2. Andrew, S. E. et al., The relationship between trinucleotide (CAG) repeat length and clinical features of Huntington's disease. *Nat Genet,* 1993. 4(4): pp. 398–403.
3. Harper, P. S., The epidemiology of Huntington's disease. *Hum Genet,* 1992. 89(4): pp. 365–76.
4. Langbehn, D. R. et al., A new model for prediction of the age of onset and penetrance for Huntington's disease based on CAG length. *Clin Genet,* 2004. 65(4): pp. 267–77.
5. Tabrizi, S. J. et al., Biological and clinical changes in premanifest and early stage Huntington's disease in the TRACK-HD study: The 12-month longitudinal analysis. *Lancet Neurol,* 2011. 10(1): pp. 31–42.
6. Rees, E. M. et al., Cerebellar abnormalities in Huntington's disease: A role in motor and psychiatric impairment? *Mov Disord,* 2014. 29(13): pp. 1648–54.
7. Waldvogel, H. J. et al., The neuropathology of Huntington's disease. *Curr Top Behav Neurosci,* 2014.
8. Gutekunst, C. A. et al., Nuclear and neuropil aggregates in Huntington's disease: Relationship to neuropathology. *J Neurosci,* 1999. 19(7): pp. 2522–34.
9. Vonsattel, J. P. et al., Neuropathological classification of Huntington's disease. *J Neuropathol Exp Neurol,* 1985. 44(6): pp. 559–77.
10. Niccolini, F. and M. Politis, Neuroimaging in Huntington's disease. *World J Radiol,* 2014. 6(6): pp. 301–12.

11. Grinthal, A. et al., PR65, the HEAT-repeat scaffold of phosphatase PP2A, is an elastic connector that links force and catalysis. *Proc Natl Acad Sci USA*, 2010. 107(6): pp. 2467–72.
12. Martin, D. D. et al., Identification of a post-translationally myristoylated autophagy-inducing domain released by caspase cleavage of huntingtin. *Hum Mol Genet*, 2014. 23(12): pp. 3166–79.
13. Cong, X. et al., Mass spectrometric identification of novel lysine acetylation sites in huntingtin. *Mol Cell Proteomics*, 2011. 10(10): pp. M111 009829.
14. Ehrnhoefer, D. E., L. Sutton, and M. R. Hayden, Small changes, big impact: Posttranslational modifications and function of huntingtin in Huntington disease. *Neuroscientist*, 2011. 17(5): pp. 475–92.
15. Marques Sousa, C. and S. Humbert, Huntingtin: Here, there, everywhere! *J Huntingtons Dis*, 2013. 2(4): pp. 395–403.
16. Dragatsis, I., A. Efstratiadis, and S. Zeitlin, Mouse mutant embryos lacking huntingtin are rescued from lethality by wild-type extraembryonic tissues. *Development*, 1998. 125(8): pp. 1529–39.
17. Trottier, Y. et al., Cellular localization of the Huntington's disease protein and discrimination of the normal and mutated form. *Nat Genet*, 1995. 10(1): pp. 104–10.
18. Shirasaki, D. I. et al., Network organization of the huntingtin proteomic interactome in mammalian brain. *Neuron*, 2012. 75(1): pp. 41–57.
19. Kaltenbach, L. S. et al., Huntingtin interacting proteins are genetic modifiers of neurodegeneration. *PLoS Genet*, 2007. 3(5): p. e82.
20. Zuccato, C. and E. Cattaneo, Role of brain-derived neurotrophic factor in Huntington's disease. *Prog Neurobiol*, 2007. 81(5–6): pp. 294–330.
21. Zuccato, C. et al., Huntingtin interacts with REST/NRSF to modulate the transcription of NRSE-controlled neuronal genes. *Nat Genet*, 2003. 35(1): pp. 76–83.
22. Marcora, E., K. Gowan, and J. E. Lee, Stimulation of NeuroD activity by huntingtin and huntingtin-associated proteins HAP1 and MLK2. *Proc Natl Acad Sci USA*, 2003. 100(16): pp. 9578–83.
23. Zuccato, C. et al., Loss of huntingtin-mediated BDNF gene transcription in Huntington's disease. *Science*, 2001. 293(5529): pp. 493–8.
24. Miller, J. et al., Quantitative relationships between huntingtin levels, polyglutamine length, inclusion body formation, and neuronal death provide novel insight into Huntington's disease molecular pathogenesis. *J Neurosci*, 2010. 30(31): pp. 10541–50.
25. Augood, S. J., R. L. Faull, and P. C. Emson, Dopamine D1 and D2 receptor gene expression in the striatum in Huntington's disease. *Ann Neurol*, 1997. 42(2): pp. 215–21.
26. Han, I. et al., Differential vulnerability of neurons in Huntington's disease: The role of cell type-specific features. *J Neurochem*, 2010. 113(5): pp. 1073–91.
27. Juopperi, T. A. et al., Astrocytes generated from patient induced pluripotent stem cells recapitulate features of Huntington's disease patient cells. *Mol Brain*, 2012. 5: pp. 17.
28. Shin, J. Y. et al., Expression of mutant huntingtin in glial cells contributes to neuronal excitotoxicity. *J Cell Biol*, 2005. 171(6): pp. 1001–12.
29. Zuccato, C., M. Valenza, and E. Cattaneo, Molecular mechanisms and potential therapeutical targets in Huntington's disease. *Physiol Rev*, 2010. 90(3): pp. 905–81.

30. Plotkin, J. L. et al., Impaired TrkB receptor signaling underlies corticostriatal dysfunction in Huntington's disease. *Neuron*, 2014. 83(1): pp. 178–88.
31. Jiang, H. et al., Depletion of CBP is directly linked with cellular toxicity caused by mutant huntingtin. *Neurobiol Dis*, 2006. 23(3): pp. 543–51.
32. Nucifora, F. C., Jr. et al., Interference by huntingtin and atrophin-1 with cbp-mediated transcription leading to cellular toxicity. *Science*, 2001. 291(5512): pp. 2423–8.
33. Subramaniam, S. et al., Rhes, a striatal specific protein, mediates mutant-huntingtin cytotoxicity. *Science*, 2009. 324(5932): pp. 1327–30.
34. Baiamonte, B. A. et al., Attenuation of Rhes activity significantly delays the appearance of behavioral symptoms in a mouse model of Huntington's disease. *PLoS One*, 2013. 8(1): p. e53606.
35. Lu, B. and J. Palacino, A novel human embryonic stem cell-derived Huntington's disease neuronal model exhibits mutant huntingtin (mHTT) aggregates and soluble mHTT-dependent neurodegeneration. *FASEB J*, 2013. 27(5): pp. 1820–9.
36. Gong, B. et al., Time-lapse analysis of aggregate formation in an inducible PC12 cell model of Huntington's disease reveals time-dependent aggregate formation that transiently delays cell death. *Brain Res Bull*, 2008. 75(1): pp. 146–57.
37. Saudou, F. et al., Huntingtin acts in the nucleus to induce apoptosis but death does not correlate with the formation of intranuclear inclusions. *Cell*, 1998. 95(1): pp. 55–66.
38. Arrasate, M. et al., Inclusion body formation reduces levels of mutant huntingtin and the risk of neuronal death. *Nature*, 2004. 431(7010): pp. 805–10.
39. Fernandez-Nogales, M. et al., Huntington's disease is a four-repeat tauopathy with tau nuclear rods. *Nat Med*, 2014. 20(8): pp. 881–5.
40. Vattakatuchery, J. J. and R. Kurien, Acetylcholinesterase inhibitors in cognitive impairment in Huntington's disease: A brief review. *World J Psychiatry*, 2013. 3(3): pp. 62–64.
41. Videnovic, A., Treatment of huntington disease. *Curr Treat Options Neurol*, 2013. 15(4): pp. 424–38.
42. Hayden, M. R. et al., Tetrabenazine. *Nat Rev Drug Discov*, 2009. 8(1): pp. 17–8.
43. Takahashi, K. et al., Induction of pluripotent stem cells from adult human fibroblasts by defined factors. *Cell*, 2007. 131(5): pp. 861–72.
44. Takahashi, K. and S. Yamanaka, Induction of pluripotent stem cells from mouse embryonic and adult fibroblast cultures by defined factors. *Cell*, 2006. 126(4): pp. 663–76.
45. Moretti, A. et al., Pluripotent stem cell models of human heart disease. *Cold Spring Harb Perspect Med*, 2013. 3(11).
46. An, M. C. et al., Genetic correction of Huntington's disease phenotypes in induced pluripotent stem cells. *Cell Stem Cell*, 2012. 11(2): pp. 253–63.
47. Kim, H. S. et al., Genomic editing tools to model human diseases with isogenic pluripotent stem cells. *Stem Cells Dev*, 2014. 23(22): pp. 2673–86.
48. Soldner, F. et al., Generation of isogenic pluripotent stem cells differing exclusively at two early onset Parkinson point mutations. *Cell*, 2011. 146(2): pp. 318–31.
49. Difiglia, M., T. Pasik, and P. Pasik, Ultrastructure of Golgi-impregnated and gold-toned spiny and aspiny neurons in the monkey neostriatum. *J Neurocytol*, 1980. 9(4): pp. 471–92.

50. Difiglia, M., P. Pasik, and T. Pasik, Early postnatal development of the monkey neostriatum: A Golgi and ultrastructural study. *J Comp Neurol*, 1980. 190(2): pp. 303–31.
51. DiFiglia, M., P. Pasik, and T. Pasik, A Golgi study of neuronal types in the neostriatum of monkeys. *Brain Res*, 1976. 114(2): pp. 245–56.
52. Albin, R. L. et al., Excitatory amino acid binding sites in the basal ganglia of the rat: A quantitative autoradiographic study. *Neuroscience*, 1992. 46(1): pp. 35–48.
53. Reiner, A. et al., Differential loss of striatal projection neurons in Huntington disease. *Proc Natl Acad Sci USA*, 1988. 85(15): pp. 5733–7.
54. Crossman, A. R., Primate models of dyskinesia: The experimental approach to the study of basal ganglia-related involuntary movement disorders. *Neuroscience*, 1987. 21(1): pp. 1–40.
55. Hedreen, J. C. et al., Neuronal loss in layers V and VI of cerebral cortex in Huntington's disease. *Neurosci Lett*, 1991. 133(2): pp. 257–61.
56. Cudkowicz, M. and N. W. Kowall, Degeneration of pyramidal projection neurons in Huntington's disease cortex. *Ann Neurol*, 1990. 27(2): pp. 200–4.
57. Sotrel, A. et al., Morphometric analysis of the prefrontal cortex in Huntington's disease. *Neurology*, 1991. 41(7): pp. 1117–23.
58. Kim, E. H. et al., Cortical interneuron loss and symptom heterogeneity in Huntington disease. *Ann Neurol*, 2014. 75(5): pp. 717–27.
59. Faideau, M. et al., In vivo expression of polyglutamine-expanded huntingtin by mouse striatal astrocytes impairs glutamate transport: A correlation with Huntington's disease subjects. *Hum Mol Genet*, 2010. 19(15): pp. 3053–67.
60. Tong, X. et al., Astrocyte Kir4.1 ion channel deficits contribute to neuronal dysfunction in Huntington's disease model mice. *Nat Neurosci*, 2014. 17(5): pp. 694–703.
61. Kwan, W. et al., Mutant huntingtin impairs immune cell migration in Huntington disease. *J Clin Invest*, 2012. 122(12): pp. 4737–47.
62. Tai, Y. F. et al., Microglial activation in presymptomatic Huntington's disease gene carriers. *Brain*, 2007. 130(Pt 7): pp. 1759–66.
63. Pavese, N. et al., Microglial activation correlates with severity in Huntington disease: A clinical and PET study. *Neurology*, 2006. 66(11): pp. 1638–43.
64. Sapp, E. et al., Early and progressive accumulation of reactive microglia in the Huntington disease brain. *J Neuropathol Exp Neurol*, 2001. 60(2): pp. 161–72.
65. Nayak, A. et al., Huntington's disease: An immune perspective. *Neurol Res Int*, 2011. 2011: 563784.
66. Martin, B. et al., Therapeutic perspectives for the treatment of Huntington's disease: Treating the whole body. *Histol Histopathol*, 2008. 23(2): pp. 237–50.
67. Zielonka, D. et al., Skeletal muscle pathology in Huntington's disease. *Front Physiol*, 2014. 5: p. 380.
68. Magnusson-Lind, A. et al., Skeletal muscle atrophy in R6/2 mice—Altered circulating skeletal muscle markers and gene expression profile changes. *J Huntingtons Dis*, 2014. 3(1): pp. 13–24.
69. Lindenberg, K. S. et al., Two-point magnitude MRI for rapid mapping of brown adipose tissue and its application to the R6/2 mouse model of Huntington disease. *PLoS One*, 2014. 9(8): p. e105556.
70. Her, L. S. et al., The differential profiling of ubiquitin-proteasome and autophagy systems in different tissues before the onset of Huntington's disease models. *Brain Pathol*, 2014.

71. Trager, U. et al., HTT-lowering reverses Huntington's disease immune dysfunction caused by NFkappaB pathway dysregulation. *Brain*, 2014. 137(Pt 3): pp. 819–33.
72. Beal, M. F. et al., Replication of the neurochemical characteristics of Huntington's disease by quinolinic acid. *Nature*, 1986. 321(6066): pp. 168–71.
73. Reynolds, G. P. et al., Brain quinolinic acid in Huntington's disease. *J Neurochem*, 1988. 50(6): pp. 1959–60.
74. Boegman, R. J., Y. Smith, and A. Parent, Quinolinic acid does not spare striatal neuropeptide Y-immunoreactive neurons. *Brain Res*, 1987. 415(1): pp. 178–82.
75. A novel gene containing a trinucleotide repeat that is expanded and unstable on Huntington's disease chromosomes. The Huntington's Disease Collaborative Research Group. *Cell*, 1993. 72(6): pp. 971–83.
76. Mangiarini, L. et al., Exon 1 of the HD gene with an expanded CAG repeat is sufficient to cause a progressive neurological phenotype in transgenic mice. *Cell*, 1996. 87(3): pp. 493–506.
77. Hickey, M. A. et al., Improvement of neuropathology and transcriptional deficits in CAG 140 knock-in mice supports a beneficial effect of dietary curcumin in Huntington's disease. *Mol Neurodegener*, 2012. 7: p. 12.
78. Menalled, L. et al., Systematic behavioral evaluation of Huntington's disease transgenic and knock-in mouse models. *Neurobiol Dis*, 2009. 35(3): pp. 319–36.
79. Lee, C. Y., J. P. Cantle, and X. W. Yang, Genetic manipulations of mutant huntingtin in mice: New insights into Huntington's disease pathogenesis. *FEBS J*, 2013. 280(18): pp. 4382–94.
80. Pouladi, M. A., A. J. Morton, and M. R. Hayden, Choosing an animal model for the study of Huntington's disease. *Nat Rev Neurosci*, 2013. 14(10): pp. 708–21.
81. Carter, R. L. et al., Reversal of cellular phenotypes in neural cells derived from Huntington's disease monkey-induced pluripotent stem cells. *Stem Cell Reports*, 2014. 3(4): pp. 585–93.
82. Zhang, S. et al., A genomewide RNA interference screen for modifiers of aggregates formation by mutant Huntingtin in *Drosophila*. *Genetics*, 2010. 184(4): pp. 1165–79.
83. Ehrlich, M. E., Huntington's disease and the striatal medium spiny neuron: Cell-autonomous and non-cell-autonomous mechanisms of disease. *Neurotherapeutics*, 2012. 9(2): pp. 270–84.
84. Bradford, J. et al., Expression of mutant huntingtin in mouse brain astrocytes causes age-dependent neurological symptoms. *Proc Natl Acad Sci USA*, 2009. 106(52): pp. 22480–5.
85. Brown, T. B., A. I. Bogush, and M. E. Ehrlich, Neocortical expression of mutant huntingtin is not required for alterations in striatal gene expression or motor dysfunction in a transgenic mouse. *Hum Mol Genet*, 2008. 17(20): pp. 3095–104.
86. Thomas, E. A. et al., In vivo cell-autonomous transcriptional abnormalities revealed in mice expressing mutant huntingtin in striatal but not cortical neurons. *Hum Mol Genet*, 2011. 20(6): pp. 1049–60.
87. Kim, S. H. et al., Forebrain striatal-specific expression of mutant huntingtin protein in vivo induces cell-autonomous age-dependent alterations in sensitivity to excitotoxicity and mitochondrial function. *ASN Neuro*, 2011. 3(3): p. e00060.

88. Gu, X. et al., Pathological cell-cell interactions are necessary for striatal pathogenesis in a conditional mouse model of Huntington's disease. *Mol Neurodegener,* 2007. 2: p. 8.
89. Gu, X. et al., Pathological cell-cell interactions elicited by a neuropathogenic form of mutant Huntingtin contribute to cortical pathogenesis in HD mice. *Neuron,* 2005. 46(3): pp. 433–44.
90. Wang, N. et al., Neuronal targets for reducing mutant huntingtin expression to ameliorate disease in a mouse model of Huntington's disease. *Nat Med,* 2014. 20(5): pp. 536–41.
91. Trettel, F. et al., Dominant phenotypes produced by the HD mutation in STHdh(Q111) striatal cells. *Hum Mol Genet,* 2000. 9(19): pp. 2799–809.
92. Bradley, C. K. et al., Derivation of Huntington's disease-affected human embryonic stem cell lines. *Stem Cells Dev,* 2011. 20(3): pp. 495–502.
93. Niclis, J. et al., Human embryonic stem cell models of Huntington disease. *Reprod Biomed Online,* 2009. 19(1): pp. 106–13.
94. Malik, N. and M. S. Rao, A review of the methods for human iPSC derivation. *Methods Mol Biol,* 2013. 997: pp. 23–33.
95. Cyranoski, D., Stem-cell pioneer banks on future therapies. *Nature,* 2012. 488(7410): pp. 139.
96. Park, I. H. et al., Disease-specific induced pluripotent stem cells. *Cell,* 2008. 134(5): pp. 877–86.
97. Induced pluripotent stem cells from patients with Huntington's disease show CAG-repeat-expansion-associated phenotypes. *Cell Stem Cell,* 2012. 11(2): pp. 264–78.
98. Crane, J. D. et al., Massage therapy attenuates inflammatory signaling after exercise-induced muscle damage. *Sci Transl Med,* 2012. 4(119): p. 119ra13.
99. Camnasio, S. et al., The first reported generation of several induced pluripotent stem cell lines from homozygous and heterozygous Huntington's disease patients demonstrates mutation related enhanced lysosomal activity. *Neurobiol Dis,* 2012. 46(1): pp. 41–51.
100. Chae, J. I. et al., Quantitative proteomic analysis of induced pluripotent stem cells derived from a human Huntington's disease patient. *Biochem J,* 2012. 446(3): pp. 359–71.
101. Jeon, I. et al., Neuronal properties, in vivo effects, and pathology of a Huntington's disease patient-derived induced pluripotent stem cells. *Stem Cells,* 2012. 30(9): pp. 2054–62.
102. Sundberg, M. et al., Improved cell therapy protocols for Parkinson's disease based on differentiation efficiency and safety of hESC-, hiPSC-, and non-human primate iPSC-derived dopaminergic neurons. *Stem Cells,* 2013. 31(8): pp. 1548–62.
103. Kriks, S. et al., Dopamine neurons derived from human ES cells efficiently engraft in animal models of Parkinson's disease. *Nature,* 2011. 480(7378): pp. 547–51.
104. Hargus, G. et al., Differentiated Parkinson patient-derived induced pluripotent stem cells grow in the adult rodent brain and reduce motor asymmetry in Parkinsonian rats. *Proc Natl Acad Sci USA,* 2010. 107(36): pp. 15921–6.
105. Swistowski, A. et al., Efficient generation of functional dopaminergic neurons from human induced pluripotent stem cells under defined conditions. *Stem Cells,* 2010. 28(10): pp. 1893–904.

106. Cooper, O. et al., Differentiation of human ES and Parkinson's disease iPS cells into ventral midbrain dopaminergic neurons requires a high activity form of SHH, FGF8a and specific regionalization by retinoic acid. *Mol Cell Neurosci*, 2010. 45(3): pp. 258–66.
107. Delli Carri, A. et al., Human pluripotent stem cell differentiation into authentic striatal projection neurons. *Stem Cell Rev*, 2013. 9(4): pp. 461–74.
108. Delli Carri, A. et al., Developmentally coordinated extrinsic signals drive human pluripotent stem cell differentiation toward authentic DARPP-32+ medium-sized spiny neurons. *Development*, 2013. 140(2): pp. 301–12.
109. Consortium, H. D. I., Induced pluripotent stem cells from patients with Huntington's disease show CAG-repeat-expansion-associated phenotypes. *Cell Stem Cell*, 2012. 11(2): pp. 264–78.
110. Shin, E. et al., GABAergic neurons from mouse embryonic stem cells possess functional properties of striatal neurons in vitro, and develop into striatal neurons in vivo in a mouse model of Huntington's disease. *Stem Cell Rev*, 2012. 8(2): pp. 513–31.
111. Zhang, N. et al., Characterization of human Huntington's disease cell model from induced pluripotent stem cells. *PLoS Curr*, 2010. 2: p. RRN1193.
112. Aubry, L. et al., Striatal progenitors derived from human ES cells mature into DARPP32 neurons in vitro and in quinolinic acid-lesioned rats. *Proc Natl Acad Sci USA*, 2008. 105(43): pp. 16707–12.
113. Roybon, L. et al., Human stem cell-derived spinal cord astrocytes with defined mature or reactive phenotypes. *Cell Rep*, 2013. 4(5): pp. 1035–48.
114. Espuny-Camacho, I. et al., Pyramidal neurons derived from human pluripotent stem cells integrate efficiently into mouse brain circuits in vivo. *Neuron*, 2013. 77(3): pp. 440–56.
115. Shaltouki, A. et al., Efficient generation of astrocytes from human pluripotent stem cells in defined conditions. *Stem Cells*, 2013. 31(5): pp. 941–52.
116. Shi, Y., P. Kirwan, and F. J. Livesey, Directed differentiation of human pluripotent stem cells to cerebral cortex neurons and neural networks. *Nat Protoc*, 2012. 7(10): pp. 1836–46.
117. Shi, Y. et al., Human cerebral cortex development from pluripotent stem cells to functional excitatory synapses. *Nat Neurosci*, 2012. 15(3): pp. 477–86, S1.
118. Calamini, B., D. C. Lo, and L. S. Kaltenbach, Experimental models for identifying modifiers of polyglutamine-induced aggregation and neurodegeneration. *Neurotherapeutics*, 2013. 10(3): pp. 400–15.
119. Schulte, J. et al., High-content chemical and RNAi screens for suppressors of neurotoxicity in a Huntington's disease model. *PLoS One*, 2011. 6(8): p. e23841.
120. Miller, J. P. et al., A genome-scale RNA-interference screen identifies RRAS signaling as a pathologic feature of Huntington's disease. *PLoS Genet*, 2012. 8(11): p. e1003042.
121. Zhang, N. et al., Inhibition of lipid signaling enzyme diacylglycerol kinase epsilon attenuates mutant huntingtin toxicity. *J Biol Chem*, 2012. 287(25): pp. 21204–13.
122. Lancaster, M. A. et al., Cerebral organoids model human brain development and microcephaly. *Nature*, 2013. 501(7467): pp. 373–9.
123. Kadoshima, T. et al., Self-organization of axial polarity, inside-out layer pattern, and species-specific progenitor dynamics in human ES cell-derived neocortex. *Proc Natl Acad Sci USA*, 2013. 110(50): pp. 20284–9.

124. Mali, P. et al., RNA-guided human genome engineering via Cas9. *Science*, 2013. 339(6121): pp. 823–6.
125. Sanjana, N. E. et al., A transcription activator-like effector toolbox for genome engineering. *Nat Protoc*, 2012. 7(1): pp. 171–92.
126. Czosek, R. J. et al., Cardiac rhythm devices in the pediatric population: Utilization and complications. *Heart Rhythm*, 2012. 9(2): pp. 199–208.
127. An, M. C. et al., Polyglutamine disease modeling: Epitope based screen for homologous recombination using CRISPR/Cas9 system. *PLoS Curr*, 2014: p. 6. doi:10.1371/currents.hd.0242d2e7ad72225efa72f6964589369a.
128. Astashkina, A., B. Mann, and D. W. Grainger, A critical evaluation of in vitro cell culture models for high-throughput drug screening and toxicity. *Pharmacol Ther*, 2012. 134(1): pp. 82–106.
129. Schindler, R., Use of cell culture in pharmacology. *Annu Rev Pharmacol*, 1969. 9: pp. 393–406.
130. Cai, S., Y. S. Chan, and D. K. Shum, Induced pluripotent stem cells and neurological disease models. *Sheng Li Xue Bao*, 2014. 66(1): pp. 55–66.
131. Xu, X. H. and Z. Zhong, Disease modeling and drug screening for neurological diseases using human induced pluripotent stem cells. *Acta Pharmacol Sin*, 2013. 34(6): pp. 755–64.
132. Crook, J. M. and N. R. Kobayashi, Human stem cells for modeling neurological disorders: Accelerating the drug discovery pipeline. *J Cell Biochem*, 2008. 105(6): pp. 1361–6.
133. Sarantos, M. R. et al., Pizotifen activates ERK and provides neuroprotection in vitro and in vivo in models of Huntington's disease. *J Huntingtons Dis*, 2012. 1(2): pp. 195–210.
134. Doumanis, J. et al., RNAi screening in *Drosophila* cells identifies new modifiers of mutant huntingtin aggregation. *PLoS One*, 2009. 4(9): p. e7275.
135. Marsh, J. L., J. Pallos, and L. M. Thompson, Fly models of Huntington's disease. *Hum Mol Genet*, 2003. 12 Spec No 2: pp. R187–93.
136. Crane, A. T., J. Rossignol, and G. L. Dunbar, Use of genetically altered stem cells for the treatment of Huntington's disease. *Brain Sci*, 2014. 4(1): pp. 202–19.
137. Fink, K. D. et al., Intrastriatal transplantation of adenovirus-generated induced pluripotent stem cells for treating neuropathological and functional deficits in a rodent model of Huntington's disease. *Stem Cells Transl Med*, 2014. 3(5): pp. 620–31.
138. Mu, S. et al., Transplantation of induced pluripotent stem cells improves functional recovery in Huntington's disease rat model. *PLoS One*, 2014. 9(7): p. e101185.
139. Cyranoski, D., Next-generation stem cells cleared for human trial. *Nature*, 2014. doi:10.1038/nature.2014.15897. http://www.nature.com/news/next-generation-stem-cells-cleared-for-human-trial-1.15897.
140. Marchetto, M. C. et al., Non-cell-autonomous effect of human SOD1 G37R astrocytes on motor neurons derived from human embryonic stem cells. *Cell Stem Cell*, 2008. 3(6): pp. 649–57.
141. McKinstry, S. U. et al., Huntingtin is required for normal excitatory synapse development in cortical and striatal circuits. *J Neurosci*, 2014. 34(28): pp. 9455–72.
142. Milnerwood, A. J. and L. A. Raymond, Corticostriatal synaptic function in mouse models of Huntington's disease: Early effects of huntingtin repeat length and protein load. *J Physiol*, 2007. 585(Pt 3): pp. 817–31.

7

Applications of Pluripotent Stem Cells in the Therapy and Modeling of Diabetes and Metabolic Diseases

Suranjit Mukherjee and Shuibing Chen

CONTENTS

7.1 Introduction

The opportunities and the hopes promised by the field of stem cell research and stem cell therapy are numerous and diverse. From disease modeling to tissue regeneration therapy, stem cell biology opens many avenues toward treatment of numerous disorders in completely novel ways. The applications of reprogramming and stem cell therapy also present a unique opportunity to understand and treat metabolic diseases, specifically diabetes mellitus.

Diabetes mellitus is a group of metabolic disorders defined by the inability to maintain homeostatic glucose levels, or euglycemia, as a result of insufficient insulin production. Clinically, diabetes mellitus is characterized by a blood glucose level greater than or equal to 126 mg/dL two hours after a glucose tolerance test or a glycated hemoglobin (HbA1c) level greater than or equal to 6.5%. This hyperglycemia results in symptoms such as frequent urination, increased thirst, and increased hunger. If untreated, diabetic patients will develop many complications, including acute complications, such as diabetic ketoacidosis and nonketotic hyperosmolar coma, and long-term complications, including heart disease, stroke, kidney failure, foot ulcers, and damage to the eyes. Diabetes mellitus itself can be manifested in three categories: type 1 diabetes mellitus (T1DM), type 2 diabetes mellitus (T2DM), and monogenic forms of diabetes mellitus (maturity-onset diabetes mellitus of the young [MODY]). The numerous organ systems involved in the pathology of these various forms of diabetes are shown in Figure 7.1.

T1DM, also called *juvenile diabetes* or *insulin-dependent diabetes*, is an autoimmune disorder where the β cells of the pancreas are destroyed by the immune system as a result of self-antigen recognition. As a result, T1DM

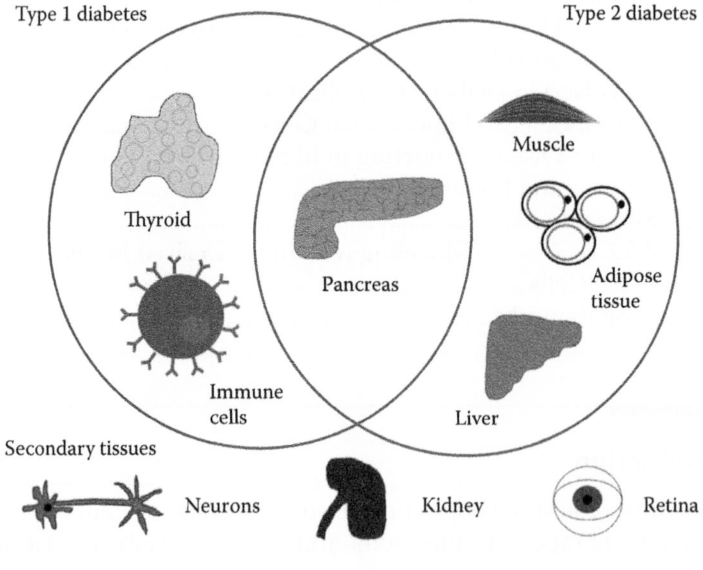

FIGURE 7.1
(See color insert.) Organs involved in the development of type 1 and type 2 diabetes.

patients cannot produce enough insulin in response to glucose and must self-administer insulin to maintain euglycemia. Although several factors, such as virus or environmental toxins, have been shown to be involved in the pathogenesis of T1DM, the trigger of T1DM is unknown. T1DM accounts for approximately 5% of all diabetic cases and is characterized with a peak age of onset around the mid-teen years.

T2DM, on the other hand, accounts for 90–95% of all diabetes mellitus cases. T2DM patients first develop insulin resistance, in which the cells of the muscle, the liver, and the adipose tissue do not correctly respond to the presence of insulin, thus leading to hyperglycemia. As T2DM progresses, the β cells of the pancreas attempt to compensate with an overproduction of insulin, leading to β cell dysfunction and, eventually, apoptosis for late-stage diabetics (although the degree and the timing of insulin resistance and β cell dysfunction vary from patient to patient). T2DM is strongly associated with obesity. Numerous studies using both *in vitro* and *in vivo* based models suggest that insulin resistance is highly related to defects and deregulation in the insulin signaling and lipid regulatory pathways, along with obesity-associated inflammation (1,2).

Monogenic forms of diabetes mellitus, also called MODY, are group of inherited autosomal dominant diseases caused by the mutation of proteins involved in pancreatic development and/or pancreatic β cell function. There are six common types of MODYs that are summarized in Table 7.1.

Multiple factors including genetic factors, environmental factors, and lifestyle contribute to the progression of diabetes mellitus. Several rodent models of obesity and diabetes mellitus, such as the *ob/ob* and *db/db* mouse, high-fat diet mice, the Zucker diabetic fatty rat, and non-obese diabetic mice, have provided valuable insight into the progression of diabetes mellitus. However, there are fundamental differences between rodents and humans. For example, rodent pancreatic β cells secrete both insulin 1 and insulin 2, while human pancreatic β cells only secrete insulin. Thus, rodent models fail to fully recapitulate the key aspects of human biology that are critical to the progression of diabetes mellitus. As a consequence, there is a need to develop novel methods and systems to study diabetes mellitus. iPSCs from diabetic patients represent an original and invaluable resource from which new knowledge and perspective on the pathogenesis of diabetes can be gathered. Utilizing the process of cellular reprogramming pioneered by Yamanaka, scientists now have the ability to generate multiple diabetic iPSCs from individual patients that can then be differentiated into different cell types relevant to diabetes mellitus. In the following sections, we will discuss how stem cell biology and cellular reprogramming can be applied to model T1DM, T2DM, and MODY and how directed differentiation of hESCs can present therapeutic opportunities for those suffering from the disease.

TABLE 7.1

Summary of Genetics and Symptoms of MODYs

Type	Gene	Protein Function	Symptoms
MODY1	Hepatocyte nuclear factor 4 alpha (HNF4α) (3)	A key transcription factor involved in liver and pancreatic β cell development and function	Gradual hyperglycemia. Patients require medical intervention with insulin or other mediations to achieve euglycemia.
MODY2	Glucokinase (4)	A key kinase involved in glucose-stimulated insulin secretion of pancreatic β cells	Mild nonprogressive hyperglycemia, with dietary interventions helping to ameliorate elevated glucose levels.
MODY3	Hepatocyte nuclear factor 1 alpha (5)	A key transcription factor involved in pancreatic β cell development and function	Persistent hyperglycemia, with approximately 40% of patients requiring insulin therapy (6).
MODY4	Pancreatic and duodenal homeobox 1 (*PDX1*)	A key transcription factor involved in pancreatic β cell development and function	A significant defect in insulin secretion and, correspondingly, persistent hyperglycemia.
MODY5	Hepatocyte nuclear factor 1 β gene (*HNF1β*) (7)	A key transcription factor involved in pancreatic β cell development and function	Low birth weight and commonly requires insulin therapy. Patients have nondiabetic renal dysfunction, with patients exhibiting kidney dysfunction due to general renal failure with cyst formation prior to exhibiting glucose intolerance.
MODY6	NEUROD1 (8)	A key transcription factor involved in pancreatic β cell development	Classic symptoms of glucose intolerance and have a profile typical of T2DM patients.

7.2 Primary Organs Involved in the Pathogenesis of Diabetes Mellitus

7.2.1 Pancreas

The pancreas is one of the primary organs featured in the pathogenesis of diabetes mellitus, including T1DM, T2DM, and MODY. In T1DM, pancreatic β cells are destroyed by the immune system as a result of self-antigen recognition of β cell factors. In T2DM, pancreatic β cells continually attempt to counter the peripheral insulin resistance by increasing insulin output, leading to β cell dysfunction and apoptosis.

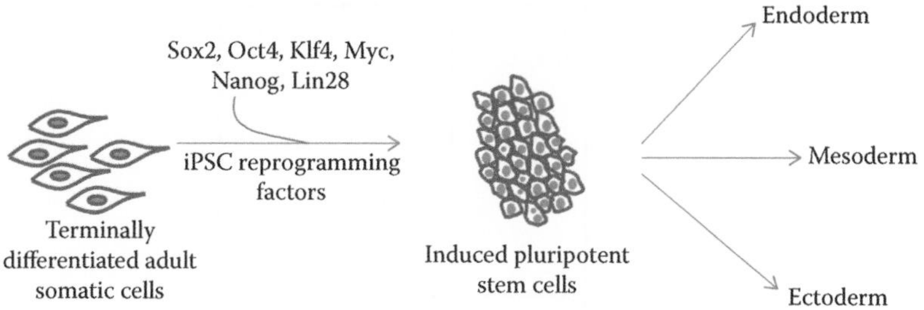

FIGURE 1.2
A schematic representation of somatic cell reprogramming.

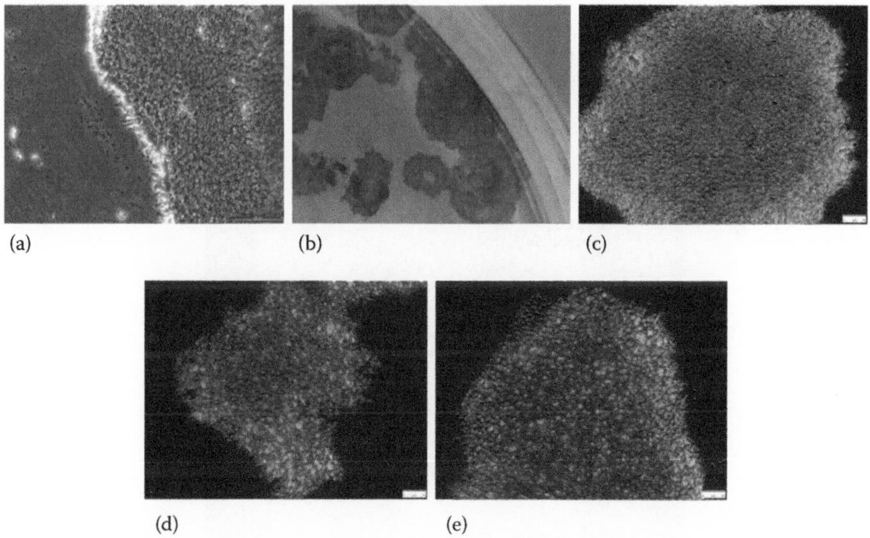

FIGURE 2.3
Characterization of iPSCs by ICC: (a) bright-field image of a healthy iPSC colony and (b) alkaline phosphatase-stained iPSC colony showing positive AP staining of undifferentiated pluripotent cells. Immunofluorescence staining images showing pluripotency markers such as (c) SSEA4 (green) and OCT3/4 (red), (d) TRA 1-81 (green) and SOX2 (red), and (e) TRA 1-60 (green) and NANOG (red).

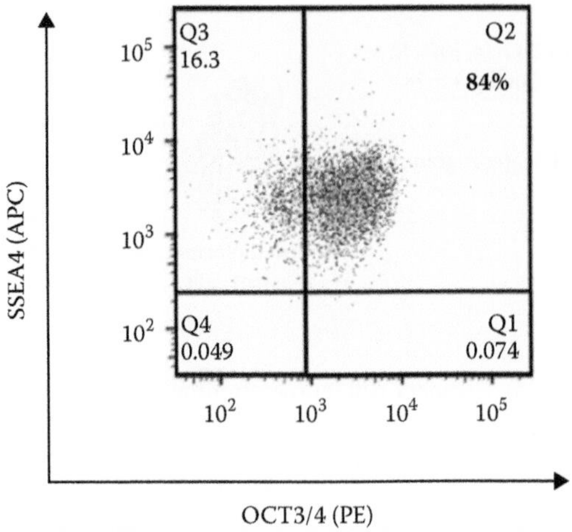

FIGURE 2.4
Flow cytometric analysis of iPSCs for the presence of pluripotency markers SSEA4 and OCT4.

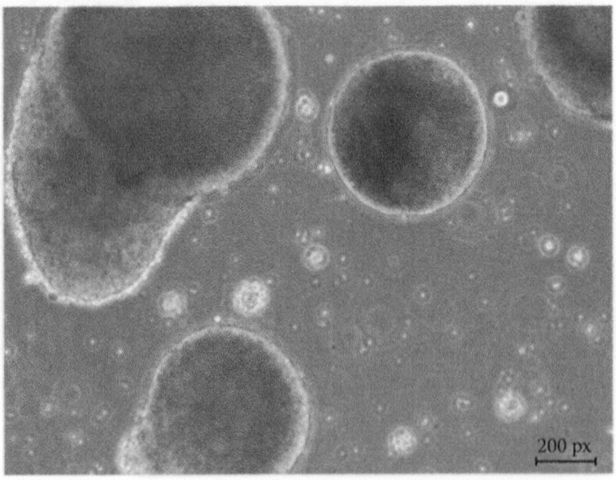

FIGURE 2.5
Embryoid bodies are clumps of spherical aggregates that contain differentiated derivatives representing all three germ layers (ectoderm, endoderm, mesoderm).

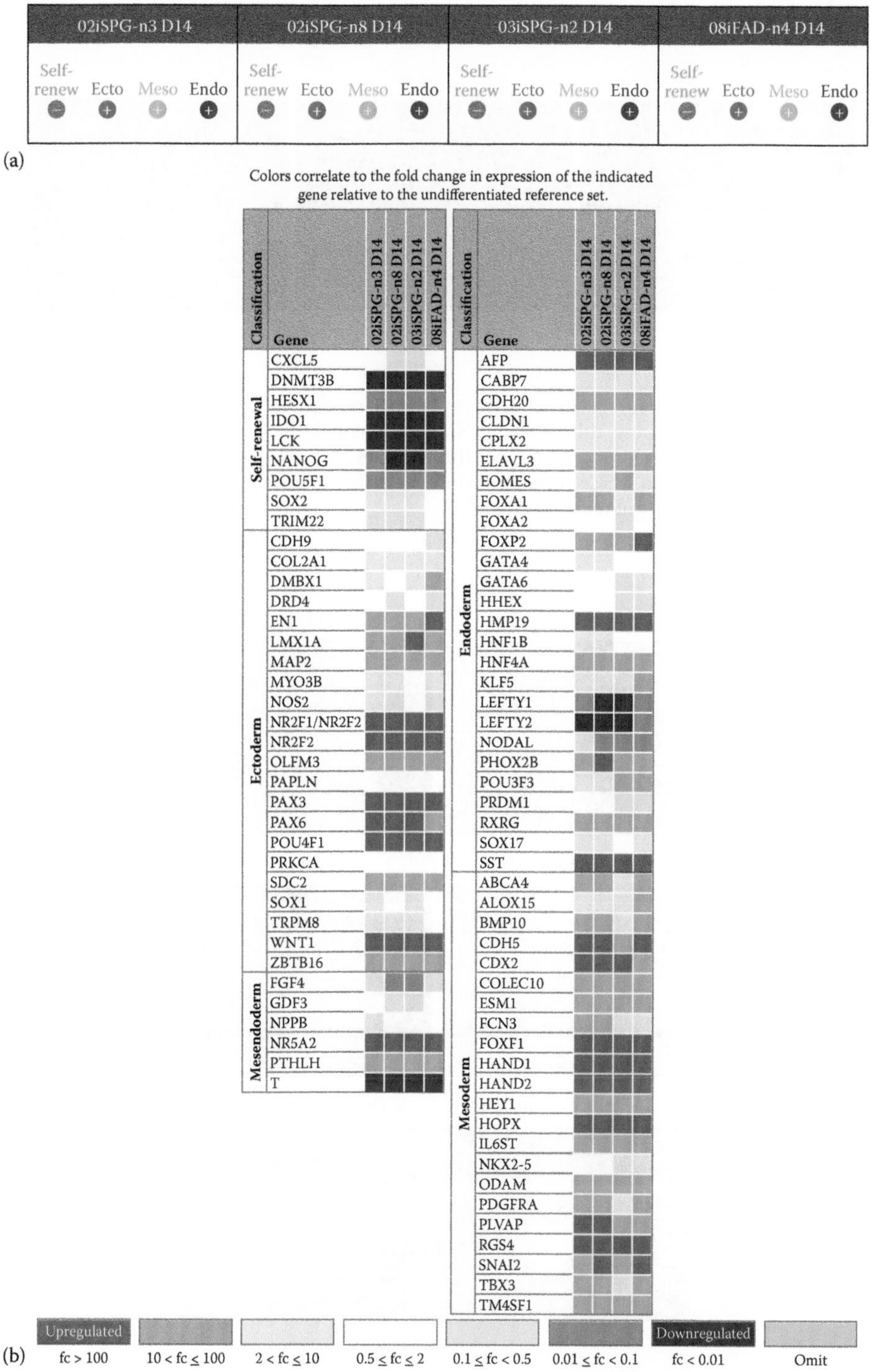

FIGURE 2.6
TaqMan hPSC Scorecard Assay. Panel (a) shows in a nutshell the pluripotency of EB indicating their ability to form the various germ layers. Panel (b) shows the expression levels of the genes that are involved in the formation of the three germ layers. Upregulated genes are shown in red and the downregulated genes in blue.

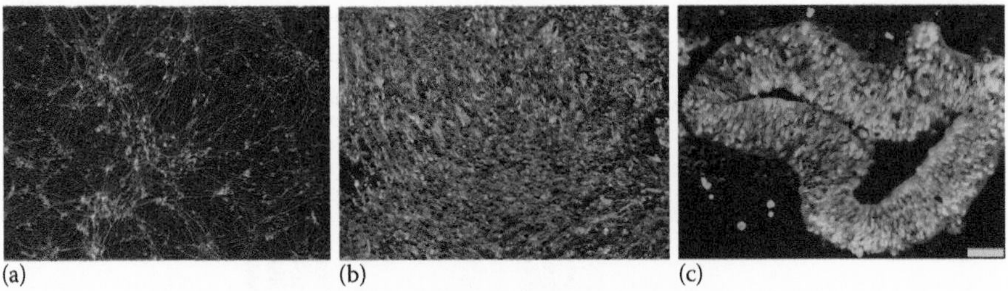

FIGURE 2.7
iPSCs differentiated into specific cell types: (a) differentiation of iPSCs to motor neurons (neurectoderm) expressing TuJ1 (red) and Nkx6.1 (green); (b) iPSCs differentiated into chondrocytes showing collagen type I (green) through mesoderm specification; (c) generation of foregut organoids from iPSCs through endoderm differentiation. Organoid expresses Sox2 (green), a posterior foregut marker on day 20.

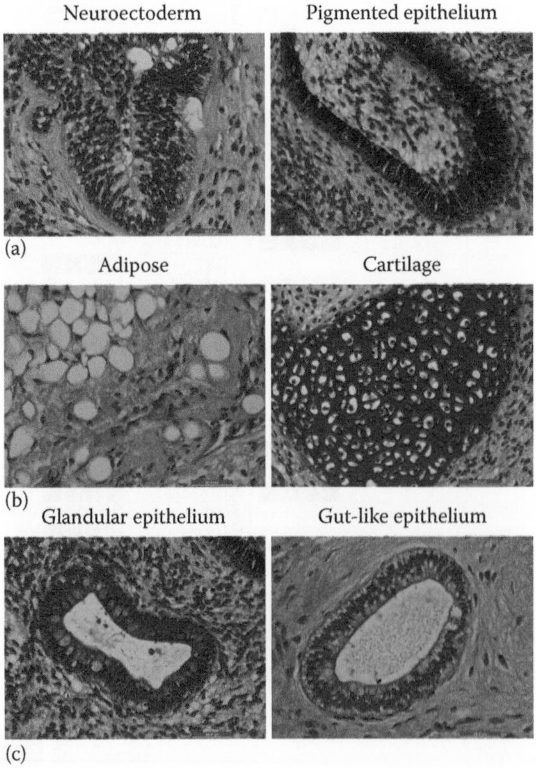

FIGURE 2.8
Panels showing teratoma staining exhibiting propensity to form all three germ layers, namely, (a) the ectoderm, (b) the mesoderm, and (c) the endoderm.

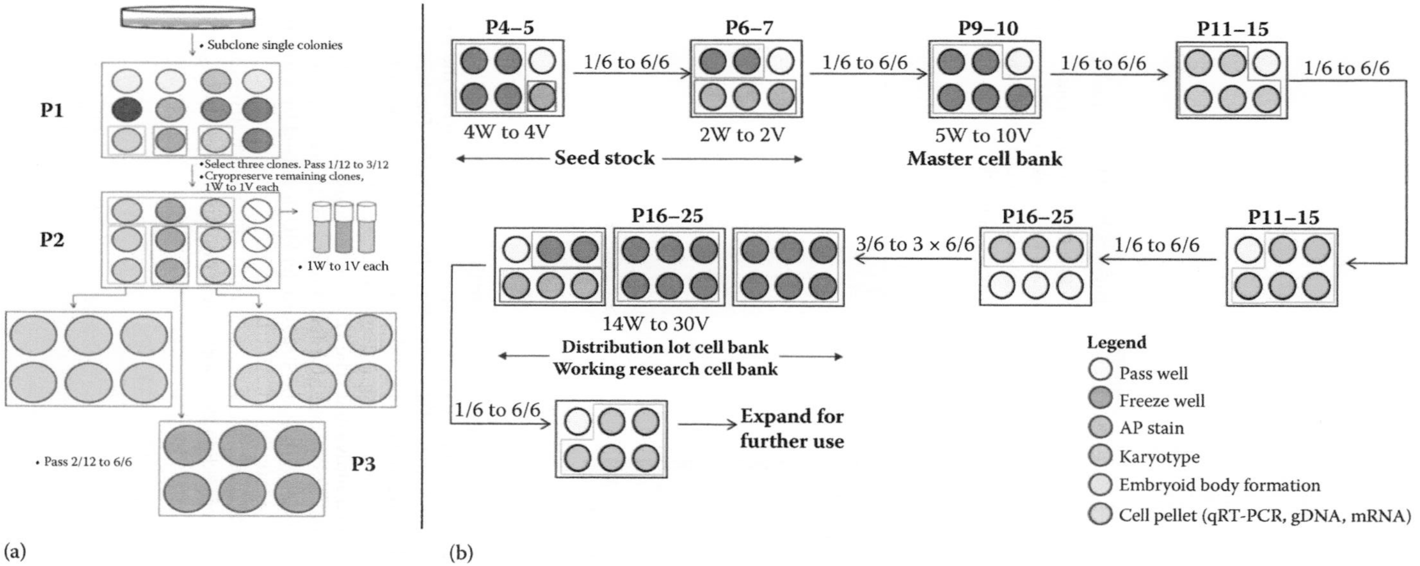

FIGURE 2.9
(a) Schematic representation of single clone isolation and selection. The 12 colors at passage 1 (P1) indicates 12 single clones. Three clones of the initial 12 are selected for further expansion and characterization. One well of a 12-well plate (1/12) is expanded to 3 wells of a 12-well plate (3/12), after which 2/12 is expanded into 6 wells of a 6-well plate (6/6). The remaining clones are cryopreserved with 1 well (1W) distributed into one vial (1V) each. Clones are selected based on cell morphology, growth rate, and rate of differentiation. (b) Schematic representation of single clone expansion, characterization, and banking. This process is repeated for each clone selected at P2 (a). Typically, 1 well of a 6-well plate (1/6) is passaged into 6 wells of a 6-well plate (6/6).

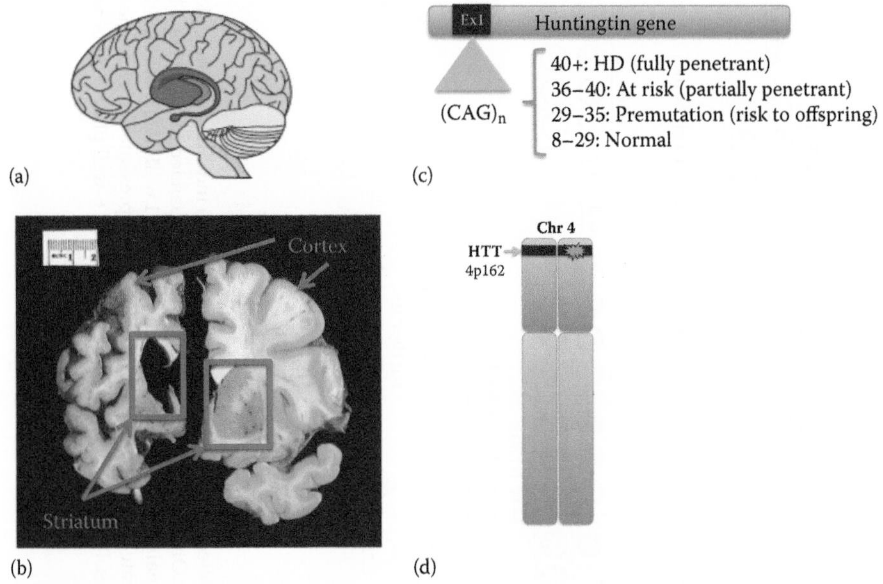

FIGURE 6.1
(a) The striatum is located below the cortex in the forebrain in a structure called the *basal ganglia*. Shown in pink are the caudate nucleus and putamen, which are part of the striatum. In orange is the thalamus, which separates the cortex from the midbrain. The basal ganglia play an important role in controlling movement and behavior, and patients with HD experience disturbances in both those actions. (b) Postmortem brain tissue from (*left*) a Huntington's disease patient and (*right*) a normal individual. The HD brain has substantial atrophy in the striatum due to the loss of striatal MSNs and also has atrophy in the cortex due to loss of cortical projection neurons that innervate in the striatum. (c) HD is caused by a mutation in the Huntingtin (HTT) gene. The mutation is a CAG trinucleotide repeat expansion located in exon 1 of HTT. Normal individuals have 29 repeats or less while those with 40 repeats or more will get HD. (d) The HTT gene is located on the short arm of human chromosome 4.

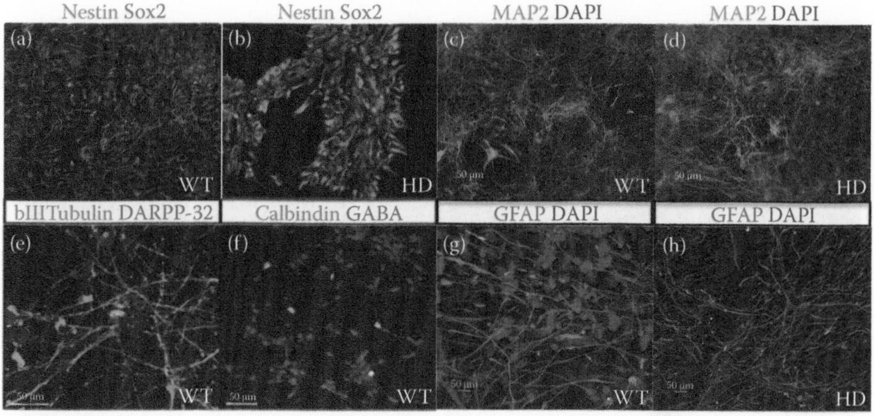

FIGURE 6.3
Immunocytochemical staining of neural cell types derived from HD and genetically corrected wild type (WT) iPSCs. iPSCs can differentiate into (a, b) NSCs, (c–f) neurons, and (g, h) astrocytes. Striatal neurons expressing DARPP-32 and GABA can also be derived from (e, f) iPSCs.

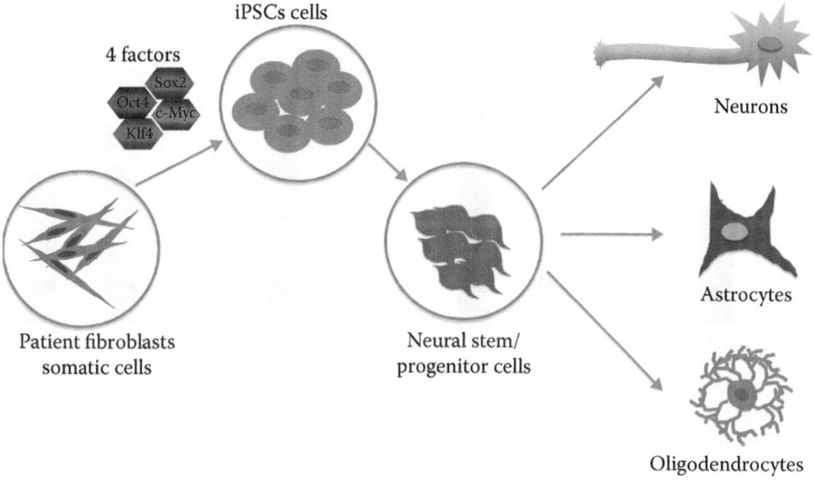

FIGURE 6.4
Human skin cells can be reprogrammed back to a PSC state by ectopically expressing the transcription factors Oct4, Sox2, KLF4, and c-Myc. These cells are called iPSCs and have identical characteristics and differentiation ability compared to hESCs. iPSCs can differentiate into any cell type in the body. In the case of HD, iPSCs can be differentiated into NSCs or neural progenitor cells, which can then be differentiated into mature brain cell types including neurons, astrocytes, and oligodendrocytes.

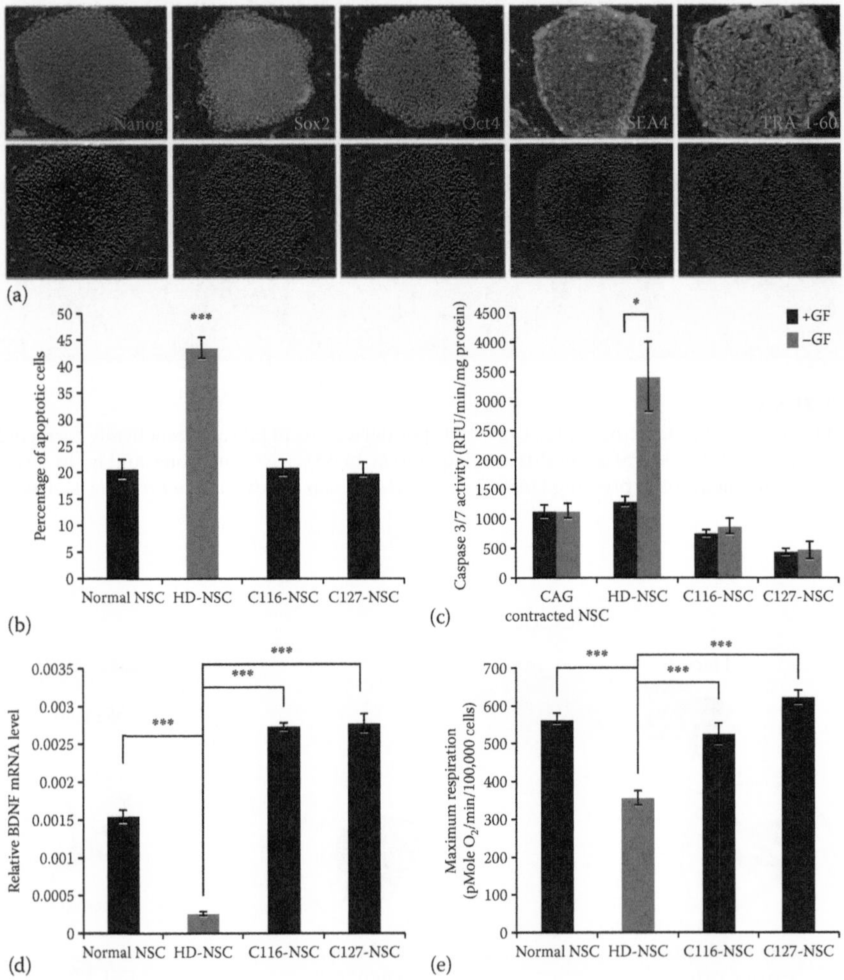

FIGURE 6.5
Using iPSCs to model HD. (a) HD iPSCs with 72 CAG repeats were genetically corrected using homologous recombination and characterized. Corrected iPSCs expressed hallmark pluripotency markers including Nanog, Sox2, Oct4, SSEA4, and TRA-1-60. (b–e) NSCs derived from HD iPSCs exhibit HD phenotypes including increased cellular apoptosis, elevated Caspase 3/7 activity, reduced BDNF expression, and reduced mitochondrial respiration compared to wild-type and genetically corrected (C116 or C127) NSCs. (Reprinted from *Cell Stem Cell*, 11, An, M. C. et al., Genetic correction of Huntington's disease phenotypes in induced pluripotent stem cells, 253–63, Copyright 2012, with permission from Elsevier.)

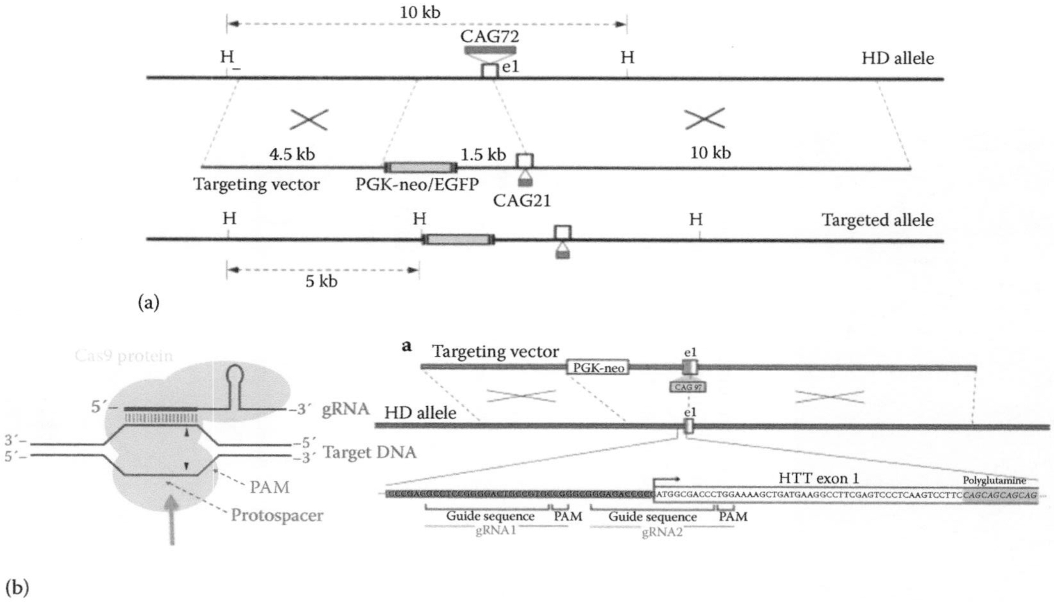

FIGURE 6.6
Generation of genetically corrected iPSC lines from HD iPSCs. (a) Using the traditional method of homologous recombination, HD IPSCs with 72 CAG repeats were corrected to have only 21 repeats. (b) Using the CRISPR/Cas9 genome targeting technique, the HD mutation can be targeted and corrected, or longer repeat sequences can be inserted. Specific gRNA sequences are selected to target exon 1 of the HTT gene, and donor sequences containing either the corrected allele or a longer mutant allele can be used. This method has a much higher efficiency than homologous recombination. (From An, M. C. et al. *PLoS Curr.* 6, 2014; reprinted from *Cell Stem Cell*, 11, An, M. C. et al., Genetic correction of Huntington's disease phenotypes in induced pluripotent stem cells, 253–63, Copyright 2012, with permission from Elsevier.)

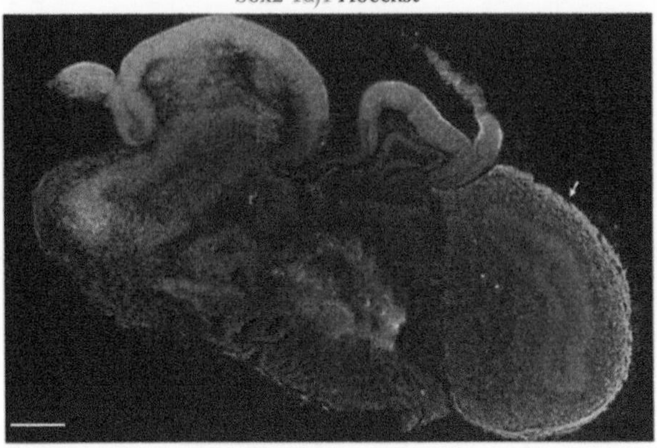

FIGURE 6.7
Human cerebral organoids can be generated in culture. These organoids are a 3D representation of cortical development in the human brain and contain most of the commonly found brain cells including neurons and astrocytes. (Reprinted by permission from Macmillan Publishers Ltd. *Nature*, Lancaster, M. A. et al., 501(7467): pp. 373–9, copyright 2013.)

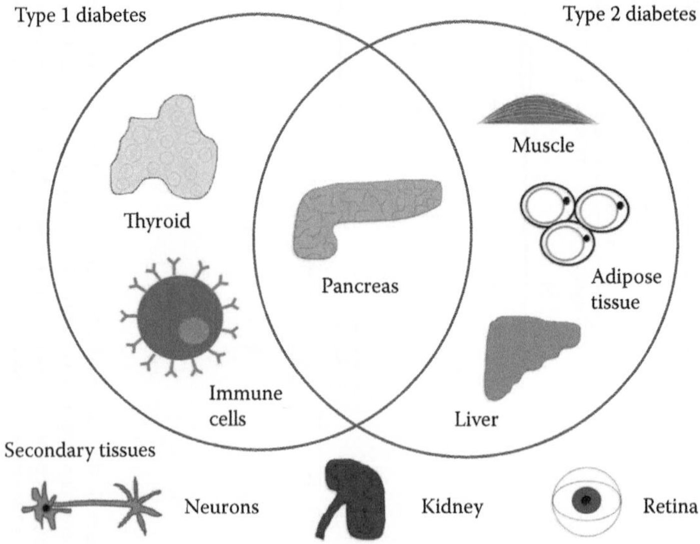

FIGURE 7.1
Organs involved in the development of type 1 and type 2 diabetes.

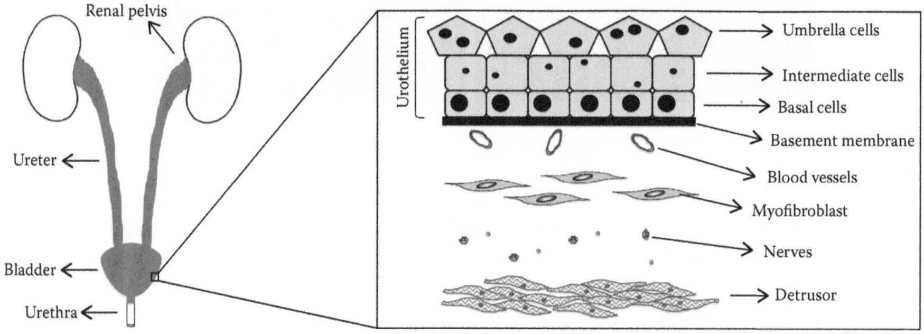

FIGURE 9.1
Diagram showing that the urothelium expands to cover the renal pelvis, the ureters, and the bladder (red). Bladder urothelium: A three-layered epithelium is apparent, consisting of large, binucleated superficial cells overlying an intermediate and a basal cell layer. This is separated by a basement membrane from a suburothelial layer that contains blood vessels, nerves, and myofibroblasts.

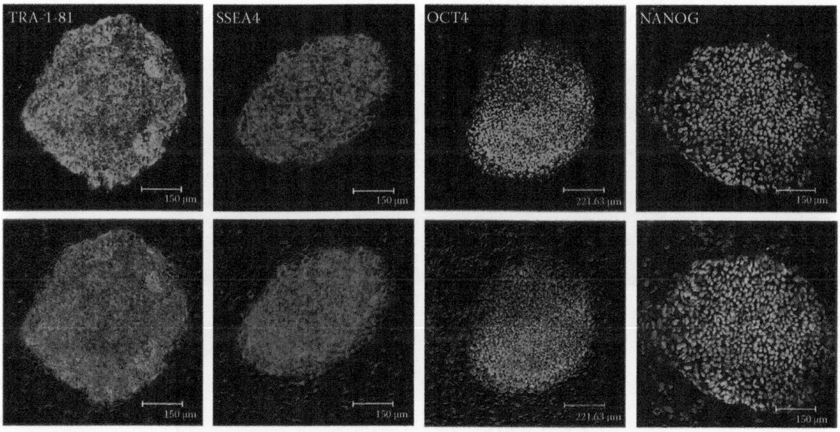

FIGURE 9.3
Immunofluorescence of generated UT-iPSCs for the expression of specific hES cell surface markers, podocalyxin (TRA-1-81) and stage-specific embryonic antigen 4 (SSEA4), and nuclear transcription factors octamer-binding transcription factor 4 (OCT4) and homeobox transcription factor nanog (NANOG). Nuclei were stained with DAPI (blue). MEF feeder cells served as the negative control.

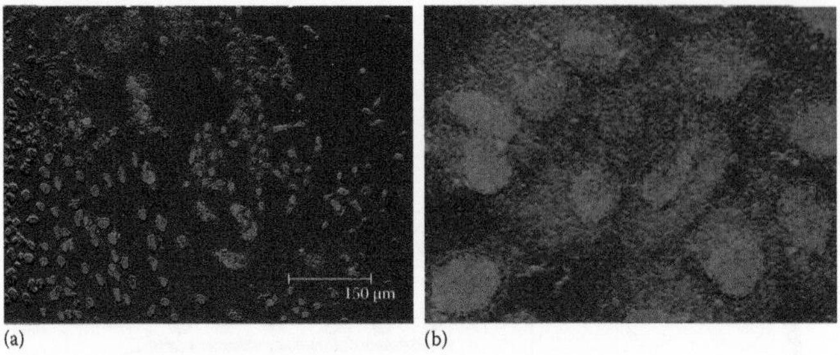

FIGURE 9.4
Immunofluorescence of differentiated cells derived from UT-iPSCs treated with conditioned medium. Positive staining for UPIb (red) juxtaposed with an area of UPIb-negative staining, (a) with DAPI nuclear counterstain (blue). (b) High magnification of uroplaisin 1B (UPIb) immunostaining.

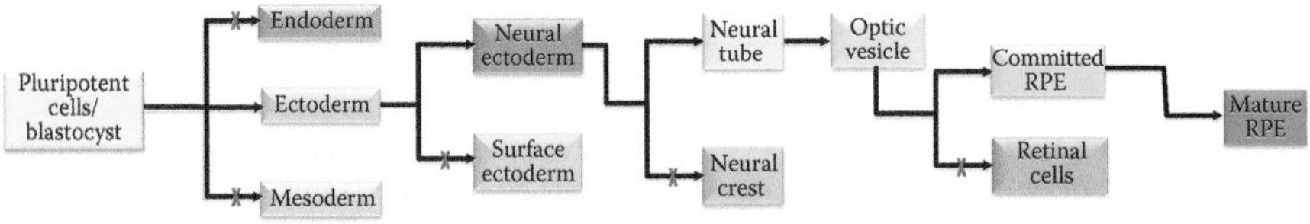

FIGURE 10.1
In vitro differentiation of PSCs (ESCs and iPSCs) follows the same developmental stages as the *in vivo* differentiation of a blastocyst. Spontaneous differentiation of ESCs or iPSCs *in vitro* induces the formation of all three germ layers—endoderm, ectoderm, and mesoderm. Developmental pathways can be manipulated *in vitro* by suppressing TGF, activin, and canonical WNT pathways and by activating IGF to specifically induce optic neuroectoderm from PSCs. RPE cells are induced from optic neuroectoderm by activating TGF and canonical WNT pathways.

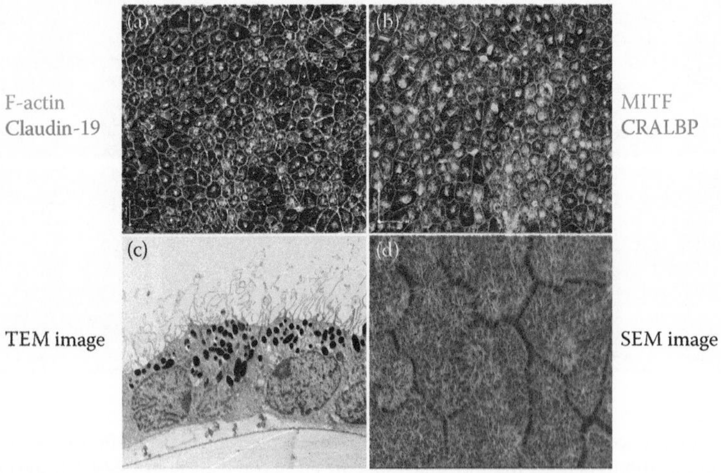

FIGURE 10.2
Fully differentiated iPSC-derived RPE are characterized by the following: (a) the presence of tight junctions between neighboring cells that are marked by claudin-19 (red) immunostaining. Actin cytoskeleton (F-actin, green) in differentiated RPE cells aligns along the cell boundaries; (b) the expression of transcription factor MITF (green) and visual cycle enzyme cellular retinaldehyde-binding protein (CRALBP, red); (c) the presence of extensive apical processes and apically located melanosomes (RPE pigment) as seen by a transmission electron microscope (TEM) image; (d) the presence of apical processes on the entire RPE monolayer as confirmed by a scanning electron microscope (SEM) image.

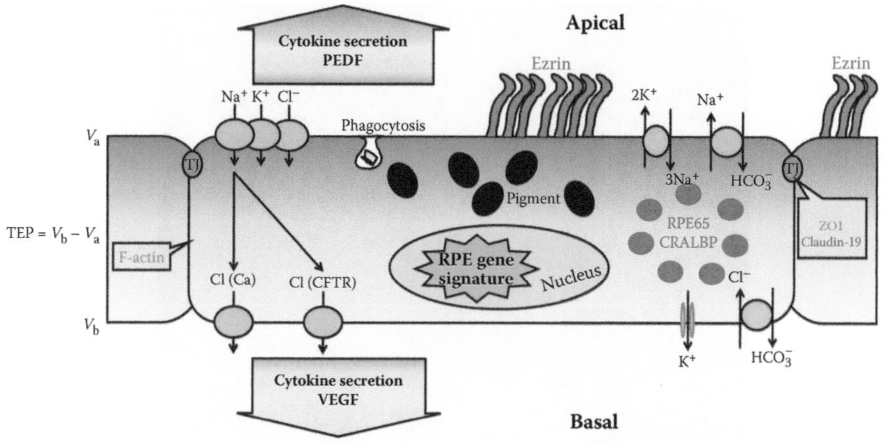

FIGURE 10.3
Mature RPE cells have a characteristic gene signature that is different from more fetal-like RPE cells. Mature RPE cells express higher levels of visual cycle enzymes RPE65 and CRALBP. Actin cytoskeleton (F-actin immunostaining) organizes around cell boundaries. Tight junctions (TJs) form between neighboring RPE cells and can be visualized with zonula occludens-1 and claudin-19 immunostaining. Because of these TJs, RPE cells in a monolayer exhibit transepithelial resistance of several hundred ohms per square centimeter. All the cells in the monolayer have apically located actin-based processes that can be labeled with ezrin. Cells secrete PEDF predominantly towards the apical side and VEGF predominantly towards the basal side. Functional RPE cells are able to phagocytose photoreceptor outer segments. Polarized RPE cells in an electrically intact monolayer express specific ion channels on the apical and the basal sides. These ion channels maintain apical (V_a) and basolateral (V_b) membrane resting potentials, whereby the basolateral membrane potential is slightly more positive compared to the apical membrane resulting in a transepithelial potential of 2–10 mV. Some of these ion channels (especially the chloride and potassium channels) also help drive vectorial fluid flow of 5–10 μl × cm^{-2} × h^{-1} from the apical toward the basolateral side.

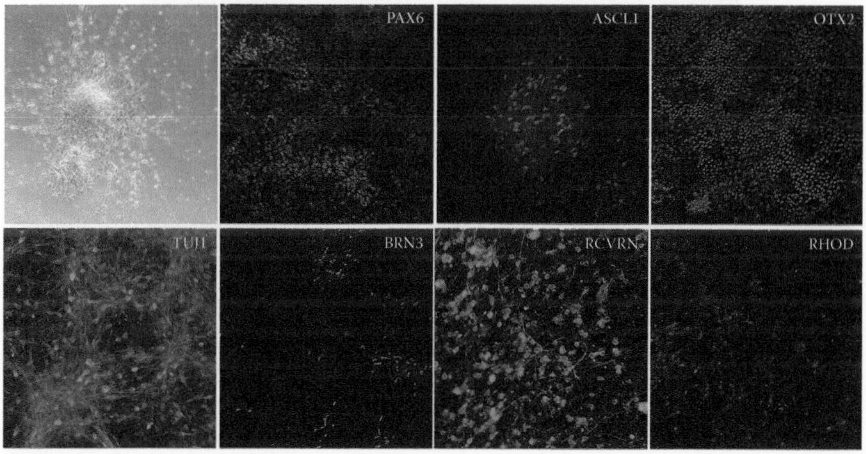

FIGURE 11.2
Characterization of 2D differentiated retinal cells from PSCs.

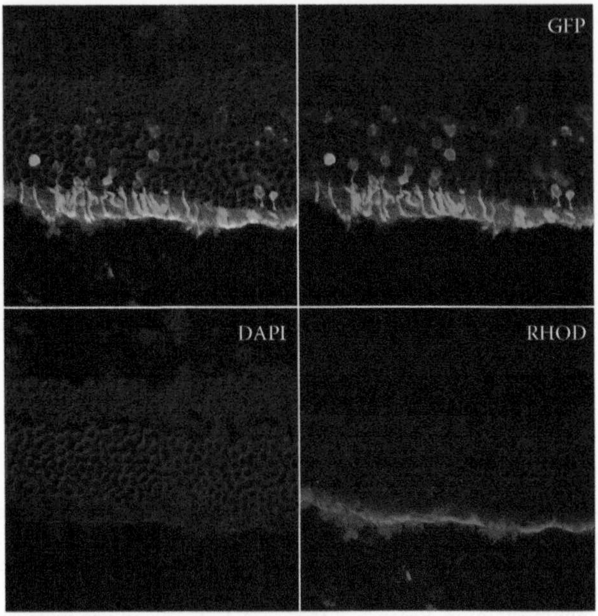

FIGURE 11.3
Characterization of 3D differentiated retinal cells from PSCs.

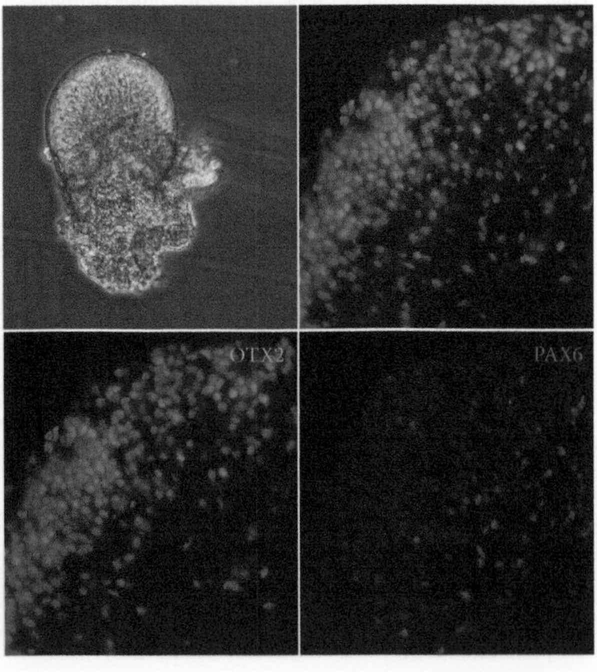

FIGURE 11.4
Integration of stem cell-derived retinal photoreceptors in a mouse retina.

7.2.1.1 Pancreatic Development

Developing protocols and methods to differentiate human PSCs into functional pancreatic β cells would provide novel means to study human β cell development, perform disease modeling, and provide a nearly inexhaustible supply of β cells that can be used for regenerative and therapeutic purposes for diabetic patients. Most studies attempting to generate β cells utilize cues from pancreatic development to create stepwise differentiation protocols. Pancreatic β cells trace their origins to definitive endoderm, where the transcription factors SRY-related HMG-box 17 and forkhead box A2 are used as markers of definitive endoderm. Following definitive endoderm specification, gut tube endoderm specification follows, which is marked by HNF4α and HNF1β expressions. Pancreatic specification then occurs around E8.5 in mouse and around three weeks postfertilization in humans. Lineage tracing studies have determined that all pancreatic cell types arise from pancreatic and duodenal homeobox 1 ($PDX1^{+}$) progenitors (9), with PDX1 expression becoming progressively restricted toward β cells during development in both mouse and human. As a result, its expression during pancreatic progenitor specification is the key when deriving mature β cells *in vitro*. Endocrine specification follows pancreatic specification with the transcription factors NK6 homeobox 1 (NKX6.1) and Neurogenin 3, being distinguishing markers for endocrine progenitors, the progenitors of all the endocrine cells in the pancreas (10,11). Once the endocrine progenitor cells become fully mature β cells, they are functionally characterized by their monohormonal insulin status and the capacity to secrete insulin upon glucose stimulation.

7.2.1.2 Therapeutic Opportunities with hESC-Derived Pancreatic β Cells

Generating human β cells *in vitro* will provide a therapeutic purpose in delivering glucose-responsive insulin-producing cells to treat diabetic patients. Early studies were able to demonstrate the key step of endoderm specification of hESCs through activin A and WNT signaling. The second stage of the stepwise differentiation is the specification toward pancreatic progenitors. Based on our knowledge of pancreatic development, fibroblast growth factor (FGF) 7 or FGF10, cyclopamine–KAAD or SANT-1 (hedgehog inhibitors), retinoic acid, dorsomorphin (a bone morphogenetic protein [BMP] inhibitor), SB431542 (a TGFβ inhibitor), and/or Noggin (a BMP inhibitor) have been tested in this step. A combination treatment with these chemicals and growth factors produces a heterogeneous population containing around 40–60% $PDX1^{+}$ cells (a pancreatic marker) for *in vitro* differentiation from hESCs/iPSCs. In addition, (–)-indolactam V (a protein kinase C [PKC] activator) was identified from an unbiased high-content chemical screen and identified as a top candidate to increase both the number and the percentage of pancreatic progenitors (12). The follow-up work from two groups shows that (–)-indolactam V promotes the generation of pancreatic progenitors from iPSCs derived from T1DM patients (13) and healthy patients, respectively. The current differentiation protocol containing FGF7/10, cyclopamine–KAAD, retinoic acid, Noggin, and PKC activators gives rise to a heterogeneous population with 60–80% $PDX1^+$ cells. It has been shown that this heterogeneous population is able to differentiate into glucose-responding cells after transplantation into immunodeficient mice and protects them from streptozotocin-induced hyperglycemia. Analysis of grafts from these transplants demonstrated that the progenitors matured into monohormonal insulin-positive β cells that expressed PDX1, NKX6.1, MAF bZIP transcription factor A (MAFA), and c-peptide (14,15). More recently, the same group reported that the purified pancreatic progenitors show a similar differentiation potential as the heterogeneous population. Building on these results allows for the study of the components constituting an *in vivo* environment that support β cell maturation.

Although *in vivo* transplantation promotes the generation of mature pancreatic β cells, transplanting embryonic progenitors poses oncogenic concerns, making the strategy poor for human therapeutic purposes. As a result, deriving mature β cells *in vitro* that could be subsequently transplanted would serve as a more suitable option. Pagliuca et al. reported the generation of mature monohormonal β cells derived from hESCs and iPSCs (stem cell-derived β cells [SC-βs]) that expressed many mature β cell markers, repeated calcium flux in the presence of glucose, and secretion of insulin at levels comparable to adult human islets in static glucose-stimulated insulin secretion assays. Moreover, human insulin was detected in serum of mice that received SC-βs as early as two weeks after transplantation compared to stem cell-derived pancreatic progenitors. The SC-βs showed the ability to significantly reduce hyperglycemia in Akira mouse models of diabetes mellitus as early as 18 days posttransplantation compared to hESC- and iPSC-derived pancreatic progenitors (16). A second published study from BetaLogics also demonstrated the derivation of insulin-producing cells from hESCs that expressed mature β cell markers including MAFA, which reversed diabetes in streptozotocin-treated mice within 40 days of transplant. However, the insulin-positive cells derived from the protocol did not mirror the same temporal and acute responses in calcium signaling as human islets, as well as a blunted response to potassium chloride in dynamic glucose-stimulated insulin secretion assays (17).

7.2.1.3 Modeling Diabetes Mellitus Using hESCs/iPSCs

Deriving robust protocols to differentiate PSCs into β cells would also assist in the disease modeling of diabetes mellitus. While the number of diabetic iPSC lines currently available is limited, published studies have demonstrated the derivation of iPSCs from patients with various forms of diabetes mellitus (13,18,19). The subsequent differentiation of these iPSCs into insulin-secreting cells could yield valuable insight into the dysfunctions in insulin secretion, pancreatic cell development, glucose sensing, and more. For example, the insulin-secreting cells derived from iPSCs of MODY2 patients require higher glucose levels to stimulate insulin secretion than the cells derived from a healthy control (20). Alternatively, the introduction of the various single-nucleotide polymorphisms (SNPs) that occur in the genes related to diabetes mellitus via clustered regularly interspaced short palindromic repeat (CRISPR) or transcription activator-like effector nuclease (TALEN) technologies can be utilized to characterize the impact of specific mutations on β cell function. Understanding the mechanisms of dysfunction can allow for the screening of new drug candidates to target and correct various pancreatic β cell defects.

7.2.2 Adipose Tissue

The adipose tissue organ is another pivotal player in the pathogenesis of diabetes and insulin resistance, specifically obesity-related T2DM. Adipocytes are characterized by two distinct types in mice and humans: white adipocytes and brown adipocytes. White adipocytes are primarily involved in energy storage, taking glucose from circulation and converting it to triacylglycerides to be used as needed by the organism. Brown adipocytes are the mirror opposite, taking the chemical energy stored in their lipid droplets and converting it to heat to primarily maintain core body temperature. In 2012, a third and novel type of adipocyte was identified in mice called *beige adipocytes* (21). These adipocytes maintain many of the functional properties of brown adipocytes but emerge in white fat depots specifically upon cold exposure or high-fat diet challenge. Human brown adipose tissue has also been shown to express markers that are characteristic of beige cells over classical brown adipose tissue (22). Developmentally, adipocytes arise from mesenchymal cells that originate from the mesoderm, although it has been reported that adipocytes can be derived from neuroectodermal neural crest cells (23,24). At the molecular level, adipocyte development and differentiation are governed by the master regulatory transcription factor peroxisome proliferator-activated receptor gamma (PPARγ), along with members of the CCAAT/enhancer-binding protein (CEBP) family (25,26). The transcription factors PR domain containing 16 (PRDM16) and PPARγ coactivator 1 alpha (PGC1α) have been shown to be critical to lineage commitment and function of brown adipocytes (27).

In the pathogenesis of obesity-linked T2DM, the adipose tissue becomes increasingly insulin resistant, resulting in less glucose uptake and increased hydrolysis of stored lipids. These free fatty acids then enter circulation and deposit in the muscle and the liver, leading to steatosis and tissue dysfunction. Attempts to elucidate the mechanism underlying adipose tissue expansion and insulin resistance have revealed that in mouse models of obesity, the expansion of adipocytes in response to increasing triacylglyceride content leads to a lipotoxicity that recruits macrophages and other inflammatory cell types, setting off a series of cytokine events that are linked to insulin resistance (2). GWASs have also helped shed light on some of the genetic components behind predispositions to obesity, with the identification of SNPs in PPARγ and polymorphisms in the *FTO* gene, being highly correlated to obesity and T2DM (28,29).

The adipocyte also plays a critical role in nonobese cases of insulin resistance. Many monogenic forms of lipodystrophy result in the complete or partial absence of subcutaneous adipose tissue that leads to varying degrees of insulin resistance and hypertriglyceridemia in patients among other symptoms. Mutations mapped to the laminin A/C gene as well as PPARγ are associated with familial partial lipodystrophy (30,31). Mutations mapped to the enzyme 1-acylglycerol-3-phosphate *O*-acyltransferase 2 and the protein seipin are linked to the autosomal recessive Berardinelli–Seip congenital lipodystrophy (BSCL) syndrome (32,33). While genome linkage studies have implicated these loci and genes as the cause of these forms of lipodystrophy and insulin resistance, little work has been done to identify the underlying mechanism of these mutations that lead to adipocyte dysfunction.

7.2.2.1 Disease Modeling Using hESC-Derived Adipocytes

Thus, the fact that iPSCs are derived from patients with T2DM or forms of lipodystrophy holds enormous potential to reveal mechanisms and phenotypes that result in human adipocyte dysfunction via disease modeling. Currently, only a limited number of protocols to differentiate human PSCs to human adipocytes have been reported. Early studies for hES cell-derived adipocytes exhibited limited analysis on expression of maturity markers such as CEBPα as well as minimal characterization of functional properties, such as adiponectin and leptin secretion and glucose uptake in the presence of insulin (34,35). More recently, published studies have emerged demonstrating robust generation of white and brown adipocytes from hESCs using both lentiviral and growth factor cocktail-based differentiation strategies (36,37). The differentiation of white adipocytes from hESCs and iPSCs was achieved using lentiviral-based expression of PPARγ. At the end of the differentiation period, lipid-filled white adipocytes emerged, expressing maturity markers such as CEBPα, fatty acid binding protein 4, hormone-sensitive lipase, and lipoprotein lipase. The white adipocytes also demonstrated functional activities, including glycerol release, glucose uptake in the presence of insulin, and leptin and adiponectin secretion. Using the same lentiviral-based approach, brown adipocytes were derived from hESCs and iPSCs via overexpression of PPARγ, CEBPβ, and/or PRDM16. The derived brown adipocytes express maturity markers, such as uncoupling protein 1 (UCP1), PGC1α, and cytochrome c1 (CYC1), as well as functional properties, such as elevated oxygen consumption. The drawback of this lentiviral-based differentiation protocol is that it cannot be used to study the defects related to genes upstream of the PPARγ signaling pathway. The second reported protocol for brown adipocyte differentiation used embryoid bodies in a hemopoietin growth factor cocktail containing KIT ligand, FLT3 ligand, interleukin-6, vascular endothelial growth factor, insulin-like growth factor-2, and BMP4 and BMP7 in two different phases. The reported brown adipocytes expressed maturity markers such as UCP1, PGC1α, CIDEA, CYC1, ELOVL3, and proliferator-activated receptor-α. The transplantation of these brown adipocytes into mice demonstrated improved fasting blood glucose and triglyceride levels when compared to vehicle control.

With the development of protocols that yield adipocytes similar to primary adipose tissue, numerous questions regarding the mechanisms of monogenic forms of insulin resistance such as familial lipodystrophies can be revealed via patient-specific iPSCs. Using gene editing tools such as ZFNs, TALENs, and CRISPRs, it is possible to systematically explore the role of numerous genes, SNPs, and polymorphisms from GWASs of obesity and diabetes in the context of human adipocyte biology (38).

7.2.3 Hepatocytes and Skeletal Muscle

The hepatocyte is another primary cell type that plays an important role in the pathogenesis of T2DM and insulin resistance. The liver is a site of high-glycogen storage, where glucagon made by pancreatic α cells drives the conversion of glycogen to glucose via glycogenolysis in order to maintain euglycemia. On the other hand, insulin made by pancreatic β cells halts the breakdown of glycogen and terminates hepatic glucose output. In the cases of insulin resistance, insulin is no longer able to suppress hepatic glucose output, thus exacerbating the hyperglycemia already present in the patient. In addition, hepatic steatosis plays a key role in the progression of insulin resistance in the liver (39).

The skeletal muscle is another organ system involved in insulin resistance along with adipose and hepatic tissue and may even be considered the primary site of insulin resistance (40). Early studies revealed the defects of insulin receptor substrate 1 phosphorylation and AKT activation in the insulin signaling pathway of skeletal muscle biopsies of T2DM patients (41). Skeletal muscle can also atrophy of the extremities of diabetic patients suffering from diabetic neuropathy (discussed in the following), a condition whose molecular basis has not been fully understood.

7.2.3.1 Disease Modeling in hESC-Derived Hepatocytes and Skeletal Muscles

Functional hepatocytes derived from hESCs/iPSCs can be used to set up disease modeling assays to unravel mechanisms leading to hepatic insulin resistance. Several protocols have been reported for the differentiation of hESCs and/or iPSCs into hepatocyte-like cells. Most reported protocols utilize the differentiation from hESCs/iPSCs to definitive endoderm by activating TGFβ signaling through the use of activin A, followed by the addition of BMP4/FGF2 to specify hepatic lineage commitment and hepatocyte factors, such as hepatocyte growth factor (HGF), for hepatocyte maturation (42). An additional study reported a two-step protocol that required only activin A for endoderm specification, followed by culture with HGF for hepatocyte formation and maturation in the presence of oncostatin (43). However, these early protocols suffered from the lack of defined media conditions, using fetal bovine serum as well as feeders for cell culture. A recent study addressed this issue by creating a hepatocyte differentiation protocol using fully defined media conditions, which used a phosphoinositide 3-kinase inhibitor to increase endoderm differentiation followed by the use of FGF10 and retinoic acid to specify hepatic specification (44). The hepatocyte-like cells generated from these studies express key markers, such as alpha feto protease, HNF4α, and cytochrome P450 enzymes, as well as functional properties, such as glycogen storage, low-density lipoprotein uptake, and metabolism of xenobiotics. Transplantation of hESC/iPSC-derived hepatocytes has even demonstrated rescue of hepatic function in an immunodeficient mouse model of hepatic failure (45).

Using the hepatocytes derived from iPSCs of T2DM, MODY1, or MODY2, patients can help reveal new disease mechanisms within the context of the liver. For example, MODY1 patients carry mutations in the HNF4α gene that plays critical roles in β cell as well as hepatic development. MODY2 patients are characterized by mutations in the glucokinase gene and exhibit impaired hepatic glycogen synthesis concomitant with augmented hepatic gluconeogenesis. Modeling these mutations within hiPSC-derived hepatocytes provides opportunities to explore how HNF4α and glucokinase affect hepatocyte development and glucose metabolism.

Muscle atrophy is a common consequence of diabetic neuropathy. From a therapeutic standpoint, myotubes derived from hESCs/iPSCs can provide a possible tissue transplant option for diabetic patients who suffer from tissue damage from late-stage neuropathy. However, robust protocols to derive mature myotubes from hESCs/iPSCs are still missing, but one study has demonstrated the engraftment and the expansion of hESC-derived myogenic mesenchymal cells in an immunodeficient mouse injury model (46).

7.3 Organs Involved in Diabetic Complications

Although the pathogenesis of diabetes mellitus mainly involves the organs previously discussed in Section 7.2, the complications that arise from diabetes mellitus affect other systems throughout the body as well. In this section, we will discuss the secondary organ and tissue systems that are affected as a result of hyperglycemia and the potential applications of iPSCs to model or treat diabetic complications.

7.3.1 Neural Cells and Endothelial Cells

Diabetic neuropathy is a complication that arises due to nerve damage as a result of continual exposure to hyperglycemic condition. Diabetic neuropathy can be manifested as peripheral neuropathy, where nerve damage impacts the sensitivity of the extremities, such as the hands and the feet. Symptoms and complications include a sense of tingling in the feet and the hands, as well as loss of sensitivity. Reduced blood flow and loss of sensitivity in the limbs lead to cuts and wounds going unnoticed, which might result in ulcers and infections that require amputation. As a consequence, greater than half of all nontraumatic amputations occurring below the waist in the United States occur due to diabetic complications every year (47). Autonomic neuropathy impacts nerves regulating the autonomic nervous system, with symptoms including urinary tract infections, nausea, constipation, and erectile dysfunction in men.

Together with nerve damage, damage to endothelial cells lining blood vessels helps exacerbate the microvascular and macrovascular complications, which lead to diabetic foot ulcer and other conditions. Damage to macrovessels can also lead to ischemic stroke and coronary artery disease, while microvascular damage leads to diabetic retinopathy, nephropathy, and cardiomyopathy.

7.3.1.1 Disease Modeling in hESC-Derived Neural and Endothelial Cells

Little is known so far about the mechanisms behind the nerve and the endothelial cell damage that occurs in the presence of hyperglycemic conditions and how that damage leads to the eventual clinical complications. However, protocols currently exist for the differentiation of human PSCs to endothelial cells (48–50) and numerous neural cell types (51–53), but further protocols will be needed to derive all the neural types encompassed by diabetic neuropathy for future studies examining the processes underlying diabetic neuropathy.

7.3.2 Retinal Cells

Around 40–45% of diabetic patients in the United States have some form of diabetic retinopathy, where hyperglycemic conditions lead to damage of the blood vessels feeding the retina. Under diabetic conditions, microvessels surrounding the eye undergo swelling and edema, followed by neovascularization, which can lead to blood leak and eventual retinal detachment and blindness.

7.3.2.1 Disease Modeling with hESC-Derived Retinal Cells

In diabetic retinopathy, little is known about the mechanisms underlying the deterioration of the cells of the retina, since the eye is complicated in structure and difficult to reach in patients. hESCs/iPSCs from diabetic patients once again provide opportunities to study and treat this complicated condition. hESC- and swine ESC-derived retinal pigment epithelial cells have been shown to integrate into rodent and pig retinas, providing the ability to study the role and the contribution of this cell type in the context of a diabetes setting (54,55). Further derivation of other retinal cell types from hESCs will be needed to provide a more comprehensive picture of what occurs during diabetic retinopathy in humans.

7.3.3 Renal Cells

In diabetic nephropathy, damages to microvessels of the kidney glomerulus lead to the leak of proteins into the urine and defective filtration of the blood, culminating in complete renal failure. While drugs are currently on the market to help slow the progression of renal failure, kidney transplants are eventually needed for end-stage renal failure. Much like the difficulty in directly accessing and studying retinal degradation in diabetic patients, the location of the kidney glomerulus makes it difficult for it to be directly studied without the use of invasive procedures on patients. Therefore, there is a hope to use iPSC-based disease modeling using hESC/iPSC-derived kidney cell types such as mesangial cells and podocytes to better understand diabetic renal failure, as well as to provide transplant opportunities for patients in need of improved kidney function.

7.4 Concluding Remarks

The growing prevalence and impact of obesity and diabetes mellitus at the global level have led to a strong need to find new ways to better understand and treat the numerous manifestations of metabolic disorder. hESCs/iPSCs derived from diabetic patients provide a new potential window of insight into the complex etiology of diabetes mellitus. The power of PSCs to become any cell in the body under the correct developmental cues allows researchers to derive numerous cell and tissue types relevant to the pathology of diabetes. This resource can then subsequently be applied toward treatment and therapeutic purposes, such as the derivation of pancreatic β cells, myotubes, retinal cells, and renal cells for transplant into late-stage diabetic patients who suffer from organ failure and tissue damage. iPSCs from diabetic patients, on the other hand, can be used to model many monogenic forms of diabetes arising from SNPs and polymorphisms in metabolically relevant genes and cell types such as adipocytes, hepatocytes, and skeletal myotubes. The advent of genome editing technologies such as CRISPR and TALEN makes genome editing and characterization of these polymorphisms under isogenic conditions scalable and amenable to high-throughput formats. The possibility of therapeutic options custom-tailored to one's genetic makeup is now closer than ever. Differentiation protocols for metabolically relevant cell and tissue types continue to improve and fall under ever more defined culture conditions, and while their scalability and safety will have to be rigorously tested before transplants of pancreatic β cells or any other kind of cell or tissue occur, the opportunity for the future therapies of diabetes mellitus seems more tangible than ever.

References

1. Kahn CR. Banting Lecture: Insulin action, diabetogenes, and the cause of type II diabetes. *Diabetes*. 1994, August;43:1066–1084.
2. Xu H, Barnes GT, Yang Q et al. Chronic inflammation in fat plays a crucial role in the development of obesity-related insulin resistance. *J Clin Invest*. 2003;112:1821–1830.
3. Yamagata K, Furuta H, Oda N et al. Mutations in the hepatocyte nuclear factor-4 alpha gene in maturity-onset diabetes of the young (MODY1). *Nature*. 1996;384:458–460.
4. Froguel P, Zouali H, Vionnet N et al. Familial hyperglycemia due to mutations in glucokinase. *N Engl J Med*. 1993;328:697–702.
5. Yamagata K, Oda N, Kaisaki PJ et al. Mutations in the hepatocyte nuclear factor-1 alpha gene in maturity-onset diabetes of the young (MODY3). *Nature*. 1996;384:455–458.
6. Fajans SS, Bell GI, Polonsky KS. Molecular mechanisms and clinical pathophysiology of maturity-onset diabetes of the young. *N Engl J Med*. 2001;345:971–980.
7. Horikawa Y, Hara M, Hinokio Y et al. Mutation in hepatocyte nuclear factor-1 β gene (TCF2) associated with MODY. *Nat Genet*. 1997;17:384–385.
8. Malecki MT, Jhala US, Antonellis A et al. Mutations in NEUROD1 are associated with the development of type 2 diabetes mellitus. *Nat Genet*. 1999, November;23:323–328.
9. Offield MF, Jetton TL, Labosky PA et al. PDX-1 is required for pancreatic outgrowth and differentiation of the rostral duodenum. *Development*. 1996;122:983–995.
10. Gradwohl G, Dierich A, LeMeur M et al. Neurogenin3 is required for the development of the four endocrine cell lineages of the pancreas. *Proc Natl Acad Sci*. 2000;97:1607–1611.
11. Herrera PL, Nepote V, Delacour A. Pancreatic cell lineage analyses in mice. *Endocrine*. 2002;19:267–277.

12. Chen S, Borowiak M, Fox JL et al. A small molecule that directs differentiation of human ESCs into the pancreatic lineage. *Nat Chem Biol.* 2009;5:258–265.
13. Maehr R, Chen S, Snitow M et al. Generation of pluripotent stem cells from patients with type 1 diabetes. *Proc Natl Acad Sci USA.* 2009;106:15768–15773.
14. Kroon E, Martinson LA, Kadoya K et al. Pancreatic endoderm derived from human embryonic stem cells generates glucose-responsive insulin-secreting cells in vivo. *Nat Biotechnol.* 2008;26:443–452.
15. Rezania A, Bruin JE, Riedel MJ et al. Maturation of human embryonic stem cell-derived pancreatic progenitors into functional islets capable of treating pre-existing diabetes in mice. *Diabetes.* 2012;61:2016–2029.
16. Pagliuca FW, Millman JR, Gürtler M et al. Generation of functional human pancreatic β cells in vitro. *Cell.* 2014;159:428–439.
17. Rezania A, Bruin JE, Arora P et al. Reversal of diabetes with insulin-producing cells derived in vitro from human pluripotent stem cells. *Nat Biotechnol.* 2014;32:1121–1134.
18. Fujikura J, Nakao K, Sone M et al. Induced pluripotent stem cells generated from diabetic patients with mitochondrial DNA A3243G mutation. *Diabetologia.* 2012;55:1689–1698.
19. Kudva YC, Ohmine, S, Greder LV et al. Transgene-free disease-specific induced pluripotent stem cells from patients with type 1 and type 2 diabetes. *Stem Cells Transl Med.* 2012;1:451–461.
20. Hua H, Shang L, Martinez H et al. iPSC-derived β cells model diabetes due to glucokinase deficiency. *J Clin Invest.* 2013;123:3146–3153.
21. Wu J, Boström P, Sparks LM et al. Beige adipocytes are a distinct type of thermogenic fat cell in mouse and human. *Cell.* 2012;150:366–376.
22. Sharp LZ, Shinoda K, Ohno H et al. Human BAT possesses molecular signatures that resemble beige/brite cells. *PLoS One.* 2012;7:e49452.
23. Billon N, Iannarelli P, Monteiro MC et al. The generation of adipocytes by the neural crest. *Development.* 2007;134:2283–2292.
24. Lemos DR, Paylor B, Chang C et al. Functionally convergent white adipogenic progenitors of different lineages participate in a diffused system supporting tissue regeneration. *Stem Cells.* 2012;30:1152–1162.
25. Tontonoz P, Hu E, Graves RA et al. mPPAR gamma 2: Tissue-specific regulator of an adipocyte enhancer. *Genes Dev.* 1994;8:1224–1234.
26. Siersbaek R, Nielsen R, Mandrup S. PPARgamma in adipocyte differentiation and metabolism—Novel insights from genome-wide studies. *FEBS Lett.* 2010;584:3242–3249.
27. Harms M, Seale P. Brown and beige fat: Development, function and therapeutic potential. *Nat Med.* 2013;19:1252–1263.
28. Muller YL, Bogardus C, Beamer BA et al. A functional variant in the peroxisome proliferator-activated receptor gamma2 promoter is associated with predictors of obesity and type 2 diabetes in Pima Indians. *Diabetes.* 2003;52:1864–1871.
29. Frayling TM, Timpson NJ, Weedon MN et al. A common variant in the FTO gene is associated with body mass index and predisposes to childhood and adult obesity. *Science.* 2007;316:889–894.
30. Shackleton S, Lloyd DJ, Jackson SNJ et al. LMNA, encoding lamin A/C, is mutated in partial lipodystrophy. *Nat Genet.* 2000;24:153–156.

31. Savage DB, Tan GD, Acerini CL et al. Human metabolic syndrome resulting from dominant negative mutations in the nuclear receptor peroxisome proliferator activated receptor gamma. *Diabetes*. 2003, April;52:910–917.
32. Agarwal AK, Arioglu E, De Almeida S et al. AGPAT2 is mutated in congenital generalized lipodystrophy linked to chromosome 9q34. *Nat Genet*. 2002;31:21–23.
33. Magré J, Delépine M, Khallouf E et al. Identification of the gene altered in Berardinelli-Seip congenital lipodystrophy on chromosome 11q13. *Nat Genet*. 2001;28:365–370.
34. Xiong C, Xie C-Q, Zhang L et al. Derivation of adipocytes from human embryonic stem cells. *Stem Cells Dev*. 2005;14:671–675.
35. Van Harmelen V, Aström G, Strömberg A et al. Differential lipolytic regulation in human embryonic stem cell-derived adipocytes. *Obesity (Silver Spring)*. 2007;15:846–852.
36. Ahfeldt T, Schinzel RT, Lee Y-K et al. Programming human pluripotent stem cells into white and brown adipocytes. *Nat Cell Biol*. 2012;14:209–219.
37. Nishio M, Yoneshiro T, Nakahara M et al. Production of functional classical brown adipocytes from human pluripotent stem cells using specific hemopoietin cocktail without gene transfer. *Cell Metab*. 2012;16:394–406.
38. Ding Q, Lee Y, Schaefer EAKAK et al. A TALEN genome-editing system for generating human stem cell-based disease models. *Stem Cell*. 2013:1–14.
39. Samuel VT, Petersen KF, Shulman GI. Lipid-induced insulin resistance: Unravelling the mechanism. *Lancet*. 2010;375:2267–2277.
40. DeFronzo RA, Tripathy D. Skeletal muscle insulin resistance is the primary defect in type 2 diabetes. *Diabetes Care*. 2009;32:S157–S163.
41. Dresner A, Laurent D, Marcucci M et al. Effects of free fatty acids on glucose transport and IRS-1-associated phosphatidylinositol 3-kinase activity. *J Clin Invest*. 1999;103:253–259.
42. Brolén G, Sivertsson L, Björquist P et al. Hepatocyte-like cells derived from human embryonic stem cells specifically via definitive endoderm and a progenitor stage. *J Biotechnol*. 2010;145:284–294.
43. Takata A, Otsuka M, Kogiso T et al. Direct differentiation of hepatic cells from human induced pluripotent stem cells using a limited number of cytokines. *Hepatol Int*. 2011:890–898.
44. Touboul T, Hannan NRF, Corbineau S et al. Generation of functional hepatocytes from human embryonic stem cells under chemically defined conditions that recapitulate liver development. *Hepatology*. 2010;51:1754–1765.
45. Chen Y-F, Tseng C-Y, Wang H-W et al. Rapid generation of mature hepatocyte-like cells from human induced pluripotent stem cells by an efficient three-step protocol. *Hepatology*. 2012;55:1193–1203.
46. Awaya T, Kato T, Mizuno Y et al. Selective development of myogenic mesenchymal cells from human embryonic and induced pluripotent stem cells. *PLoS One*. 2012;7:e51638.
47. Centers for Disease Control and Prevention. National Diabetes Statistics Report: Estimates of diabetes and its burden in the epidemiologic estimation methods. *Cent Dis Control*. 2014;2009–2012.
48. Joo HJ, Kim H, Park S-W et al. Angiopoietin-1 promotes endothelial differentiation from embryonic stem cells and induced pluripotent stem cells. *Blood*. 2011;118:2094–2104.

49. Li Z, Hu S, Ghosh Z et al. Functional characterization and expression profiling of human induced pluripotent stem cell- and embryonic stem cell-derived endothelial cells. *Stem Cells Dev.* 2011;20:1701–1710.
50. Adams WJ, Zhang Y, Cloutier J et al. Functional vascular endothelium derived from human induced pluripotent stem cells. *Stem Cell Reports.* 2013;1:105–113.
51. Zeng H, Guo M, Martins-Taylor K et al. Specification of region-specific neurons including forebrain glutamatergic neurons from human induced pluripotent stem cells. *PLoS One.* 2010;5:e11853.
52. Swistowski A, Peng J, Liu Q et al. Efficient generation of functional dopaminergic neurons from human induced pluripotent stem cells under defined conditions. *Stem Cells.* 2010;28:1893–1904.
53. Boulting GL, Kiskinis E, Croft GF et al. A functionally characterized test set of human induced pluripotent stem cells. *Nat Biotechnol.* 2011;29.
54. Hambright D, Park K, Brooks M et al. Long-term survival and differentiation of retinal neurons derived from human embryonic stem cell lines in un-immunosuppressed mouse retina. *Mol Vis.* 2012;18:920–936.
55. Zhou L, Wang W, Liu Y et al. Differentiation of induced pluripotent stem cells of swine into rod photoreceptors and their integration into the retina. *Stem Cells.* 2011;29:972–980.

8

Role of iPSCs in Disease Modeling: Gaucher Disease and Related Disorders

Daniel K. Borger, Elma Aflaki, and Ellen Sidransky

CONTENTS

8.1 Introduction

When the development of induced pluripotent stem cells (iPSCs) was first announced by Takahashi and Yamanaka (1) in 2006, there was great enthusiasm, especially in the lay press, over the possibility of a therapeutic revolution for previously untreatable neurologic injuries and disorders. However, from early on, it was clear that the development and the implementation of stem cell therapies would take time, and major obstacles would first need to be confronted. Currently, due to issues regarding the possibility of tumorigenicity and immunogenicity, these cells still have yet to fully deliver on that initial therapeutic promise (2,3). However, despite the hurdles that remain in applying iPSC technology directly to treatment of disease and injury, these cells have already proven to be an invaluable tool for developing cellular models for a variety of diseases, especially for diseases that previously had few options available for *in vitro* studies.

TABLE 8.1

Lysosomal Storage Diseases with Published hiPSC Models

Disease	Mutated Gene	Accumulated Substrate(s)	Affected Cells/ Tissues	Signs/Symptoms	Approved Therapies	References
Gaucher disease	*GBA1*	Glucosylceramide Glucosylsphingosine	Macrophages Neurons (types 2 and 3)	Hepatosplenomegaly, bone lesions, neurological dysfunction (types 2 and 3)	Enzyme replacement Substrate reduction	(7–15)
Mucopolysaccharidosis type I (Hurler syndrome)	*IDUA*	Heparan sulfate Dermatan sulfate	Connective tissue Neurons	Coarse facial features, organomegally, short stature, intellectual disability	Stem cell therapy Enzyme replacement	(16)
Mucopolysaccharidosis type IIIB	*NAGLU*	Heparan sulfate	Connective tissue Neurons	Developmental delay, behavioral problems	Substrate reduction	(17)
Neural ceroid lipofuscinoses (Batten disease)	*TPP1*, *CLN3*, others	Lipidated protein ATP synthase, subunit c	Neurons	Developmental delay, seizures, neurodegeneration	Bone marrow transplantation	(18)
Pompe disease	*GAA*	Glycogen	Myocytes Cardiomyocytes	Cardiomegaly, cardiomyopathy, muscle weakness	Enzyme replacement	(19,20)

To date, iPSC-derived cell models have been generated for more than 40 diseases (4). These include both clear-cut Mendelian disorders and multifactorial diseases, such as schizophrenia (5) and type 1 diabetes (6). But of the various diseases that have been modeled, one class of inherited diseases has been a particular focus of iPSC-based research: lysosomal storage diseases (LSDs). To date, five LSDs have been modeled using hiPSCs (Table 8.1), with additional murine iPSC lines derived from mouse models of five other LSDs (21,22). And despite the short time that these models have existed, they have already begun a major revolution in our understanding and treatment of these very rare diseases.

8.2 LSDs: Etiology and Treatment

LSDs are a class of approximately 50 metabolic disorders caused by inherited deficiencies in various lysosomal proteins. The majority of LSDs are the result of mutations in metabolic enzymes active in the lysosomal lumen, although several LSDs are caused by defects in lysosomal transport or vesicular trafficking (23–25). While the clinical and pathological features of these diseases vary widely, they all share intracellular accumulation of specific unmetabolized substrates (Table 8.1). The lack of functional enzyme and the subsequent substrate accumulation interfere with normal lysosomal function in affected cells, which in turn impact a diverse array of cellular activities, ultimately resulting in a range of visceral, musculoskeletal, immunological, and neurological symptoms (23,26). However, the precise mechanisms linking mutant enzyme, lysosomal dysfunction, and ultimate disease phenotype have not been fully elucidated in many of the LSDs.

Treatment options for LSDs are generally limited and are, in some cases, restricted to palliative care and physical therapy (26). Enzyme replacement therapy (ERT), a treatment modality whereby the deficient enzyme is provided via intravenous infusions, is now available for several of the LSDs. While ERT is effective in treating the visceral and musculoskeletal manifestations of these diseases, generally, the infused enzymes are too large to cross the blood–brain barrier. As a result, ERT is ineffective at reversing brain involvement in neuronopathic LSDs. Even when ERT is an effective treatment option, it usually involves biweekly infusions for the remainder of the patients' life, making it inconvenient and extremely expensive (more than US$ 200,000 per year). While, individually, the LSDs are very rare, collectively, they affect between 1 in 4000 and 1 in 8000 live births (27–29), meaning that the treatment of LSDs can still constitute a substantial burden on health care systems (30). These realities make the development of less expensive, neurologically active, and orally available small-molecule drugs an attractive goal for those researching LSDs.

More recently, an alternative therapeutic approach known as substrate reduction therapy (SRT) has been met with some success. Rather than supplement or otherwise augment the activity of the deficient enzyme as in ERT, SRT uses orally administered small-molecule inhibitors to reduce the synthesis of accumulated substrates. As small molecules may pass through the blood–brain barrier, the hope is that these drugs may impact neurological manifestations of the disease (26). However, to date, SRT has demonstrated only mixed success in managing neurological symptoms of LSDs. One SRT drug, miglustat, has currently been approved for the treatment of both Gaucher disease (GD) and Niemann–Pick disease type C (31,32). And while miglustat may be effective at slowing the neurological decline in Niemann–Pick disease type C, it only impacted the visceral symptoms in GD (32,33). Eliglucerase, another SRT recently approved for GD, has been shown to reverse disease manifestations in type 1 GD, but does not penetrate into the brain (34,35). More SRT drugs are currently undergoing trials, but many in the field have begun to shift paradigms and look at new classes of drugs such as pharmacological chaperones (26).

8.3 Gaucher Disease

GD is an autosomal recessive disorder caused by mutations in the *GBA1* gene, which codes for the lysosomal enzyme glucocerebrosidase (GCase). GCase is responsible for the hydrolysis of glucosylceramide and glucosylsphingosine into ceramide and sphingosine, respectively, and glucose. Loss of GCase activity in patients with GD leads to the accumulation of glucosylceramide and glucosylsphingosine within affected tissues. GD presents with a wide spectrum of clinical manifestations and is generally divided into three types: non-neuronopathic GD (type 1); acute neuronopathic GD (type 2), a disorder presenting perinatally or during the first year of life; and chronic neuronopathic GD (type 3), which encompasses many distinct phenotypes, all with some form of brain involvement. In all cases, GD affects the macrophages of the reticuloendothelial system, while the neuronopathic forms also affect the neurons of the central nervous system.

GD is one of the most common LSDs, and therefore, it has long been a major focus of LSD research. As a result, GD was the first LSD for which both ERT and SRT were developed (31,36). More recently, the discovery that mutations in *GBA1* constitute the single most common genetic risk factor for Parkinson disease (PD) and related disorders served to thrust what many considered an arcane disease into the spotlight (37,38). In part, due to its high frequency relative to other LSDs and, in part, due to its now well-established association with PD, there have been 23 unique GD iPSC lines generated to date, far more than any other LSDs (Table 8.2). Hence, this exploration of the role of iPSCs in LSD research will focus primarily on GD.

TABLE 8.2

Generation and Differentiation of Gaucher Disease iPSC Lines

Reference	Year	Type (No. of Patients)	Genotypes (No. of Patients)	Reprogramming Method (Integrative/ Nonintegrative)	Differentiation Target	Novel Drug Testing
Park et al. (7); *Cell*	2008	Type 1 (1)	N370S/c.84dupG	Retrovirus (integrative)	N/A	No
Mazzulli et al. (8); *Cell*	2011	Type 1 (1)	N370S/c.84dupG	Retrovirus (integrative)	Dopaminergic neurons	No
Panicker et al. (9); *PNAS*	2012	Type 1 (1) Type 2 (1) Type 3 (1)	N370S/N370S L444P/Rec*Ncil* L444P/L444P	Lentivirus (integrative, Cre-excised)	Macrophages Neurons	Yes
Tiscornia et al. (10); *Hum Mol Genet*	2013	Type 2 (1)	L444P/G202R	Nucleofection (integrative)	Macrophages Dopaminergic neurons	Yes
Panicker et al. (11); *Stem Cells*	2014	Type 1 (2) Type 2 (2) Type 3 (2)	N370S/N370S (2) L444P/G202R (1), W184/D409H (1) L444P/L444P (2)	Sendai virus (nonintegrative)	Macrophages	Yes
Aflaki et al. (12); *Sci Transl Med*	2014	Type 1 (4) Type 2 (1)	N370S/N370S (3), N370S/c.84dupG (1) IVS2+1G>A/ L444P(1)	Lentivirus (integrative, Cre-excised)	Macrophages	Yes
Schöndorf et al. (13); *Nat Commun*	2014	Type 1 (2) Type 3 (2)	N370S/N370S (2) L444P/L444P (2)	Retrovirus (integrative)	Dopaminergic neurons	No
Sun et al. (14); *PLOS One*	2015	Type 2 (3)	L444P/P415R G325R/C342G L444P;E326K/ L444P;E326K	Lentivirus (integrative) Nucleofection	Dopaminergic neurons	No
Sgambato et al. (15); *Stem Cells Transl Med*	2015		No novel lines generated		Hematopoietic progenitor cells	No

8.4 Non-iPSC Models of GD

Considerable research in the field of GD, and LSDs in general, has been focused on a continuing search for useful cell and animal models. A variety of GD mouse models have been generated over the last four decades (39), and various cell models have been derived from these mice (40,41). While some of these murine cell models show promise for applications such as drug screening, the genetic and phenotypic differences between the mouse models and the human disease mean that they cannot fully replace human cell models.

However, the search for appropriate human cell models of GD has been particularly challenging. Fibroblasts, commonly used to study other LSDs, do not store glucocsylceramide and glucosylsphingosine and, hence, are not appropriate for evaluating the consequences or the modulation of glycolipid storage (42). A number of alternative human cell models of GD have therefore been developed. Hein et al. (43) approximated Gaucher macrophages in culture by treating THP-1 cells, a monocytic leukemia cell line, with conduritol β-epoxide (CBE), an irreversible inhibitor of GCase. These cells were then differentiated to macrophage-like cells using phorbol 12-myristate 13-acetate (43). A similar model for neuronopathic GD was developed by treating SH-SY5Y cells, a human neuroblastoma cell line, with CBE (44). These CBE-induced models demonstrated reduced GCase activity and concomitant glycolipid accumulation and have the advantage of being relatively cheap and easy to generate (43,44). However, they employed cancer cell lines, which complicated any attempts to generalize findings in these cells to normal (i.e., noncancerous) macrophages or neurons. Furthermore, while induced models of GD using CBE are useful for exploring the impact of glycolipid storage on cellular function, they express wild-type GCase that is inactivated by CBE, rather than the mutant forms of the enzyme seen in patients. This makes these models ineffective for the development and the testing of drugs that may act as chaperones or otherwise augment GCase activity.

Attempts to generate human cell models bearing common *GBA1* mutations began with culturing lymphoblasts and fibroblasts from patients with GD (45,46). More recently, macrophages derived from peripheral blood-derived monocytic cells (PBMCs) of patients with GD have been used as a cellular model (12). These PBMC-derived macrophages are useful, in that they represent the primary cell type affected in GD (macrophages) and provide a genetic model of GD. However, these cells are subject to the typical shortcomings of many human primary cell models. They have a relatively short lifespan in culture, in part, because they do not propagate. This in turn means that utilizing these cells requires a relatively steady source of fresh patient blood.

Considering the shortcomings of these models, it is perhaps no surprise that researchers have quickly co-opted iPSC technology to generate new GD models. iPSCs can be derived from small and readily available tissue samples, most often skin or blood samples. This makes them a feasible option, even when studying infants, as is the case in type 2 GD. iPSCs can be passaged many times, meaning that consistent access to patients is unnecessary. Finally, these cells may be differentiated to cell types that cannot be easily obtained from living patients such as neurons, which are important in both neuronopathic GD and PD. However, the creation and the maintenance of these models require major commitments of time, money, and expertise.

8.5 iPSC Models of GD

Over the past several years, a number of laboratories have developed iPSC-derived cell models of GD. Table 8.2 compares the different forms of GD used as sources for reprogrammed cells, the patient genotypes, the reprogramming methods, and the cell types derived. Patients with type 1 GD are the most common source of cells for current iPSC models, and these lines have been differentiated to macrophages and neurons exclusively, corresponding to tissues associated with GD and parkinsonism. Researchers generating these GD models have worked to address and overcome a variety of challenges associated with developing iPSC-derived cell models. And while this field is still in its early days, important insights into both the etiology of and the therapeutics for GD and PD have already been gained through the use of these models.

8.5.1 Reprogramming

GD was among the first diseases for which iPSCs were generated. Not long after the initial announcement of the development of human iPSCs, Park et al. (7) demonstrated the potential of these cells for developing *in vitro* disease models by successfully generating 10 different iPSC lines from tissue samples of patients with various inherited diseases. Among these patients was one 20-year-old male patient with type 1 GD. This study utilized a reprogramming methodology employed in other early iPSC studies; an integrative retroviral vector was used to deliver the four reprogramming factors—*Oct3/4, Klf4, Sox2,* and *c-Myc*—while relying on the spontaneous silencing of these integrated transgenes after reprogramming.

Notably, the GD iPSC line generated by Park et al. (7) demonstrated only limited silencing of these transgenes after reprogramming, highlighting one of the problems inherent in using integrative vectors to initiate reprogramming. While the use of integrative vectors can be mutagenic or can lead to de novo silencing of genes near the insertion site (47), perhaps a more concerning issue for the development of cell models of disease lies in the fact that continued expression of reprogramming factors may interfere with subsequent differentiation, impact the phenotype of terminally differentiated cells, and predispose iPSCs to genomic instability (2,48,49). Some subsequent studies have avoided this issue by using either *loxP*-flanked Cre-excisable integrative vectors or nonintegrative vectors such as Sendai virus or transfection with episomal plasmids (Table 8.2). While Park et al. (7) did not attempt to differentiate these cells, this study did serve as a vital proof of principle by demonstrating that the reprogramming of somatic cells from patients with inheritable diseases could successfully yield iPSC lines bearing the causative mutations.

8.5.2 Metabolic Impediments to Reprogramming and Differentiation

Another major consideration when attempting to generate iPSC-derived cell models from patients with LSDs, and inborn errors of metabolism in general, is whether the disruption of metabolism and the subsequent cellular dysfunction negatively impact reprogramming or differentiation. Cells must undergo major alterations in metabolic flux in order to achieve pluripotency (50). Furthermore, autophagy, a metabolic process that is known to be defective in several LSDs (51), has been implicated in the reprogramming process (52,53). Therefore, the impaired metabolism of various substrates seen in LSDs may impede cellular reprogramming.

Outside of GD, Huang et al. (19) reported difficulties in reprogramming fibroblasts from patients with Pompe disease, an LSD caused by mutations in the gene *GAA*, the gene coding for the lysosomal glycogenolytic enzyme alpha-glucosidase. *GAA* mutations lead to lysosomal glycogen accumulation in cardiomyocytes and skeletal muscle (19) (Table 8.1). Ultimately, Huang et al. (19) found that the rescue of acid alpha-glucosidase activity via lentiviral delivery of wild-type *GAA* was a prerequisite for successful reprogramming of these cells. While subsequent studies have reported success in reprogramming fibroblasts from patients with Pompe disease (20,54), this finding suggests that mutations in lysosomal proteins can significantly impact the efficiency of the reprogramming process. Park et al. (7), Mazzulli et al. (8), Panicker et al. (9,11), Tiscornia et al. (10), Aflaki et al. (12), Schöndorf et al. (13), and Sun et al. (14) did not report any such difficulty in the generation of iPSCs from the patient with GD, nor has any subsequent group. However, Park et al. (7) did not differentiate their GD iPSCs, leaving open the question as to whether *GBA1* mutations may interfere with the differentiation process.

Mazzulli et al. (8) were the first to publish the results of differentiating GD patient-derived iPSCs. In an effort to probe the link between PD and GD, they generated two iPSC lines from fibroblasts of the same 20-year-old male patient with GD used by Park et al. (7). They then attempted to differentiate these iPSCs into dopaminergic neurons. After 35 days of differentiation, they found that ~80% of both GD and control cells expressed the general neuronal marker β-III tubulin (TuJ1). Of these TuJ1$^+$ cells, ~10% were found to express dopaminergic neuron marker tyrosine hydroxylase (TH). These results matched earlier attempts to differentiate dopaminergic neurons (55) and suggested that GD patient-derived iPSCs can be successfully differentiated. However, no assays were performed to determine if these neurons were fully functional, that is, whether they synthesized dopamine or were capable of generating action potentials. In two subsequent studies, Panicker et al. (9) and Tiscornia et al. (10) used different differentiation protocols but were also able to differentiate significant populations of both control and GD dopaminergic neurons, again as determined by TH expression. In all three studies, no differences in the efficiency of the dopaminergic neuron differentiation process were noted between control and GD lines, despite the fact that two of the lines were derived from type 2 GD patients bearing severe *GBA1* mutations (L444P/Rec*Ncil* and L444P/G202R).

In addition to dopaminergic neurons, Panicker et al. (9) and Tiscornia et al. (10) also differentiated GD iPSCs into macrophages, by way of monocytes. Both studies used CD14 and CD163 as markers of macrophage differentiation, with Panicker et al. (9) including CD68 and Tiscornia et al. (10) including CD11b and CD33. In all cases, control and GD iPSC lines produced macrophages at similar efficiencies. To further confirm that reduced GCase activity does not impact the iPSC differentiation process, Tiscornia et al. (10) used a lentiviral vector to transduce their type 2 GD iPSC line with wild-type *GBA1*, which restored GCase activity to control levels. They then compared the differentiation efficiency of both transduced and nontransduced lines and found that both produced cells with similar marker patterns upon differentiation to macrophages (10). Taken together, these results suggest that the metabolic defect in GD has little to no effect on either the reprogramming or the differentiation processes.

8.5.3 Disease Phenocopying and Lessons in Etiology

Another major issue faced by attempts to model diseases using iPSC-derived cells lies in determining the extent to which these cells phenocopy diseased cells in patients, as the value of iPSC-derived cell models rests upon the assertion that these cells mimic the behavior of their *in vivo* counterparts. For neurons, this is exceedingly difficult to demonstrate for the precise reason that iPSCs are so important; primary human neurons are, for good reason, nearly impossible to obtain. In the initial three iPSC-derived neuron papers discussed earlier, GD neurons were shown to have lower GCase protein levels and greatly reduced GCase enzyme activity when compared to control iPSC-derived dopaminergic neurons, which would be expected in cells with mutant GCase.

As previously noted, the first papers to differentiate GD iPSC lines to dopaminergic neurons lacked functional assays to evaluate if the neurons were fully functional. Sun et al. (14) were the first to evaluate the resting and action potentials of GD iPSC-derived dopaminergic neurons, doing so in the context of type 2 GD. Using a patch clamp, they found that type 2 neurons demonstrated a less negative resting potential, which in turn limited the amplitude of action potentials (14). However, whether this accurately reflects the neurological dysfunction seen in type 2 GD remains to be seen.

Several groups have also further characterized GD iPSC-derived dopaminergic neurons in attempts to link decreased GCase activity with the observed predisposition toward PD and related disorders. Mazzulli et al. (8) investigated α-synuclein (SNCA) degradation in their type 1 GD dopaminergic neurons. They used immunofluorescence and Western blotting to show that these neurons exhibit higher levels of SNCA and decreased lysosomal proteolysis. While these findings could not be confirmed directly in human neurons, Western blots performed on autopsied cerebral cortex samples from patients with type 1 GD revealed elevated SNCA levels, and the same results were recapitulated by knocking down GCase in mouse cortical neurons (8). Schöndorf et al. (13) demonstrated similar increases in SNCA protein levels in type 1 and type 3 GD dopaminergic neurons. Furthermore, tying in with the decreased proteolytic activity observed by Mazzulli et al. (8), Schöndorf et al. (13) found a buildup of autophagosomes and decreased autophagosome–lysosome fusion in GD dopaminergic neurons, suggesting that lysosomal dysfunction in GD leads to an expansion of the macroautophagic compartment.

Compared to iPSC-derived dopaminergic neurons, it has been much easier to determine the degree to which iPSC macrophages phenocopy their *in vivo* counterparts, as biomarkers of macrophage activity can be observed in blood, and primary macrophages can be derived from patients. In their second iPSC paper, Panicker et al. (9) differentiated iPSCs from all three types of GD to macrophages and then evaluated the cytokine profiles of these cells. They focused primarily on type 2 GD macrophages and found that immunological challenge with lipopolysaccharide, a common component of Gram-negative bacteria cell walls, leads to greatly increased expression of proinflammatory cytokines such as interleukin (IL)-10, IL-6, IL-1β, and tumor necrosis factor-α. They also determined that all GD macrophage lines exhibited elevated chitotriosidase activity. These factors are generally found at high levels in the blood of patients with GD (56,57), supporting the findings that GD macrophages are extremely sensitive to proinflammatory signals.

Aflaki et al. (12) were the first to directly demonstrate extensive phenocopying of iPSC-derived cells and their patient-derived counterparts. In this study, the authors used a battery of functional tests to compare the phenotype of iPSC-derived macrophages to peripheral blood-derived macrophages of the same genotype and, in some cases, from the same individual. They showed that iPSC-derived and peripheral blood-derived macrophages bearing the same *GBA1* mutations have similar GCase activity and, as a result, exhibit similar glycolipid storage after undergoing phagocytosis of lipid-rich erythrocyte membranes. Furthermore, this work showed that both iPSC-derived and patient blood-derived macrophages demonstrate similar defects in reactive oxygen species (ROS) production and chemotaxis.

8.5.4 Drug Testing in iPSC Models

In addition to providing a new means for determining the cellular basis of GD, iPSC cell models have also provided a new platform for testing new drugs. While the relatively high cost to generate a small number of cells currently limits their direct use in high-throughput screening, they can serve as a valuable proving ground for drugs discovered using these methods. Panicker et al. (9,11) and Tiscornia et al. (10) both evaluated various small-molecule inhibitors of GCase that act as pharmacological chaperones and saw improvement in the clearance of erythrocytes and the reduction in the secretion of proinflammatory factors by macrophages. Furthermore, Aflaki et al. (12) noted improvement in every metric (GCase activity, lipid storage, chemotaxis, and ROS production) upon treatment of macrophages with a novel noninhibitory chaperone. These results demonstrate that iPSC-derived cells provide a wide variety of benchmarks by which to measure response to new therapies.

8.6 Conclusion

Although still a rather fledgling field, iPSCs have already rapidly changed the approach to modeling a variety of LSDs, particularly GD. These cells have allowed us to probe disease mechanisms that were previously very challenging to study and have provided a new powerful tool for evaluating potential therapeutics. But some issues involved with the use of iPSCs are still unresolved. Questions remain as to which phenomena observed in these cells are physiologically relevant and which may be artifacts of the reprogramming and differentiation processes. Furthermore, the cost in time and materials currently limits the applications that are feasible for iPSC-derived models. However, the potential of iPSC-derived cell models for elucidating disease pathogenesis and in facilitating the development of novel drugs should not be underestimated.

Acknowledgments

This work was supported by the Intramural Research Program of the National Human Genome Research Institute and the National Institutes of Health.

References

1. Takahashi, K. & Yamanaka, S. (2006). Induction of pluripotent stem cells from mouse embryonic and adult fibroblast cultures by defined factors. *Cell*, 126, 663–676.
2. Okita, K., Ichisaka, T. & Yamanaka, S. (2007). Generation of germline-competent induced pluripotent stem cells. *Nature*, 448, 313–317.
3. Zhao, T., Zhang, Z. N., Rong, Z. & Xu, Y. (2011). Immunogenicity of induced pluripotent stem cells. *Nature*, 474, 212–215.
4. Robinton, D. A. & Daley, G. Q. (2013). The promise of induced pluripotent stem cells in research and therapy. *Nature*, 481, 295–305.
5. Brennand, K. J., Simone, A., Jou, J., Gelboin-Burkhart, C., Tran, N., Sangar, S. et al. (2011). Modelling schizophrenia using human induced pluripotent stem cells. *Nature*, 473, 221–225.
6. Maehr, R., Chen, S., Snitow, M., Ludwig, T., Yagasaki, L., Goland, R. et al. (2009). Generation of pluripotent stem cells from patients with type 1 diabetes. *Proc Natl Acad Sci USA*, 106, 15768–15773.
7. Park, I. H., Arora, N., Huo, H., Maherali, N., Ahfeldt, T., Shimamura, A. et al. (2008). Disease-specific induced pluripotent stem cells. *Cell*, 134, 877–886.
8. Mazzulli, J. R., Xu, Y. H., Sun, Y., Knight, A. L., McLean, P. J., Caldwell, G. A. et al. (2011). Gaucher disease glucocerebrosidase and alpha-synuclein form a bidirectional pathogenic loop in synucleinopathies. *Cell*, 146, 37–52.
9. Panicker, L. M., Miller, D., Park, T. S., Patel, B., Azevedo, J. L., Awad, O. et al. (2012). Induced pluripotent stem cell model recapitulates pathologic hallmarks of Gaucher disease. *Proc Natl Acad Sci USA*, 109, 18054–18059.
10. Tiscornia, G., Vivas, E. L., Matalonga, L., Berniakovich, I., Barragan Monasterio, M., Eguizabal, C. et al. (2013). Neuronopathic Gaucher's disease: Induced pluripotent stem cells for disease modelling and testing chaperone activity of small compounds. *Hum Mol Genet*, 22, 633–645.
11. Panicker, L. M., Miller, D., Awad, O., Bose, V., Lun, Y., Park, T. S. et al. (2014). Gaucher iPSC-derived macrophages produce elevated levels of inflammatory mediators and serve as a new platform for therapeutic development. *Stem Cells*, 32, 2338–2349.
12. Aflaki, E., Stubblefield, B. K., Maniwang, E., Lopez, G., Moaven, N., Goldin, E. et al. (2014). Macrophage models of Gaucher disease for evaluating disease pathogenesis and candidate drugs. *Sci Transl Med*, 6, 240ra273.
13. Schöndorf, D. C., Aureli, M., McAllister, F. E., Hindley, C. J., Mayer, F., Schmid, B. et al. (2014). iPSC-derived neurons from GBA1-associated Parkinson's disease patients show autophagic defects and impaired calcium homeostasis. *Nat Commun*, 5, 4028.
14. Sun, Y., Florer, J., Mayhew, C. N., Jia, Z., Zhao, Z., Xu, K. et al. (2015). Properties of neurons derived from induced pluripotent stem cells of Gaucher disease type 2 patient fibroblasts: Potential role in neuropathology. *PLoS One*, 10, e0118771.
15. Sgambato, J. A., Park, T. S., Miller, D., Panicker, L. M., Sidransky, E., Lun, Y. et al. (2015). Gaucher disease-induced pluripotent stem cells display decreased erythroid potential and aberrant myelopoiesis. *Stem Cells Transl Med*, 4, 878–886.

16. Tolar, J., Park, I. H., Xia, L., Lees, C. J., Peacock, B., Webber, B. et al. (2011). Hematopoietic differentiation of induced pluripotent stem cells from patients with mucopolysaccharidosis type I (Hurler syndrome). *Blood*, 117, 839–847.
17. Lemonnier, T., Blanchard, S., Toli, D., Roy, E., Bigou, S., Froissart, R. et al. (2011). Modeling neuronal defects associated with a lysosomal disorder using patient-derived induced pluripotent stem cells. *Hum Mol Genet*, 20, 3653–3666.
18. Lojewski, X., Staropoli, J. F., Biswas-Legrand, S., Simas, A. M., Haliw, L., Selig, M. K. et al. (2014). Human iPSC models of neuronal ceroid lipofuscinosis capture distinct effects of TPP1 and CLN3 mutations on the endocytic pathway. *Hum Mol Genet*, 23, 2005–2022.
19. Huang, H. P., Chen, P. H., Hwu, W. L., Chuang, C. Y., Chien, Y. H., Stone, L. et al. (2011). Human Pompe disease-induced pluripotent stem cells for pathogenesis modeling, drug testing and disease marker identification. *Hum Mol Genet*, 20, 4851–4864.
20. Raval, K. K., Tao, R., White, B. E., De Lange, W. J., Koonce, C. H., Yu, J. et al. (2015). Pompe disease results in a Golgi-based glycosylation deficit in human induced pluripotent stem cell-derived cardiomyocytes. *J Biol Chem*, 290, 3121–3136.
21. Meng, X. L., Shen, J. S., Kawagoe, S., Ohashi, T., Brady, R. O. & Eto, Y. (2010). Induced pluripotent stem cells derived from mouse models of lysosomal storage disorders. *Proc Natl Acad Sci USA*, 107, 7886–7891.
22. Ogawa, Y., Tanaka, M., Tanabe, M., Suzuki, T., Togawa, T., Fukushige, T. et al. (2013). Impaired neural differentiation of induced pluripotent stem cells generated from a mouse model of Sandhoff disease. *PLoS One*, 8, e55856.
23. Boustany, R. M. (2013). Lysosomal storage diseases-the horizon expands. *Nat Rev Neurol*, 9, 583–598.
24. Ward, D. M., Griffiths, G. M., Stinchcombe, J. C. & Kaplan, J. (2000). Analysis of the lysosomal storage disease Chediak-Higashi syndrome. *Traffic*, 1, 816–822.
25. Feng, L., Novak, E. K., Hartnell, L. M., Bonifacino, J. S., Collinson, L. M. & Swank, R. T. (2002). The Hermansky-Pudlak syndrome 1 (HPS1) and HPS2 genes independently contribute to the production and function of platelet dense granules, melanosomes, and lysosomes. *Blood*, 99, 1651–1658.
26. Parenti, G., Andria, G. & Ballabio, A. (2015). Lysosomal storage diseases: From pathophysiology to therapy. *Annu Rev Med*, 66, 471–486.
27. Meikle, P. J., Ranieri, E., Ravenscroft, E. M., Hua, C. T., Brooks, D. A. & Hopwood, J. J. (1999). Newborn screening for lysosomal storage disorders. *SE Asian J Trop Med Public Health*, 30, 104–110.
28. Pinto, R., Caseiro, C., Lemos, M., Lopes, L., Fontes, A., Ribeiro, H. et al. (2004). Prevalence of lysosomal storage diseases in Portugal. *Eur J Hum Genet*, 12, 87–92.
29. Poorthuis, B. J., Wevers, R. A., Kleijer, W. J., Groener, J. E., de Jong, J. G., van Weely, S. et al. (1999). The frequency of lysosomal storage diseases in The Netherlands. *Hum Genet*, 105, 151–156.
30. Beutler, E. (2006). Lysosomal storage diseases: Natural history and ethical and economic aspects. *Mol Genet Metab*, 88, 208–215.
31. Cox, T., Lachmann, R., Hollak, C., Aerts, J., van Weely, S., Hrebicek, M. et al. (2000). Novel oral treatment of Gaucher's disease with N-butyldeoxynojirimycin (OGT 918) to decrease substrate biosynthesis. *Lancet*, 355, 1481–1485.

32. Patterson, M. C., Vecchio, D., Prady, H., Abel, L. & Wraith, J. E. (2007). Miglustat for treatment of Niemann–Pick C disease: A randomised controlled study. *Lancet Neurol*, 6, 765–772.
33. Schiffmann, R., Fitzgibbon, E. J., Harris, C., DeVile, C., Davies, E. H., Abel, L. et al. (2008). Randomized, controlled trial of miglustat in Gaucher's disease type 3. *Ann Neurol*, 64, 514–522.
34. Poole, R. M. (2014). Eliglustat: First global approval. *Drugs*, 74, 1829–1836.
35. Shayman, J. A. (2010). Eliglustat tartrate: Glucosylceramide synthase inhibitor treatment of type 1 Gaucher disease. *Drugs Future*, 35, 613–620.
36. Barton, N. W., Brady, R. O., Dambrosia, J. M., Di Bisceglie, A. M., Doppelt, S. H., Hill, S. C. et al. (1991). Replacement therapy for inherited enzyme deficiency—Macrophage-targeted glucocerebrosidase for Gaucher's disease. *N Engl J Med*, 324, 1464–1470.
37. Sidransky, E., Nalls, M. A., Aasly, J. O., Aharon-Peretz, J., Annesi, G., Barbosa, E. R. et al. (2009). Multicenter analysis of glucocerebrosidase mutations in Parkinson's disease. *N Engl J Med*, 361, 1651–1661.
38. Nalls, M. A., Duran, R., Lopez, G. et al. (2013). A multicenter study of glucocerebrosidase mutations in dementia with Lewy bodies. *JAMA Neurol*, 70, 727–735.
39. Farfel-Becker, T., Vitner, E. B. & Futerman, A. H. (2011). Animal models for Gaucher disease research. *Dis Model Mech*, 4, 746–752.
40. Farfel-Becker, T., Vitner, E., Dekel, H., Leshem, N., Enquist, I. B., Karlsson, S. & Futerman, A. H. (2009). No evidence for activation of the unfolded protein response in neuronopathic models of Gaucher disease. *Hum Mol Genet*, 18, 1482–1488.
41. Kacher, Y. & Futerman, A. H. (2009). Impaired IL-10 transcription and release in animal models of Gaucher disease macrophages. *Blood Cells Mol Dis*, 43, 134–137.
42. Saito, M. & Rosenberg, A. (1985). The fate of glucosylceramide (glucocerebroside) in genetically impaired (lysosomal beta-glucosidase deficient) Gaucher disease diploid human fibroblasts. *J Biol Chem*, 260, 2295–2300.
43. Hein, L. K., Meikle, P. J., Hopwood, J. J. & Fuller, M. (2007). Secondary sphingolipid accumulation in a macrophage model of Gaucher disease. *Mol Genet Metab*, 92, 336–345.
44. Prence, E. M., Chaturvedi, P. & Newburg, D. S. (1996). In vitro accumulation of glucocerebroside in neuroblastoma cells: A model for study of Gaucher disease pathobiology. *J Neurosci Res*, 43, 365–371.
45. Beutler, E. & Kuhl, W. (1970). Detection of the defect of Gaucher's disease and its carrier state in peripheral-blood leucocytes. *Lancet*, 1, 612–613.
46. Beutler, E., Kuhl, W., Trinidad, F., Teplitz, R. & Nadler, H. (1970). Detection of Gaucher's disease and its carrier state from fibroblast cultures. *Lancet*, 2, 369.
47. Remus, R., Kammer, C., Heller, H., Schmitz, B., Schell, G. & Doerfler, W. (1999). Insertion of foreign DNA into an established mammalian genome can alter the methylation of cellular DNA sequences. *J Virol*, 73, 1010–1022.
48. Ramos-Mejia, V., Montes, R., Bueno, C., Ayllon, V., Real, P. J., Rodriguez, R. & Menendez, P. (2012). Residual expression of the reprogramming factors prevents differentiation of iPSC generated from human fibroblasts and cord blood CD34+ progenitors. *PLoS One*, 7, e35824.

49. Ramos-Mejia, V., Munoz-Lopez, M., Garcia-Perez, J. L. & Menendez, P. (2010). iPSC lines that do not silence the expression of the ectopic reprogramming factors may display enhanced propensity to genomic instability. *Cell Res*, 20, 1092–1095.
50. Panopoulos, A. D., Yanes, O., Ruiz, S., Kida, Y. S., Diep, D., Tautenhahn, R. et al. (2012). The metabolome of induced pluripotent stem cells reveals metabolic changes occurring in somatic cell reprogramming. *Cell Res*, 22, 168–177.
51. Settembre, C., Fraldi, A., Rubinsztein, D. C. & Ballabio, A. (2007). Lysosomal storage diseases as disorders of autophagy. *Autophagy*, 4, 113–114.
52. Wang, S., Xia, P., Ye, B., Huang, G., Liu, J. & Fan, Z. (2013). Transient activation of autophagy via Sox2-mediated suppression of mTOR is an important early step in reprogramming to pluripotency. *Cell Stem Cell*, 13, 617–625.
53. Pan, H., Cai, N., Li, M., Liu, G.-H. & Izpisua Belmonte, J. C. (2013). Autophagic control of cell 'stemness'. *EMBO Mol Med*, 5, 327–331.
54. Higuchi, T., Kawagoe, S., Otsu, M., Shimada, Y., Kobayashi, H., Hirayama, R. et al. (2014). The generation of induced pluripotent stem cells (iPSCs) from patients with infantile and late-onset types of Pompe disease and the effects of treatment with acid-α-glucosidase in Pompe's iPSCs. *Mol Genet Metab*, 112, 44–48.
55. Seibler, P., Graziotto, J., Jeong, H., Simunovic, F., Klein, C. & Krainc, D. (2011). Mitochondrial Parkin recruitment is impaired in neurons derived from mutant PINK1 induced pluripotent stem cells. *J Neurosci*, 31, 5970–5976.
56. Allen, M. J., Myer, B. J., Khokher, A. M., Rushton, N. & Cox, T. M. (1997). Pro-inflammatory cytokines and the pathogenesis of Gaucher's disease: Increased release of interleukin-6 and interleukin-10. *QJM*, 90, 19–25.
57. Hollak, C. E., van Weely, S., van Oers, M. H. & Aerts, J. M. (1994). Marked elevation of plasma chitotriosidase activity: A novel hallmark of Gaucher disease. *J Clin Invest*, 93, 1288–1292.

9

Role of Induced Pluripotent Stem Cells in Urological Disease Modeling and Repair

Mohammad Moad, Emma L. Curry, Craig N. Robson, and Rakesh Heer

CONTENTS

9.1 Introduction

Many forms of urinary tract disease are associated with a loss in functionally normal cells. These include congenital and neuropathic disorders as well as malignancies and acquired injuries (Atala, 2008; Bolland and Southgate, 2008). Patients suffering from these maladies can endure multiple reconstructive surgeries to replace malformed, damaged, or cancerous urinary tract tissues. Currently, the main strategies proposed for urological disease modeling and repair involve reconstruction with autologous nonurologic tissues, heterologous tissues, or artificial materials. However, reconstructive procedures using these nonnative tissues and foreign body materials can result in serious and common complications (Atala, 2008; Ludlow et al., 2012). Therefore, cellular systems simulating native urological tissue characteristics are in urgent need to enable the development of disease-specific models for investigations of new therapeutic targets as well as for surgical treatment. For these applications, pluripotent stem cells (PSCs) are envisioned to be a valuable new resource, not only because they can be used to derive nearly any replacement cells in the body, but also because they provide a model for differentiation, which can be easily manipulated to better understand the mechanisms of urological oncology (Pastor-Navarro et al., 2010; Yu and Estrada, 2010). Specifically, it has become clear that urothelial cells can be efficiently produced from PSCs, including both human embryonic stem cells (hESCs) and induced pluripotent stem cells (iPSCs) *in vitro* (Moad et al., 2013; Kang et al., 2014; Osborn et al., 2014).

Human induced pluripotent stem cells (hiPSCs) present a unique and potential source of cells for cell therapy in urological tissues, as they have the ability to propagate themselves through self-renewal, differentiate into multiple lineages and, importantly, overcome the ethical barriers that have limited hESC research since oocytes and embryos are not required. In addition, autologous iPSCs can be derived from patients, and such patient-specific iPSC lines would be more compatible with the immune system, thus providing an attractive potential cell source for cellular therapy. Furthermore, iPSCs proliferate readily and are a potentially inexhaustible source of various cell types that could be used in regenerative medicine, drug discovery, disease modeling, and pharmaceutical applications. The aim of this chapter is to discuss the role of iPSCs in urological disease modeling and repair.

9.2 Clinical Need for Bladder Reconstruction

The normal function of the lower urinary tract may be diseased for many reasons, and people from all races, ages, and ethnic groups suffer from poor quality of life even with access to the best medical care. Some conditions are primary congenital abnormalities and become lifelong health concerns of the genitourinary system. The birth abnormalities include hypospadias, where the urethral opening develops proximal to its normal location, and bladder exstrophy, where the bladder develops on the outer surface of the abdomen. These conditions have a considerable impact on social and psychological development. Secondary conditions are much more common and include bladder malignancies, trauma, and neuropathy and are a leading cause of urinary symptoms (including incontinence), kidney failure, pelvic pain, urinary tract infection, and, in some cases, death (Atala, 2008; Ferlay et al., 2008).

In particular, carcinoma of the bladder is a significant, common, and serious healthcare problem throughout the world. Recent statistics estimate that over 380,000 people are diagnosed with the disease worldwide each year. With invasive tumors, the overall survival rate is approximately 50%

(Crawford et al., 1991; de Wit, 2003), and the excision of the bladder is a common treatment and represents one of the most common reasons for urinary tract reconstruction.

Currently, using gastrointestinal segments of the patient's own intestine, also termed *enterocystoplasty*, is the most commonly performed procedure for bladder replacement or repair (Bolland and Southgate, 2008). However, because the bladder and the intestine have different functions, this procedure has been associated with multiple complications such as infection, stone formation, metabolic disturbances, and malignancy (Atala et al., 1993; Ali-El-Dein et al., 2002). Therefore, new regenerative methods and also new models (human-specific tools that accurately reflect normal physiology) to study disease initiation and progression are required.

9.3 Anatomy of the Postnatal Human Bladder

The urinary tract drains urine from the kidneys, and the lower urinary tract consists of two ureters, the bladder and the urethra, and is responsible for the storage and the evacuation of the urine. The bladder is a balloon-like organ that stores and expels urine. The normal capacity of the bladder is about 400 mL of urine. When the bladder is filled, it sends a signal through the nerves to the brain that the bladder is getting full. When the signal comes back from the brain to void, the detrusor muscle in the bladder contracts and the sphincter relaxes to push the urine through the urethra and out of the body (Lang et al., 2006; Drake, 2007; Fowler et al., 2008). The wall of the bladder has two distinct layers—an outer muscle layer and an inner epithelial lining (urothelium) (Figure 9.1).

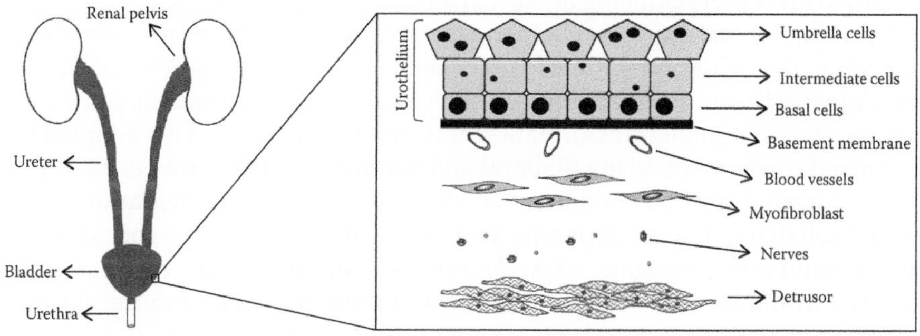

FIGURE 9.1
(See color insert.) Diagram showing that the urothelium expands to cover the renal pelvis, the ureters, and the bladder (red). Bladder urothelium: A three-layered epithelium is apparent, consisting of large, binucleated superficial cells overlying an intermediate and a basal cell layer. This is separated by a basement membrane from a suburothelial layer that contains blood vessels, nerves, and myofibroblasts.

9.3.1 Detrusor Muscle

Anatomically, the wall of the bladder consists of smooth muscular layers, called the *detrusor muscle*, which, functionally, allows for efficient bladder emptying and elasticity that permits low-pressure urine storage (Haab, 2001; Schick, 2008).

9.3.2 Urothelium

The majority of the urinary tract, including the renal pelvis, the ureters, the bladder, and the proximal urethra, is lined by specialized transitional epithelium that occurs nowhere else in the body, also known as *urothelium*. The urothelium consists of a basal, an intermediate, and a superficial cell layer (Lewis, 2000) (Figure 9.1). Although similar to epithelial cells in other types of tissues, the urothelium has unique properties. In addition to its role as a highly effective barrier between the urine and the underlying connective tissue, the urothelium modulates the movement of ions, solutes, and water across the mucosal surface of the bladder and protects the underlying tissue from pathogens (Hicks, 1975; Marceau, 1990; Limas, 1993; Baskin et al., 1997; Apodaca, 2004). Furthermore, it has been reported that the smooth muscle layers under the urothelium need an epithelial signal to differentiate from the mesenchyme (Baskin et al., 1996; Cao et al., 2008).

9.4 Urinary Bladder Replacement and Tissue Engineering Strategies

A diseased urinary tract may require either replacement or augmentation, usually with bowel segments as the current mainstay of treatment. Although this procedure can improve bladder capacity and continence, incorporating bowel into the urinary tract can be associated with several relatively common and potentially serious complications such as low-grade bacteriuria, stone formation, and malignant transformation as the intestinal lining is not adapted to prolonged contact with urine (Bolland and Southgate, 2008; Turner et al., 2011).

Experimental animal models have shown that even augmentation with de-epithelialized bowel segments is associated with fibrosis and shrinkage (Bolland and Southgate, 2008). These observations suggest that the ideal material would be the use of urothelium and the compliance and the elasticity afforded by its associated stroma containing smooth muscle.

9.4.1 Composite Enterocystoplasty

The use of native cells has been explored in a number of experimental approaches involving *ex vivo* expansion of autologous cells that would avoid rejection. The main strategies include the engraftment of urothelium onto de-epithelialized bowel (composite enterocystoplasty) or the engraftment of both urothelial and smooth muscle stromal cells into acellular biomaterials for reconstruction (Turner et al., 2011). However, preclinical models of composite enterocystoplasty show that this technique can be compromised by graft contraction and poor urothelial coverage (Fraser et al., 2004).

9.4.2 Biomaterials and Cell-Seeded Constructs

Biomaterials can be either natural or synthetic (Grise, 2002). The outcomes of the incorporation of synthetic materials such as polyglycolic acid, polyethylene, and polyvinyl into the bladder have been unfavorable due to biomechanical failure or biological incompatibility, resulting in recurrent infection, scar generation, and urinary stone formation (Elbahnasy et al., 1998). Natural tissue matrices derived from various types of tissue including small intestinal submucosa (SIS), porcine dermis (Kimuli et al., 2004), and the urinary bladder itself have been developed and investigated in both *in vitro* and *in vivo* settings. In animal models, using SIS bioscaffold for bladder regeneration resulted in rapid cellular infiltration with the resultant tissue similar to that of the native organ. However, the level of bladder damage appears to affect the success of bladder reconstruction using SIS (Zhang et al., 2006). Bladder acellular matrix grafts (BAMG) are naturally derived from dissected split thickness bladders and also from full-thickness bladders. Previous studies using BAMG in animal bladder have demonstrated the bladder regeneration potential of BAMG. However, the use of this matrix has been associated with many problems including poor vascularization, graft shrinkage, and incomplete or disorganized smooth muscle development (Bolland and Southgate, 2008).

In 2006, Atala et al. (2006) reported the first human clinical trial with engineered bladders, using autologous cells seeded onto biomaterials. However, these strategies rely on *ex vivo* cell culture to generate sufficient quantities and quality of autologous cells and patients with tissue loss and end-organ cellular damage are not ideal candidates. Given the limitations described, increasing attention has been focused on the use of stem cells that may provide a more readily expandable source of cells with the ability for sustained self-renewal (Becker and Jakse, 2007).

9.5 Tissue Engineering of Urinary Bladder Using Stem Cells

Over the past decade, an increased number of studies have investigated the utility of stem cells such as somatic urothelial stem cells, mesenchymal stem cells (MSCs), ESCs, embryonic germ cells (EGCs), amniotic fluid-derived stem cells (Yu and Estrada, 2010), and iPSCs (Moad et al., 2013) in the field of regenerative urology. In addition, several differentiation protocols have been utilized to direct stem cell differentiation to bladder tissue. The most powerful described method utilizes tissue recombinant xenografts of embryonic bladder mesenchyme (EBLM) (Baskin et al., 1996; Oottamasathien et al., 2006, 2007).

9.5.1 Embryonic Stem Cells

Interesting results have been achieved using ESCs. Oottamasathien et al. (2007) showed that mouse ESCs can differentiate into bladder cells when associated with embryonic rat bladder mesenchyme and implanted under the kidney capsule for up to 42 days. The endodermal markers forkhead box A1 (Foxa1) and forkhead box A2 (Foxa2), but not uroplakin, were first detected at day 7 after grafting. At 42 days, optimized numbers of cells resulted in pure urothelial cells with mature bladder tissues derived from the ESCs that were evident by hematoxylin and eosin staining. Maturation was evident based on the expression of uroplakin, a selective marker for urothelial cell differentiation, and the basal cell marker p63, whereas smooth muscle α-actin (SMA) was used as a marker to identify smooth muscle cells (SMCs) (Oottamasathien et al., 2006, 2007).

9.5.2 Mesenchymal Stem Cells

There are ethical and immunological debates about using ESCs in humans. In addition, the differences observed between murine ESCs and hESCs regarding molecular and developmental properties may represent an obstacle for direct translation to humans. Therefore, increasing attention has been paid to the use of adult stem cells, which are less controversial but equally promising cells, MSCs in particular, due to their versatility and their ability to differentiate into a wide range of adult tissue cell types (Caplan, 2007; da Silva Meirelles et al., 2008), including muscle (Luttun et al., 2006; Crisan et al., 2008), liver (Mimeault and Batra, 2008), lung (Nolen-Walston et al., 2008), neuronal (Duan et al., 2007), and gut tissues (Jiang et al., 2002). Utilizing the same model, Anumanthan et al. (2008) used a recombinant xenograft of MSCs with EBLM to differentiate mouse MSCs toward mature bladder cells. Histological examination showed a bladder tissue structure with the expression of uroplakin, SMA, and desmin (Oottamasathien et al., 2007). MSCs and human EGC-derived cells seeded on porcine SIS grafts were also found to enhance bladder reconstitution in animal models. Three months after augmentation, only the stem cell-seeded biohybrid displayed normal bladder structure with both urothelial and SMCs exhibiting gene expression levels similar to those of sham-operated animals (Chung et al., 2005; Frimberger et al., 2005). However, cytotoxic effects of the commercially available SIS on urothelial cells have been reported (Feil et al., 2006).

Tian et al. (2010b) reported that bone marrow mesenchymal stem cells (BMSCs) can be differentiated into urothelial cells and SMCs *in vitro* and *in vivo* when cocultured with bladder cells or conditioned media derived from bladder cell culture. Later, the same group published that BMSCs could be induced to differentiate into bladder SMCs and urothelial cells when seeded on a highly porous poly-L-lactide scaffold and treated with several key growth factors including platelet-derived growth factor B and TGFβ1 (Tian et al., 2010a; Petrovic et al., 2011). However, the clinical utility of BMSC is currently limited due to their extremely low frequency, the intricacy and the pain of the process, and the difficulty in maintaining such cells in culture (Arai et al., 2002).

9.5.3 Adipose-Derived Stem Cells

Another source of autologous adult stem cells has been obtained from stromal elements of adipose tissue (referred to as *adipose-derived stem cells*), which have been successfully used to tissue engineer the smooth muscle of the urinary bladders in rats (Zuk et al., 2001; Jack et al., 2009).

9.5.4 Urothelial Somatic Stem Cells

Recently, Zhang et al. (2008) isolated a subpopulation of cells with progenitor cell characteristics from urine samples. These cells showed the ability to differentiate *in vitro* into multiple lineages that expressed cell markers of urothelial, endothelial, smooth muscle, and interstitial cells and maintained normal karyotype even after several passages.

Although many trials have documented the great therapeutic potential of adult stem cells, experimental studies have reported that adult stem cells are able to form other cell types by fusion with them rather than by transdifferentiation, which might produce cells with karyotypic abnormalities (Terada et al., 2002; Ying et al., 2002; Sievert et al., 2007). Alternatively, PSCs with their ability to proliferate indefinitely and to differentiate into any of the cell types in the body represent a serious alternative and major avenue in regenerative medicine.

9.5.5 Direct Transdifferentiation of Somatic Stem Cells

In an attempt to overcome problems concerning appropriate cell sources for tissue regeneration of the bladder, Drewa et al. showed that stem cells from rat hair follicle seeded on a bladder acellular matrix scaffold and grafted into a surgically created defect within the anterior bladder wall were able to reconstruct both the urothelial and the muscle layers. However, they did not demonstrate any uroplakin expression and the urothelial cells showed incomplete differentiation with weak expression of cytokeratin (CK) 7. Most importantly, they were unable to control the differentiation of the hair follicle stem cells after transplantation. Muscle layers were thick in bladders reconstructed with cell-seeded grafts and very thin in acellular grafts. Again, obtaining sufficient cell numbers posed a major challenge regarding the use of these cells (Drewa, 2008; Petrovic et al., 2011).

9.5.6 Induced Pluripotent Stem Cells

A recent landmark discovery by Takahashi et al., for which he was awarded the Nobel Prize in medicine in 2012, showed that somatic cells can be transformed to an embryonic-like state termed iPSCs by the overexpression of defined factors (Takahashi and Yamanaka, 2006; Takahashi et al., 2007). These iPSCs exhibit ESC characteristics and can differentiate into all three embryonic germ layers. Disease-specific and patient-specific iPSCs have also been generated, thus providing a new method to evaluate potential therapeutics and to gain mechanistic insight into a variety of diseases (Park et al., 2008). Because the production of iPSCs does not involve the use of embryos or oocytes, it overcomes the ethical restrictions that clearly obstruct the isolation, the study, and the use of ESCs, while maintaining the general promise of ESCs, which is pluripotency, and thus, the ability to form any desired tissue (Yu and Estrada, 2010; Robinton and Daley, 2012). Such cells have great potential for use in tissue engineering, understanding the mechanisms of bladder disease, and drug screening. The potential to generate human bladder-derived iPSCs and an urothelial differentiation model was first realized by our work, and these data and subsequent work by others are discussed in the following (Moad et al., 2013).

Recently, a number of studies have investigated methods for the differentiation of hiPSCs to the urothelium. Osborn et al. (2014) used a protocol for the differentiation of hESCs to urothelial cells via definitive endoderm, which mimics the differentiation process during embryogenesis. Two human iPSC lines were differentiated to the urothelium using urothelium-specific medium without the use of matrices or cell–cell contact. The resulting cells showed the expression of the urothelial markers uroplakin 1a and 1b when analyzed by flow cytometry and immunocytochemistry (Osborn et al., 2014).

Consistent with these results, Kang et al. (2014) utilized a similar differentiation strategy with modification to allow the generation of urothelium in a serum- and feeder-free environment, which is critical for the use of iPSC-derived cells in a clinical setting. The results demonstrated the ability of iPSCs derived from human foreskin fibroblasts to differentiate into urothelial cells expressing the urothelial markers uroplakin II, CK8/18, and p63 as well as tight junction markers. Importantly, these cells lacked the expression of markers of pluripotency or other cell lineages confirming their specificity (Kang et al., 2014).

Although iPSCs were shown to be similar to ESCs with respect to gene expression of pluripotency markers and the ability to differentiate into cell types from the three embryonic germ layers both *in vitro* and in teratoma assays, differences between iPSCs and ESCs in their gene expression profiles (Chin et al., 2009), differentiation abilities (Feng et al., 2010), and persistence of donor cell gene expression (Ghosh et al., 2010; Bar-Nur et al., 2011) have been recently documented. It has been shown that iPSCs at low passages display an epigenetic memory inherited from the original somatic cell that will likely favor iPSC differentiation toward lineages related to that cell (Chin et al., 2009; Marchetto et al., 2009; Kim et al., 2010). Genome-wide expression analysis of hiPSCs and their embryo-derived counterparts showed that early- and late-passage iPSCs have different gene expression signatures, whereas late-passage iPSCs appeared to be much more similar to their embryo-derived counterparts than early-passage iPSCs. Analyzing the expression differences between early-passage iPSC lines and their related hESCs showed that most of the genes highly expressed in iPSCs against ESCs were associated with differentiation. Although extended passaging significantly reduced these transcriptional differences, late-passage iPSCs were still distinguishable from ESCs. Other studies also revealed unique DNA methylation and gene expression patterns that are inherited from a parental cell following reprogramming in both human and mouse iPSCs (Kim et al., 2010; Lister et al., 2011; Ohi et al., 2011).

Recently, Bar-Nur et al. (2011) utilized a genetic lineage tracing approach for monitoring the origin of reprogrammed cells and evaluating the differentiation potential of iPSCs derived from human beta cells (BiPSCs). Generated BiPSCs were found to retain an epigenetic memory during their expansion *in vitro* that may preferentially drive their differentiation more readily into insulin-producing cells. Beta cell induced pluripotent stem (BiPS) cell lines showed a typical ESC-like morphology, expressed most pluripotency markers at both RNA and protein levels, silenced the retroviral transgenes, maintained a normal diploid karyotype, and generated cells from all three embryonic germ layers. These observations collectively indicated that BiPSCs were truly reprogrammed pluripotent cells. Chromatin immunoprecipitation showed that epigenetic imprinting was preserved in the insulin and PDX1 gene promoters in BiPS cell lines at similar levels to those of beta cell-derived (BCD) progeny, while not detected in nonbeta pancreatic induced pluripotent stem (PiPS) cell lines, iPSCs derived from fibroblasts, or ESCs of similar passage numbers. Similar epigenetic memory was observed at the DNA methylation level with unique DNA methylation signature in BiPS cells that segregated them from BCD, PiPS cell lines, iPSCs, and ESCs.

The authors next investigated whether the observed epigenetic memory might skew the differentiation of BiPSCs into insulin-producing cells and found that differentiated cells derived from BiPSCs expressed higher levels of insulin, PDX1, and Foxa2 compared to differentiated cells derived from PiPS cell lines, iPSCs, and ESCs of similar passage numbers, suggesting preferential lineage-specific differentiation in BiPS (Bar-Nur et al., 2011). Therefore, it is essential to generate hiPSCs from different tissues and compare their safety and differentiation capacities.

Consistent with these results, iPSCs derived from murine ventricular myocytes showed a dramatically higher tendency to redifferentiate back to cardiovascular progenitors and contribute to functionally beating cardiomyocytes compared to genetically matched ESCs or iPSCs derived from tail tip fibroblasts (Xu et al., 2012). Global gene expression and DNA methylation analysis of these iPSCs showed a distinct transcriptional and epigenetic signature that may potentially be involved in directing iPSCs to ventricular myocytes fate. Later, Lee et al. (2012) extended these studies to demonstrate that the differentiation potential of iPSCs may depend on the lineage stage-specific differentiation state of donor cells. iPSC lines established from hepatic lineage cells at an early stage (hepatoblast) can differentiate more effectively toward a hepatocytic lineage compared to iPSCs established from adult hepatocyte (late stage), mouse embryonic fibroblasts, or mESCs (Lee et al., 2012). All the generated iPSC lines showed ESC-like morphology, expressed pluripotency markers, and underwent multilineage differentiation *in vitro* and *in vivo*, demonstrating their complete reprogramming. Moreover, a global gene expression analysis of hepatoblast-derived iPSCs also exhibited a unique gene expression signature, which clearly differentiated them from iPSCs established from adult hepatocytes as well as mESCs. These differences in gene expression were suggested to be responsible for the variability observed in differentiation potency.

Based on this, human urothelial-associated stromal cells isolated from urinary tract tissue have been successfully reprogrammed to an ESC-like pluripotent state (Moad et al., 2013) (Figure 9.2). These cells were validated as *de facto* iPSCs by confirming their ability for sustained self-renewal, silencing of transgenes, expression of ESC-specific genes, and pluripotent differentiation into all three germ lineages (Figure 9.3). Furthermore, within the appropriate inductive

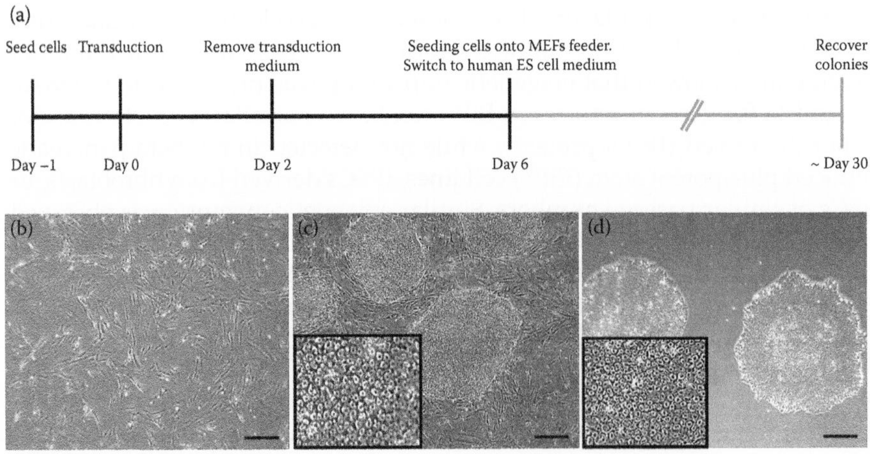

FIGURE 9.2
Generation of urinary tract induced pluripotent stem cells (UT-iPSCs). (a) Timeline for UT-iPSC generation. (b) Phase-contrast micrographs of UT-stroma cells before transduction. (c, d) Example of established UT-iPSC colonies in feeder-dependent and feeder-free defined culture, respectively. Insets show higher magnification. Scale bars: 100 μm.

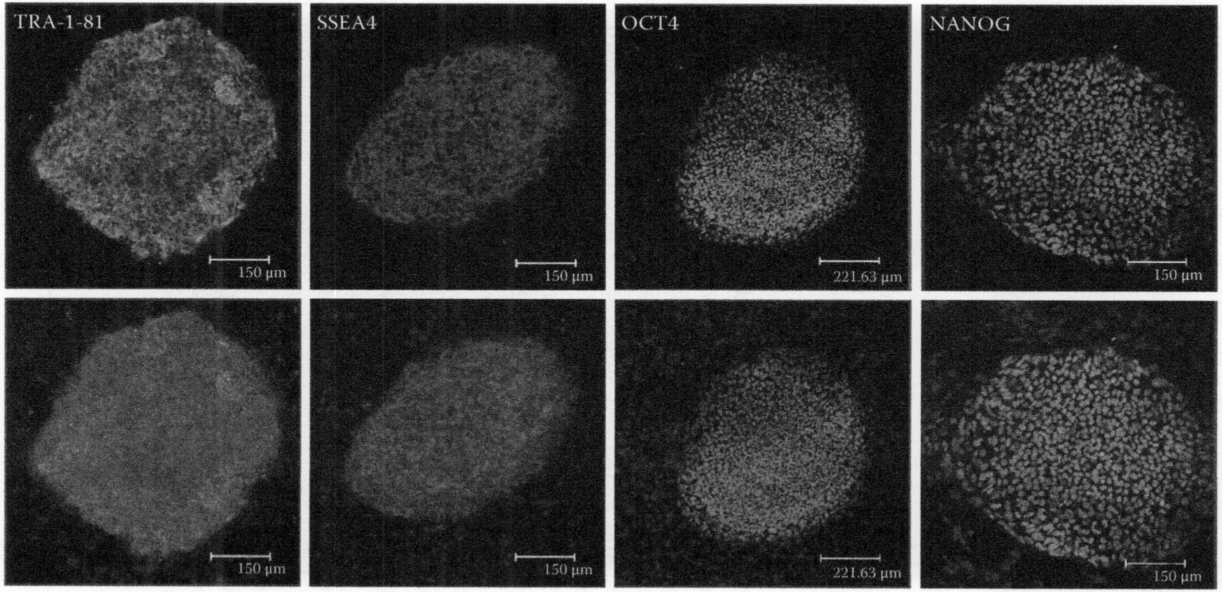

FIGURE 9.3
(See color insert.) Immunofluorescence of generated UT-iPSCs for the expression of specific hES cell surface markers, podocalyxin (TRA-1-81) and stage-specific embryonic antigen 4 (SSEA4), and nuclear transcription factors octamer-binding transcription factor 4 (OCT4) and homeobox transcription factor nanog (NANOG). Nuclei were stained with DAPI (blue). MEF feeder cells served as the negative control.

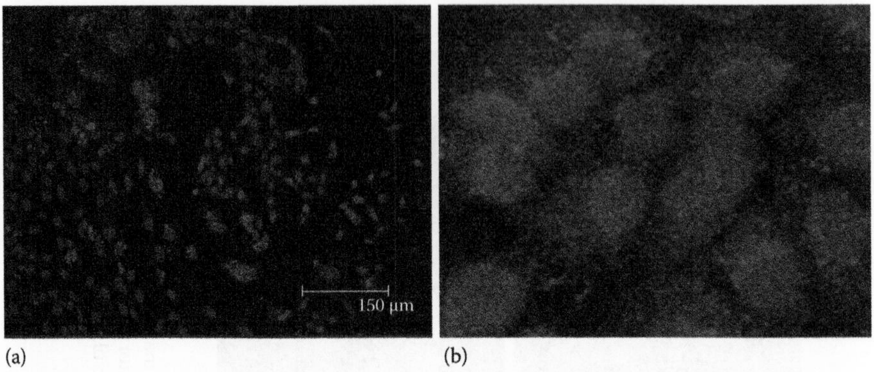

FIGURE 9.4
(See color insert.) Immunofluorescence of differentiated cells derived from UT-iPSCs treated with conditioned medium. Positive staining for UPIb (red) juxtaposed with an area of UPIb-negative staining, (a) with DAPI nuclear counterstain (blue). (b) High magnification of uroplaisin 1B (UPIb) immunostaining.

environment, urinary tract-iPSC (UT-iPSC) differentiation could be directed into bladder-specific lineages (Figure 9.4), allowing for enormous scope in future studies of tissue engineering, disease mechanisms, and drug treatments.

Future challenges include developing viral- and transgene-free reprogramming and xeno-free and feeder-free approaches in building toward clinical translation (Warren et al., 2010). Alternative approaches are now becoming established to realize this ambition, including the development of delivery protocols for nonintegrated genetic constructs (adenoviruses, plasmid transfection, doxycycline-inducible system, and excisable PiggyBac transposon) or the generation of iPSCs using recombinant proteins or small molecules rather than genes (Lyssiotis et al., 2009). Additionally, there are also developments in the use of feeder-free and albumin-free cultures that would again facilitate clinical translation (Chen et al., 2011). Because of their potential to form tumors, concern has been raised about using iPSCs in tissue engineering (Ibarretxe et al., 2012). When these cells are transplanted in the undifferentiated state, they form teratomas. Therefore, an emerging rationale to first differentiate the iPSCs, enrich for the desired cell type, and screen for the presence of undifferentiated cells and genetic mutations before transplantation is now considered the best way to ensure that their proliferation is controlled (Zhu et al., 2011).

9.6 Summary

In summary, iPSCs seem to meet virtually all the defining criteria of true PSCs but without the ethical and immunological concerns that have limited progress with hESCs. Alternative approaches are now becoming established to produce virus/integration-free iPSC lines, and new approaches using controlled 3D scaffolds and cell–cell and cell–extracellular matrix induction systems are being investigated to control the differentiation of PSCs along a favored lineage. These initiatives will further improve the prospects of clinical translation and will open up new perspectives in the treatment of diseases for patients. The results of initial studies in the experimental domain and clinical urology suggest that iPSCs may provide great benefits in regenerative strategies, specifically for urological structures. Nevertheless, the field of iPSCs remains in its infancy, and a better understanding of the reprogramming process and its effects on the cells is needed in order to develop safer and more rapid and efficient approaches for pluripotency induction. No doubt the development of these protocols for the differentiation and the purification of desired cells prior to their application will form a major emerging translational research focus in the management of urological disease.

References

Ali-El-Dein, B., El-Tabey, N., Abdel-Latif, M., Abdel-Rahim, M., and El-Bahnasawy, M. S. (2002) Late uro-ileal cancer after incorporation of ileum into the urinary tract, *J Urol*, 167, pp. 84–88.

Anumanthan, G., Makari, J. H., Honea, L., Thomas, J. C., Wills, M. L., Bhowmick, N. A., Adams, M. C. et al. (2008). Directed differentiation of bone marrow derived mesenchymal stem cells into bladder urothelium. *J Urol*, 180 (4 Suppl), pp. 1778–1783.

Apodaca, G. (2004) The uroepithelium: Not just a passive barrier, *Traffic*, 5, pp. 117–128.

Arai, F., Ohneda, O., Miyamoto, T., Zhang, X. Q., and Suda, T. (2002) Mesenchymal stem cells in perichondrium express activated leukocyte cell adhesion molecule and participate in bone marrow formation, *J Exp Med*, 195, pp. 1549–1563.

Atala, A. (2008) Bioengineered tissues for urogenital repair in children, *Pediatr Res*, 63, pp. 569–575.

Atala, A., Bauer, S. B., Hendren, W. H., and Retik, A. B. (1993) The effect of gastric augmentation on bladder function, *J Urol*, 149, pp. 1099–1102.

Atala, A., Bauer, S. B., Soker, S., Yoo, J. J., and Retik, A. B. (2006) Tissue-engineered autologous bladders for patients needing cystoplasty, *Lancet*, 367, pp. 1241–1246.

Bar-Nur, O., Russ, H. A., Efrat, S., and Benvenisty, N. (2011) Epigenetic memory and preferential lineage-specific differentiation in induced pluripotent stem cells derived from human pancreatic islet beta cells, *Cell Stem Cell*, 9, pp. 17–23.

Baskin, L. S., Hayward, S. W., Young, P., and Cunha, G. R. (1996) Role of mesenchymal-epithelial interactions in normal bladder development, *J Urol*, 156, pp. 1820–1827.

Baskin, L. S., Sutherland, R. S., Thomson, A. A., Nguyen, H. T., Morgan, D. M., Hayward, S. W., Hom, Y. K., DiSandro, M., and Cunha, G. R. (1997) Growth factors in bladder wound healing, *J Urol*, 157, pp. 2388–2395.

Becker, C., and Jakse, G. (2007) Stem cells for regeneration of urological structures, *Eur Urol*, 51, pp. 1217–1228.

Bolland, F., and Southgate, J. (2008) Bio-engineering urothelial cells for bladder tissue transplant, *Expert Opin Biol Ther*, 8, pp. 1039–1049.

Cao, M., Liu, B., Cunha, G., and Baskin, L. (2008) Urothelium patterns bladder smooth muscle location, *Pediatr Res*, 64, pp. 352–357.

Caplan, A. I. (2007) Adult mesenchymal stem cells for tissue engineering versus regenerative medicine, *J Cell Physiol*, 213, pp. 341–347.

Chen, G., Gulbranson, D. R., Hou, Z., Bolin, J. M., Ruotti, V., Probasco, M. D., Smuga-Otto, K. et al. (2011) Chemically defined conditions for human iPSC derivation and culture, *Nat Methods*, 8, pp. 424–429.

Chin, M. H., Mason, M. J., Xie, W., Volinia, S., Singer, M., Peterson, C., Ambartsumyan, G. et al. (2009) Induced pluripotent stem cells and embryonic stem cells are distinguished by gene expression signatures, *Cell Stem Cell*, 5, pp. 111–123.

Chung, S. Y., Krivorov, N. P., Rausei, V., Thomas, L., Frantzen, M., Landsittel, D., Kang, Y. M., Chon, C. H., Ng, C. S., and Fuchs, G. J. (2005) Bladder reconstitution with bone marrow derived stem cells seeded on small intestinal submucosa improves morphological and molecular composition, *J Urol*, 174, pp. 353–359.

Crawford, E. D., Saiers, J. H., Baker, L. H., Costanzi, J. H., and Bukowski, R. M. (1991) Gallium nitrate in advanced bladder carcinoma: Southwest Oncology Group study, *Urology*, 38, pp. 355–357.

Crisan, M., Casteilla, L., Lehr, L., Carmona, M., Paoloni-Giacobino, A., Yap, S., Sun, B. et al. (2008) A reservoir of brown adipocyte progenitors in human skeletal muscle, *Stem Cells*, 26, pp. 2425–2433.

da Silva Meirelles, L., Caplan, A. I., and Nardi, N. B. (2008) In search of the in vivo identity of mesenchymal stem cells, *Stem Cells*, 26, pp. 2287–2299.

de Wit, R. (2003) Overview of bladder cancer trials in the European Organization for Research and Treatment, *Cancer*, 97, pp. 2120–2126.

Drake, M. J. (2007) The integrative physiology of the bladder, *Ann R Coll Surg Engl*, 89, pp. 580–585.

Drewa, T. (2008) Using hair-follicle stem cells for urinary bladder-wall regeneration, *Regen Med*, 3, pp. 939–944.

Duan, X., Chang, J. H., Ge, S., Faulkner, R. L., Kim, J. Y., Kitabatake, Y., Liu, X. B. et al. (2007) Disrupted-In-Schizophrenia 1 regulates integration of newly generated neurons in the adult brain, *Cell*, 130, pp. 1146–1158.

Elbahnasy, A. M., Shalhav, A., Hoenig, D. M., Figenshau, R., and Clayman, R. V. (1998) Bladder wall substitution with synthetic and non-intestinal organic materials, *J Urol*, 159, pp. 628–637.

Feil, G., Christ-Adler, M., Maurer, S., Corvin, S., Rennekampff, H. O., Krug, J., Hennenlotter, J., Kuehs, U., Stenzl, A., and Sievert, K. D. (2006) Investigations of urothelial cells seeded on commercially available small intestine submucosa, *Eur Urol*, 50, pp. 1330–1337.

Feng, Q., Lu, S. J., Klimanskaya, I., Gomes, I., Kim, D., Chung, Y., Honig, G. R., Kim, K. S., and Lanza, R. (2010) Hemangioblastic derivatives from human induced pluripotent stem cells exhibit limited expansion and early senescence, *Stem Cells*, 28, pp. 704–712.

Ferlay, J., Randi, G., Bosetti, C., Levi, F., Negri, E., Boyle, P., and La Vecchia, C. (2008) Declining mortality from bladder cancer in Europe, *BJU Int*, 101, pp. 11–19.

Fowler, C. J., Griffiths, D., and de Groat, W. C. (2008) The neural control of micturition, *Nat Rev Neurosci*, 9, pp. 453–466.

Fraser, M., Thomas, D. F., Pitt, E., Harnden, P., Trejdosiewicz, L. K., and Southgate, J. (2004) A surgical model of composite cystoplasty with cultured urothelial cells: A controlled study of gross outcome and urothelial phenotype, *BJU Int*, 93, pp. 609–616.

Frimberger, D., Morales, N., Shamblott, M., Gearhart, J. D., Gearhart, J. P., and Lakshmanan, Y. (2005) Human embryoid body-derived stem cells in bladder regeneration using rodent model, *Urology*, 65, pp. 827–832.

Ghosh, Z., Wilson, K. D., Wu, Y., Hu, S., Quertermous, T., and Wu, J. C. (2010) Persistent donor cell gene expression among human induced pluripotent stem cells contributes to differences with human embryonic stem cells, *PLoS One*, 5, p. e8975.

Grise, P. (2002) The future of biomaterials in urology, *Prog Urol*, 12, pp. 1305–1309.

Haab, F., Sebe, P., Mondet, F., Ciofu, C. (2001) Functional anatomy of the bladder and urethra in females. New York: *The Urinary Sphincter*. Florida: CRC Press.

Hicks, R. M. (1975) The mammalian urinary bladder: An accommodating organ, *Biol Rev Camb Philos Soc*, 50, pp. 215–246.

Ibarretxe, G., Alvarez, A., Canavate, M. L., Hilario, E., Aurrekoetxea, M., and Unda, F. (2012) Cell reprogramming, IPS limitations, and overcoming strategies in dental bioengineering, *Stem Cells Int*, 2012, p. 365932.

Jack, G. S., Zhang, R., Lee, M., Xu, Y., Wu, B. M., and Rodriguez, L. V. (2009) Urinary bladder smooth muscle engineered from adipose stem cells and a three dimensional synthetic composite, *Biomaterials*, 30, pp. 3259–3270.

Jiang, Y., Jahagirdar, B. N., Reinhardt, R. L., Schwartz, R. E., Keene, C. D., Ortiz-Gonzalez, X. R., Reyes, M. et al. (2002) Pluripotency of mesenchymal stem cells derived from adult marrow, *Nature*, 418, pp. 41–49.

Kang, M., Kim, H. H., and Han, Y. M. (2014) Generation of bladder urothelium from human pluripotent stem cells under chemically defined serum- and feeder-free system, *Int J Mol Sci*, 15, pp. 7139–7157.

Kim, K., Doi, A., Wen, B., Ng, K., Zhao, R., Cahan, P., Kim, J. et al. (2010) Epigenetic memory in induced pluripotent stem cells, *Nature*, 467, pp. 285–290.

Kimuli, M., Eardley, I., and Southgate, J. (2004) In vitro assessment of decellularized porcine dermis as a matrix for urinary tract reconstruction, *BJU Int*, 94, pp. 859–866.

Lang, R. J., Nguyen, D. T., Matsuyama, H., Takewaki, T., and Exintaris, B. (2006) Characterization of spontaneous depolarizations in smooth muscle cells of the Guinea pig prostate, *J Urol*, 175, pp. 370–380.

Lee, S. B., Seo, D., Choi, D., Park, K. Y., Holczbauer, A., Marquardt, J. U., Conner, E. A., Factor, V. M., and Thorgeirsson, S. S. (2012) Contribution of hepatic lineage stage-specific donor memory to the differential potential of induced mouse pluripotent stem cells, *Stem Cells*, 30, pp. 997–1007.

Lewis, S. A. (2000) Everything you wanted to know about the bladder epithelium but were afraid to ask, *Am J Physiol Renal Physiol*, 278, pp. F867–F874.

Limas, C. (1993) Proliferative state of the urothelium with benign and atypical changes. Correlation with transferrin and epidermal growth factor receptors and blood group antigens, *J Pathol*, 171, pp. 39–47.

Lister, R., Pelizzola, M., Kida, Y. S., Hawkins, R. D., Nery, J. R., Hon, G., Antosiewicz-Bourget, J. et al. (2011) Hotspots of aberrant epigenomic reprogramming in human induced pluripotent stem cells, *Nature*, 471, pp. 68–73.

Ludlow, J. W., Kelley, R. W., and Bertram, T. A. (2012) The future of regenerative medicine: Urinary system, *Tissue Eng Part B Rev*, 18, pp. 218–224.

Luttun, A., Ross, J. J., Verfaillie, C., Aranguren, X. L., and Prosper, F. (2006) Differentiation of multipotent adult progenitor cells into functional endothelial and smooth muscle cells, *Curr Protoc Immunol*, Chapter 22, p. Unit 22F.9.

Lyssiotis, C. A., Foreman, R. K., Staerk, J., Garcia, M., Mathur, D., Markoulaki, S., Hanna, J. et al. (2009) Reprogramming of murine fibroblasts to induced pluripotent stem cells with chemical complementation of Klf4, *Proc Natl Acad Sci USA*, 106, pp. 8912–8917.

Marceau, N. (1990) Cell lineages and differentiation programs in epidermal, urothelial and hepatic tissues and their neoplasms, *Lab Invest*, 63, pp. 4–20.

Marchetto, M. C., Yeo, G. W., Kainohana, O., Marsala, M., Gage, F. H., and Muotri, A. R. (2009) Transcriptional signature and memory retention of human-induced pluripotent stem cells, *PLoS One*, 4, p. e7076.

Mimeault, M., and Batra, S. K. (2008) Recent progress on tissue-resident adult stem cell biology and their therapeutic implications, *Stem Cell Rev*, 4, pp. 27–49.

Moad, M., Pal, D., Hepburn, A. C., Williamson, S. C., Wilson, L., Lako, M., Armstrong, L. et al. (2013) A novel model of urinary tract differentiation, tissue regeneration, and disease: Reprogramming human prostate and bladder cells into induced pluripotent stem cells, *Eur Urol*, 64, pp. 753–761.

Nolen-Walston, R. D., Kim, C. F., Mazan, M. R., Ingenito, E. P., Gruntman, A. M., Tsai, L., Boston, R., Woolfenden, A. E., Jacks, T., and Hoffman, A. M. (2008) Cellular kinetics and modeling of bronchioalveolar stem cell response during lung regeneration, *Am J Physiol Lung Cell Mol Physiol*, 294, pp. L1158–L1165.

Ohi, Y., Qin, H., Hong, C., Blouin, L., Polo, J. M., Guo, T., Qi, Z. et al. (2011) Incomplete DNA methylation underlies a transcriptional memory of somatic cells in human iPS cells, *Nat Cell Biol*, 13, pp. 541–549.

Oottamasathien, S., Wang, Y., Williams, K., Franco, O. E., Wills, M. L., Thomas, J. C., Saba, K. et al. (2007) Directed differentiation of embryonic stem cells into bladder tissue, *Dev Biol*, 304, pp. 556–566.

Oottamasathien, S., Williams, K., Franco, O. E., Thomas, J. C., Saba, K., Bhowmick, N. A., Staack, A. et al. (2006) Bladder tissue formation from cultured bladder urothelium, *Dev Dyn*, 235, pp. 2795–2801.

Osborn, S. L., Thangappan, R., Luria, A., Lee, J. H., Nolta, J., and Kurzrock, E. A. (2014) Induction of human embryonic and induced pluripotent stem cells into urothelium, *Stem Cells Transl Med*, 3, pp. 610–619.

Park, I. H., Arora, N., Huo, H., Maherali, N., Ahfeldt, T., Shimamura, A., Lensch, M. W., Cowan, C., Hochedlinger, K., and Daley, G. Q. (2008) Disease-specific induced pluripotent stem cells, *Cell*, 134, pp. 877–886.

Pastor-Navarro, T., Beamud-Cortes, M., Fornas-Buil, E., Moratalla-Charcos, L. M., Osca-Garcia, J. M., and Gil-Salom, M. (2010) Stem cells and regenerative medicine in urology, part 2: Urothelium, urinary bladder, urethra and prostate, *Actas Urol Esp*, 34, pp. 592–597.

Petrovic, V., Stankovic, J., and Stefanovic, V. (2011) Tissue engineering of the urinary bladder: Current concepts and future perspectives, *Scientific World Journal*, 11, pp. 1479–1488.

Robinton, D. A., and Daley, G. Q. (2012) The promise of induced pluripotent stem cells in research and therapy, *Nature*, 481, pp. 295–305.

Schick, J. C. E. (2008) *Textbook of the Neurogenic Bladder*. New York: Taylor & Francis.

Sievert, K. D., Feil, G., Renninger, M., Selent, C., Maurer, S., Conrad, S., Hennenlotter, J. et al. (2007) Tissue engineering and stem cell research in urology for a reconstructive or regenerative treatment approach, *Urologe A*, 46, pp. 1224–1230.

Takahashi, K., Tanabe, K., Ohnuki, M., Narita, M., Ichisaka, T., Tomoda, K., and Yamanaka, S. (2007) Induction of pluripotent stem cells from adult human fibroblasts by defined factors, *Cell*, 131, pp. 861–872.

Takahashi, K., and Yamanaka, S. (2006) Induction of pluripotent stem cells from mouse embryonic and adult fibroblast cultures by defined factors, *Cell*, 126, pp. 663–676.

Terada, N., Hamazaki, T., Oka, M., Hoki, M., Mastalerz, D. M., Nakano, Y., Meyer, E. M., Morel, L., Petersen, B. E., and Scott, E. W. (2002) Bone marrow cells adopt the phenotype of other cells by spontaneous cell fusion, *Nature*, 416, pp. 542–545.

Tian, H., Bharadwaj, S., Liu, Y., Ma, H., Ma, P. X., Atala, A., and Zhang, Y. (2010a) Myogenic differentiation of human bone marrow mesenchymal stem cells on a 3D nano fibrous scaffold for bladder tissue engineering, *Biomaterials*, 31, pp. 870–877.

Tian, H., Bharadwaj, S., Liu, Y., Ma, P. X., Atala, A., and Zhang, Y. (2010b) Differentiation of human bone marrow mesenchymal stem cells into bladder cells: Potential for urological tissue engineering, *Tissue Eng Part A*, 16, pp. 1769–1779.

Turner, A., Subramanian, R., Thomas, D. F., Hinley, J., Abbas, S. K., Stahlschmidt, J., and Southgate, J. (2011) Transplantation of autologous differentiated urothelium in an experimental model of composite cystoplasty, *Eur Urol*, 59, pp. 447–454.

Warren, L., Manos, P. D., Ahfeldt, T., Loh, Y. H., Li, H., Lau, F., Ebina, W. et al. (2010) Highly efficient reprogramming to pluripotency and directed differentiation of human cells with synthetic modified mRNA, *Cell Stem Cell*, 7, pp. 618–630.

Xu, H., Yi, B. A., Wu, H., Bock, C., Gu, H., Lui, K. O., Park, J. H. et al. (2012) Highly efficient derivation of ventricular cardiomyocytes from induced pluripotent stem cells with a distinct epigenetic signature, *Cell Res*, 22, pp. 142–154.

Ying, Q. L., Nichols, J., Evans, E. P., and Smith, A. G. (2002) Changing potency by spontaneous fusion, *Nature*, 416, pp. 545–548.

Yu, R. N., and Estrada, C. R. (2010) Stem cells: A review and implications for urology, *Urology*, 75, pp. 664–670.

Zhang, Y., Frimberger, D., Cheng, E. Y., Lin, H. K., and Kropp, B. P. (2006) Challenges in a larger bladder replacement with cell-seeded and unseeded small intestinal submucosa grafts in a subtotal cystectomy model, *BJU Int*, 98, pp. 1100–1105.

Zhang, Y., McNeill, E., Tian, H., Soker, S., Andersson, K. E., Yoo, J. J., and Atala, A. (2008) Urine derived cells are a potential source for urological tissue reconstruction, *J Urol*, 180, pp. 2226–2233.

Zhu, W. Z., Van Biber, B., and Laflamme, M. A. (2011) Methods for the derivation and use of cardiomyocytes from human pluripotent stem cells, *Methods Mol Biol*, 767, pp. 419–431.

Zuk, P. A., Zhu, M., Mizuno, H., Huang, J., Futrell, J. W., Katz, A. J., Benhaim, P., Lorenz, H. P., and Hedrick, M. H. (2001) Multilineage cells from human adipose tissue: Implications for cell-based therapies, *Tissue Eng*, 7, pp. 211–228.

10

Induced Pluripotent Stem Cells: A Research Tool and a Potential Therapy for RPE-Associated Blinding Eye Diseases

Ruchi Sharma, Balendu Shekhar Jha, and Kapil Bharti

CONTENTS

10.1 Retinal Pigment Epithelium

The retinal pigment epithelium (RPE) is a polarized monolayer epithelium located in the back of the eye between the photoreceptors and the choroidal blood supply. The retina/RPE/choroid form a homeostatic unit in the back of the eye, and RPE is critically important for maintaining the health of this homeostatic unit. RPE is also responsible for photoreceptor development, and the formation of photoreceptor outer segments is tightly coupled to the presence of functionally normal RPE in the adjacent layer (Raymond and Jackson, 1995; Bumsted et al., 2001; Adijanto et al., 2009; Nasonkin et al., 2013). Mouse models with incomplete RPE differentiation or with dysfunctional RPE do not develop normal photoreceptor outer segments (Raymond and Jackson, 1995; Bumsted et al., 2001; Nasonkin et al., 2013). In addition to its developmental role in regulating photoreceptor differentiation, RPE maintains several photoreceptor functions by: (a) reisomerization of all-trans-retinal, a by-product of visual cycle in photoreceptors, to 11-*cis*-retinal components and transporting it back to photoreceptors (von Lintig et al., 2010); (b) phagocytosis of photoreceptor outer segments that are damaged by light-induced photooxidation of proteins and lipids (Mazzoni et al., 2014); (c) maintenance of chemical composition and volume of the subretinal space and the choroid. RPE continuously transports water and CO_2 from the subretinal space toward the choroid (Adijanto et al., 2009; Li et al., 2009); and (d) constitutively secretes of cytokines in a polarized fashion toward the retina and the choroid to regulate their development, function, and pathophysiology (Shi et al., 2008). The relevance of RPE and its functions in photoreceptor health is underscored by gene mutations and disease conditions where these functions are compromised. For instance, mutations in the enzyme retinal pigment epithelium-specific protein 65 kDa (RPE65) that causes reisomerization of all-*trans*-retinal to 11-*cis*-retinal in the RPE, lead to photoreceptor cell death and a disease called Leber congenital amaurosis (LCA) (Cideciyan, 2010). Similarly, mutations in the cell surface protein MER proto-oncogene tyrosine kinase (MERTK) that affects RPE's ability to phagocytose photoreceptor outer segments also result in photoreceptor cell death subsequently leading to another form of LCA (den Hollander et al., 2008). Furthermore, in age-related macular degeneration (AMD), one of the main factors leading to choroidal neovascularization and to compromised RPE barrier function is the increased basal VEGF secretion by the RPE (Marneros, 2013).

The polarized nature of the RPE monolayer is fundamentally important for it to perform several of its functions. This polarized architecture is achieved in part by proteinaceous tight junction complexes present on the lateral sides of each RPE cell. The tight junctions between neighboring RPE cells interact to form an electrically tight monolayer sheet and physically segregate apical and basal compartments in each RPE cell to form the outer ocular blood–retina barrier (Cunha-Vaz et al., 2011; Rodriguez-Boulan and Macara, 2014). Toward its apical side, RPE has microtubule-based apical processes that interact with photoreceptor outer segments. The basal surface of RPE has integrins that attach the entire monolayer to a special proteinaceous basal membrane called *Bruch's membrane* (Lehmann et al., 2014). Based on these anatomical details and genetic insights, one can conclude that RPE is strategically located to help photoreceptors maintain their health and integrity. Both its structural and functional features have evolved to provide this role. A deeper understanding of the RPE biology and the mechanisms that control homeostasis in the back of the eye will help illuminate potential disease-causing pathways and provide potential therapeutic interventions. Here, we summarize how iPSC technology is providing more insight into RPE biology, mechanisms of RPE-associated blinding eye diseases, and potential therapies for these diseases.

10.2 Retinal Pigment Epithelium Development

RPE is a neuroepithelial-derived, nonmigratory cell type that is formed in vertebrates around the time when morphological eye development begins with the evagination of the optic neuroepithelium toward the surface ectoderm. As the distal part of the optic neuroepithelium touches the surface ectoderm, it begins hyperproliferating and invaginating, likely due to differential proliferation in the center against the periphery of this tissue (Eiraku et al., 2011). The distal part that continues to hyperproliferate becomes the future neuroretina, whereas the proximal part that stays as a monolayer tissue becomes the nonneuronal RPE (Bharti et al., 2006). The formation of these two tissues is tightly coupled with both cell-autonomous and noncell-autonomous signals emanating from the neuroepithelium, the surrounding mesenchyme, and the overlying surface ectoderm (Fuhrmann et al., 2000; Bharti et al., 2006). The differentiation of the distal optic neuroepithelium into the neuroretina is regulated by bFGF secreted by the surface ectoderm that upregulates a homeodomain transcription factor, visual system homeobox 2 (VSX2), in the distal neuroepithelium cells (Nguyen and Arnheiter, 2000; Rowan et al., 2004; Horsford et al., 2005). Previously, we have demonstrated that VSX2 selectively downregulates an RPE-inducing basic helix-loop-helix transcription factor microphthalmia-associated transcription factor (MITF) (Horsford et al., 2005; Bharti et al., 2008). The downregulation of MITF in the distal domain is critical to its hyperproliferation and cell fate specification as a neurosensory retina (Nguyen and Arnheiter, 2000; Horsford et al., 2005). In the proximal part of the neuroepithelium, MITF keeps VSX2 expression in check, thus maintaining its monolayer identity and cell fate specification as an RPE tissue (Horsford et al., 2005). The expression of MITF and other RPE-inducing transcription factors such as the homeodomain factor orthodenticle homeobox 2 (OTX2) are regulated in the proximal neuroepithelium by canonical WNT and TGF-beta ligands that are likely secreted by the mesenchymal tissue that surrounds the proximal neuroepithelium, and likely by the developing RPE itself (Fuhrmann et al., 2000; Westenskow et al., 2009; Steinfeld et al., 2013). Both these transcription factors are also responsible for inducing pigmentation in the RPE tissue, directly regulating the expression of the key pigmentation enzymes such as tyrosinase and tyrosinase-related protein 1 (Nakayama et al., 1998; Martinez-Morales et al., 2003; Bharti et al., 2006, 2012). This complex tissue morphogenesis is further fine-tuned by feedback loops that ensure the fates of distal and proximal neuroepithelium tissues as future neuroretina and RPE, respectively. For instance, paired homeodomain transcription factor paired box 6 (PAX6) can cooperate with MITF or its paralog transcription factor EC to inhibit, in the RPE, the expression of Dickkopf-related protein 1, a canonical WNT inhibitor, and fibroblast growth factor 15 (FGF15); both of which together can respecify the fate of proximal neuroepithelium into a second retina (Bharti et al., 2012). Furthermore, PAX6 cooperates with MITF to directly regulate the expression of pigmentation genes (Raviv et al., 2014). These studies, performed mostly in mouse and chicken models, suggest a close interplay between extrinsic and intrinsic factors in regulating RPE fate specification and differentiation. These developmental studies have become a major cornerstone while designing the protocols for efficient and robust differentiation of both retinal and RPE cells from PSCs.

10.3 Pluripotent Stem Cell to RPE Differentiation

One of the first reports of RPE differentiation from PSCs goes back to 2002 (Kawasaki et al., 2002). These initial efforts were focused on monkey embryonic stem cells (ESCs). These efforts relied on the spontaneous differentiation capability of ESCs and the ease to detect RPE, likely due to its pigmented nature. Nevertheless, these mixed cultures could be used to obtain relatively pure RPE cultures that showed typical RPE-like gene expression (Haruta et al., 2004). Similar findings were subsequently reported for hESCs as well (Klimanskaya et al., 2004). ESC-derived RPE, when transplanted in Royal College of Surgeon (RCS) rats were able to transiently rescue the dying photoreceptors (Haruta et al., 2004). In RCS rats, photoreceptor cell death is associated with mutation in gene MERTK that is important for the phagocytosis of photoreceptor outer segments (Dowling and Sidman, 1962; Edwards and Szamier, 1977; D'Cruz et al., 2000). This early work formed the basis for developmentally guided approaches that are currently being used to efficiently generate RPE and retinal cells both from ESCs and induced pluripotent stem cells (iPSCs). As discussed earlier, RPE and retina are neuroepithelium-derived tissues. Therefore, a developmentally guided generation of RPE from PSCs would require a stepwise protocol starting with the generation of optic neuroepithelium (Figure 10.1). Previous work performed on vertebrate embryonic development has shown that optic neuroepithelium develops from the anterior part of the developing nervous system, the diencephalon (reviewed in the studies by Martinez-Morales et al. [2004] and Bharti et al. [2006]). The development of this anterior part of the nervous system is dependent on the activation of the insulin-like growth factor (IGF) pathway and the inhibition of canonical WNT and BMP pathways (del Barco Barrantes et al., 2003; Cavodeassi et al., 2005). Based on these observations, pioneering work from Reh, Takahashi, and Gamm laboratories developed protocols to efficiently generate optic neuroepithelium-like cells from both ESCs and iPSCs (Lamba et al., 2006; Osakada et al., 2009a; Meyer et al., 2009, 2011). These optic neuroepithelium cells can be characterized by a high expression of eye-field transcription factors (PAX6, retinal homeobox protein, LIM homeobox 2, SIX homeobox 3, and SIX6) (Zuber et al., 2003). These cells with optic neuroepithelium phenotype were differentiated into RPE lineage by removing FGF2 from the culture media (Meyer et al., 2009) and/or by adding RPE-promoting TGFβ ligand (ACTIVIN A) to the culture media (Idelson et al., 2009). With the exception of the study by Meyer et al. (2009), most of these original developmentally guided protocols were developed only using ESCs. The initial efforts with iPSCs also mostly involved the use of spontaneous differentiation methods (Buchholz et al., 2009; Carr et al., 2009; Maruotti et al., 2013), but subsequently, other developmentally guided RPE differentiation protocols have also been successfully used with hiPSCs albeit with variable efficiency (Krohne et al., 2012). It is thought that the variable differentiation efficiency with different iPSC lines is likely due to their variable ground-state pluripotency, but a comprehensive analysis comparing different iPSCs and differentiation protocols will be needed to provide a definite answer to this question. Nevertheless, in almost all cases of RPE differentiation, the resulting RPE-like cells were characterized *in vitro* for their epithelial morphology, pigmented nature, presence of RPE markers, and ability to phagocytose photoreceptor outer segments (Figure 10.2). Although these assays were limited in their potential to define the RPE character of PSC-derived cells, they provided satisfactory data for these earlier differentiation attempts. But, as the field continues to make progress toward getting RPE differentiation protocols that are robust and reproducible across multiple research laboratories, there is an intense need to develop *in vitro* functional assays that can be used to fully authenticate RPE cells made from hiPSCs. Some of these assays can include tests to verify that (a) RPE monolayer is electrically intact and exhibits transepithelial resistance of several hundred ohms per square centimeter, (b) RPE monolayer has appropriately polarized distribution of channels and receptors

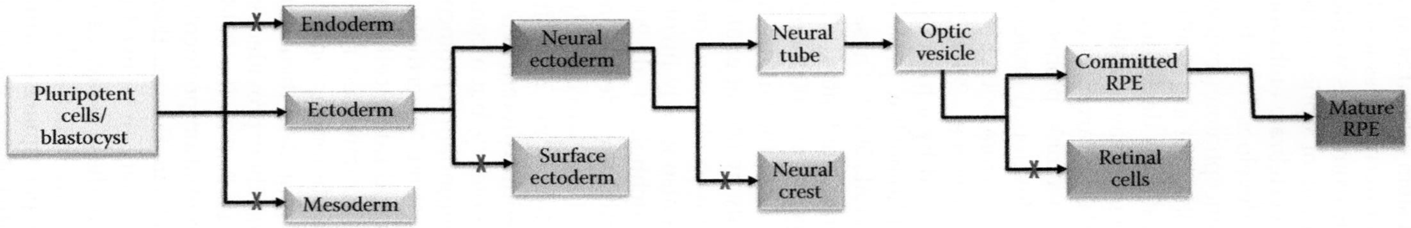

FIGURE 10.1
(See color insert.) *In vitro* differentiation of PSCs (ESCs and iPSCs) follows the same developmental stages as the *in vivo* differentiation of a blastocyst. Spontaneous differentiation of ESCs or iPSCs *in vitro* induces the formation of all three germ layers—endoderm, ectoderm, and mesoderm. Developmental pathways can be manipulated *in vitro* by suppressing TGF, activin, and canonical WNT pathways and by activating IGF to specifically induce optic neuroectoderm from PSCs. RPE cells are induced from optic neuroectoderm by activating TGF and canonical WNT pathways.

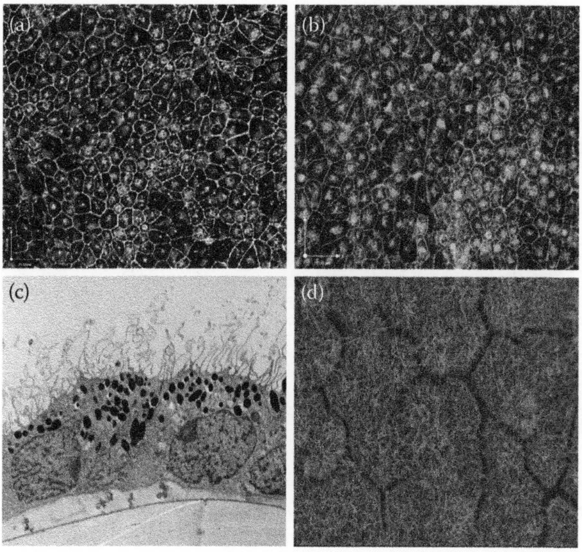

FIGURE 10.2
(See color insert.) Fully differentiated iPSC-derived RPE are characterized by the following: (a) the presence of tight junctions between neighboring cells that are marked by claudin-19 (red) immunostaining. Actin cytoskeleton (F-actin, green) in differentiated RPE cells aligns along the cell boundaries; (b) the expression of transcription factor MITF (green) and visual cycle enzyme cellular retinaldehyde-binding protein (CRALBP, red); (c) the presence of extensive apical processes and apically located melanosomes (RPE pigment) as seen by a transmission electron microscope (TEM) image; (d) the presence of apical processes on the entire RPE monolayer as confirmed by a scanning electron microscope (SEM) image.

and has resting membrane potentials of −50 to −60 mV with the apical side positive compared to the basal side resulting in a transepithelial potential of 2–10 mV, (c) RPE monolayer exhibits polarized fluid transport from the apical to the basal side (5–10 $\mu l \times cm^{-2} \times h^{-1}$), (d) RPE monolayer constitutively secretes cytokines (e.g., monocyte chemoattractant protein 1, interleukin 6, interleukin 8, pigment epithelium-derived factor [PEDF], vascular endothelial growth factor) in a polarized fashion toward the apical and basal sides, (e) RPE monolayer can regulate visual cycle (see the study by Bharti et al. [2011] for a thorough review of RPE authentication criteria) (Figure 10.3). Some of these functional properties have been analyzed and defined in recently published work from others, and our laboratory (Maeda et al., 2013; Ferrer et al., 2014; Muniz et al., 2014). It is, however, worth mentioning that currently, there is no unified protocol for generating RPE from iPSCs or ESCs. As the field progresses toward more clinical applications of ESC- and iPSC-derived RPE, there will be a critical need for fully defined differentiation protocol that are reproducible across multiple laboratories and that can be used as a benchmark of differentiation.

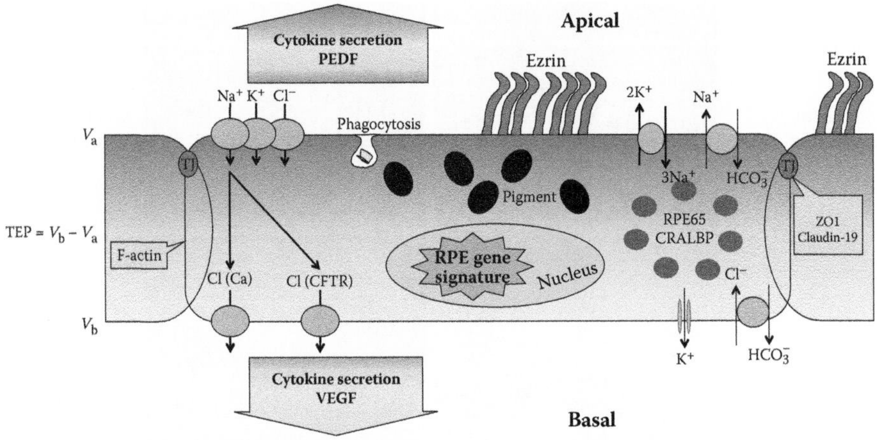

FIGURE 10.3
(See color insert.) Mature RPE cells have a characteristic gene signature that is different from more fetal-like RPE cells. Mature RPE cells express higher levels of visual cycle enzymes RPE65 and CRALBP. Actin cytoskeleton (F-actin immunostaining) organizes around cell boundaries. Tight junctions (TJs) form between neighboring RPE cells and can be visualized with zonula occludens-1 and claudin-19 immunostaining. Because of these TJs, RPE cells in a monolayer exhibit transepithelial resistance of several hundred ohms per square centimeter. All the cells in the monolayer have apically located actin-based processes that can be labeled with ezrin. Cells secrete PEDF predominantly towards the apical side and VEGF predominantly towards the basal side. Functional RPE cells are able to phagocytose photoreceptor outer segments. Polarized RPE cells in an electrically intact monolayer express specific ion channels on the apical and the basal sides. These ion channels maintain apical (V_a) and basolateral (V_b) membrane resting potentials, whereby the basolateral membrane potential is slightly more positive compared to the apical membrane resulting in a transepithelial potential of 2–10 mV. Some of these ion channels (especially the chloride and potassium channels) also help drive vectorial fluid flow of 5–10 μl × cm^{-2} × h^{-1} from the apical toward the basolateral side.

10.4 Use of iPSC-Derived RPE for *In Vitro* Disease Modeling

10.4.1 Congenital Dystrophies of the RPE

There are two main congenital ocular abnormalities (anophthalmia and microphthalmia) that cause major structural defects in the eye recognizable at birth. Both these defects also affect RPE development (Fitzpatrick and van Heyningen, 2005). However, in both cases, direct effects of human mutations on RPE pathology are unknown. In more than 50% of the known cases, both these ocular conditions occur as syndromic malformations. But it is only in a small number of cases (<25%) that the underlying genetic mutation has been identified (Slavotinek, 2011). Anophthalmia is defined as a condition where almost no eye tissue is present, and microphthalmia is a condition where the eye globes are significantly smaller in size (Fitzpatrick and van Heyningen, 2005). A thorough description of ocular and syndromic clinical features associated with these two conditions is beyond the scope of this chapter (see the study by FitzPatrick and van Heyningen [2005] for more details). Of relevance to this chapter is the idea that genes affecting early ocular development are also critical for RPE development and differentiation (Fitzpatrick and van Heyningen, 2005; Bharti et al., 2006). Two interesting examples are the genes MITF and VSX2 (formerly known as CHX10). Mouse mutations in both of these genes and human mutations in VSX2 are associated with microphthalmia and affect RPE development, but through different developmental pathways (Hodgkinson et al., 1993; Hemesath et al., 1994; Ferda Percin et al., 2000; Rowan et al., 2004; Liang and Sandell, 2008). MITF is a basic helix-loop-helix zipper transcription factor and a key regulator of RPE fate specification in mice and human (Tachibana et al., 1994; Yasumoto et al., 1998; Bharti et al., 2008). Although MITF expression in vertebrate eye begins at the optic vesicle stage of eye development, its mutations in rodent models mainly causes RPE-related phenotype leading to respecification of the dorsal RPE part into a second retina and RPE hyperproliferation and depigmentation (Nguyen and Arnheiter, 2000; Bharti et al., 2008). In humans, only heterozygous MITF mutations have been identified and are associated with pigmentary abnormalities of the eye and the hair, but no microphthalmia or RPE fate respecification is known for human mutations (Tassabehji et al., 1994; Nobukuni et al., 1996; Pingault et al., 2010). This is likely because homozygous null MITF mutations in human are lethal. To further dissect the role of MITF in human RPE cells, recently, Capowski et al. (2014) generated the first human MITF homozygous null ESC line. Similar to the findings in mouse models, MITF homozygous mutation led to reduced RPE pigmentation and reduced expression of RPE-specific genes in mutant ESC-derived RPE. Interestingly, again similar to mouse models, MITF knockout resulted in increased proliferation of cells at the RPE stage. In addition, authors identified a novel role for MITF in regulating proliferation at the optic vesicle stage as well. The exact mechanism of this differential MITF effect on cell proliferation in early eye development and in RPE cells is unknown. Authors noted a reduced expression of eye-field transcription factors that are needed for early optic vesicle patterning and proliferation (Zuber et al., 2003). Although this work was done using ESCs, it provides the first *in vitro* analysis of human cells with MITF knockout and provides insight into a previously underappreciated function of this transcription factor in early ocular specification. A comparison of MITF null ESCs with iPSCs generated from patients with specific MITF mutations will provide more details on the mechanism of RPE pathophysiology associated with its mutations.

Another example of a gene that is pivotal for early ocular morphogenesis in human and rodents is the transcription factor VSX2 (Ferda Percin et al., 2000; Rowan et al., 2004; Liang and Sandell, 2008). The expression of VSX2, a homeodomain transcription factor, begins in retinal progenitor cells at the optic vesicle stage and is a key to their fate specification as a neuroretina (Chen and Cepko, 2000; Horsford et al., 2005). Studies performed in mouse models showed that VSX2 suppresses MITF expression in retinal progenitors. In VSX2 null mice, MITF expression continues in future retinal tissue and changes its fate into a second RPE (Rowan et al., 2004; Horsford et al., 2005; Bharti et al., 2008). In a recent work, iPSCs derived from a family with microphthalmia caused by a R200Q mutation in the DNA-binding domain of VSX2 were differentiated into optic vesicle-like structures *in vitro* (Ferda Percin et al., 2000; Phillips et al., 2014). As expected, iPSCs from healthy siblings formed optic vesicles that differentiated both into RPE and retina-like tissues, but iPSCs from patients with R200Q mutation in VSX2 gene mostly formed pigmented RPE-like tissue that overexpressed transcription factors MITF. This observation confirmed the previously published work from mouse models (Rowan et al., 2004; Horsford et al., 2005; Bharti et al., 2008). Furthermore, similar to previous findings, human VSX2 also inhibited the expression of multiple MITF isoforms (A, D, and H) (Bharti et al., 2008; Phillips et al., 2014). These two studies involving MITF and VSX2 transcription factors underscore the importance of human PSCs for studying human developmental biology in a dish. Both these studies have nicely recapitulated previous observations from mouse models and have provided additional molecular insights into the roles played by these transcription factors in regulating early tissue patterning in the human eye.

Mutations in several other genes are linked to syndromic anophthalmia and microphthalmia phenotype. These include SOX2, OTX2, stimulated by retinoic acid 6 (STRA6), b-cell lymphoma 6 corepressor (BCOR6), holocytochrome c synthase (HCCS), BMP4, secreted protein acidic and cysteine rich-related modular calcium binding 1 (SMOC1), growth differentiation factor 6 (GDF6), retina and anterior neural fold homeobox (RAX), SIX6, sonic hedgehog, and PAX6 (Fitzpatrick and van Heyningen, 2005; Slavotinek, 2011). Several of these genes are known for their role in mouse RPE development, but their role in human RPE development and differentiation is not well understood. The use of iPSC technology combined with *in vitro* differentiation methodology provide a very unique possibility to dissect the individual roles played by these genes and their genetic networks in orchestrating early events of ocular morphogenesis in man.

10.4.2 Retinitis Pigmentosa

Retinitis pigmentosa (RP) is a hereditary retinal dystrophy that starts as a rod photoreceptor degeneration disease, leading to night blindness but also progressively leads to central vision loss in these patients (van Soest et al., 1999; Hartong et al., 2006). It is a rare genetic disorder with a worldwide prevalence of 1 in 4000 individuals. Its inheritance pattern is quite heterogeneous and can be caused by autosomal or X-linked mutations and can be dominant or recessive (Rivolta et al., 2002). More than 60 gene mutations are known to cause RP, which can be syndromic or nonsyndromic in phenotype (Hartong et al., 2006). Most of these genes can be categorized into subfamilies depending upon their known function in the eye or in other organs. Within the eye, there are three categories of genes whose mutations lead to RP (Hartong et al., 2006): (a) gene mutations that directly affect photoreceptor function and survival (e.g., rhodopsin, phosphodiesterase 6A [PDE6A], PDE6B, cyclic nucleotide gated channel alpha 1 [CNGA1], CNGB1, S-antigen visual arrestin, peripherin 2, retinal outer segment membrane protein 1, fascin actin-bundling protein 2, tubby-like protein 1, crumbs 1, retinitis pigmentosa 1 [RP1]); (b) gene mutations that lead to syndromic phenotype and may affect both RPE and photoreceptors (e.g., usher syndrome 1 protein network component [USH1C], Usher syndrome type 2A [USH2A], USH3A, pre-mRNA processing factor 31 [PRPF31], PRPF8, PRPF3, RP9, Bardet–Biedl syndrome 1 [BBS1], BBS2, adenosine diphosphate ribosylation factor-like GTPase 6 [ARL6], BBS4, BBS5, McKusick–Kaufman syndrome [MKKS], BBS7, tetratricopeptide repeat domain 8 [TTC8], parathyroid hormone responsive B1, retinitis pigmentosa GTPase regulator); and (c) gene mutations that predominantly affect RPE function but also lead to photoreceptor cell death (e.g., membrane-frizzled-related protein [MFRP], RPE65, retinaldehyde-binding protein 1 [RLBP1], lecithin retinol acyltransferase [LRAT], myosin 7a, carbonic anhydrase 4 [CA4], MERTK). Here, we focus primarily on genes that either specifically affect RPE function or are syndromic in nature and affect both RPE and photoreceptors.

As stated earlier, the use of patient-specific iPSCs to develop *in vitro* disease models depends on the ability to recapitulate disease cellular endophenotypes. In the case of RP, the primary clinical phenotype seen in patients is night blindness caused by the death of rod photoreceptors. But in the case of gene mutations that directly affect RPE function and are only indirectly responsible for rod photoreceptor cell death, attention needs to be given to RPE-specific cellular endophenotypes, not rod photoreceptor phenotype. For example, it is expected that RPE made from iPSCs of patients with (a) RPE65, LRAT, and RLBP1/cellular retinoic acid binding protein 1 mutations will have defects on visual cycle; (b) MERTK mutations will lead to defective phagocytosis of photoreceptor outer segments; and (c) CA4 mutations will lead to defective carbon dioxide transport and intracellular pH regulation (see the study by Hartong et al. [2006] for details) (Yang et al., 2005; den Hollander et al., 2008; Kiser et al., 2012; Lukovic et al., 2015). In fact, two recent publications show that healthy iPSC-derived RPE express all three of these visual function genes (RPE65, LRAT, and RLBP1), and *in vitro*, the visual cycle is active in these cells (Maeda et al., 2013; Muniz et al., 2014). Maeda et al. (2013) demonstrated that iPSC-derived RPE can generate the visual chromophore 11-*cis* retinal and form retinosomes (lipid-like bodies that contain metabolite intermediates of the visual cycle and are involved in regeneration of the visual chromophore). Furthermore, the transplantation of iPSC-derived RPE in the subretinal space of $LRAT^{-/-}$ and $RPE65^{-/-}$ mice increased endogenous 11-*cis* retinal, rescued visual function, and improved vision in these mice (Maeda et al., 2013). Based on these data, one can speculate that iPSC-derived RPE from patients with mutations in visual cycle genes will be insightful in further understanding this pathway and why photoreceptors in these patients actually degenerate. For instance, does the RPE with mutation in any of these genes generate any toxic intermediates of the visual cycle that lead to photoreceptor cell death? Until now, there is no report on the generation of iPSCs or RPE with mutation in any of the visual cycle genes.

Two recent publications discuss *in vitro* disease models for RPE-related RP. Li et al. (2014) focus on the MFRP gene. MFRP mutations lead to an autosomal recessive form of RP (Ayala-Ramirez et al., 2006). As the name suggests, the MFRP gene encodes for a type II transmembrane protein that is homologous to known WNT-binding frizzled proteins (Katoh, 2001; Kameya et al., 2002). Interestingly, MFRP gene encodes for a dicistronic transcript that in its 3′-untranslated region (UTR) expresses, another protein related to the complement C1q tumor necrosis factor-related protein-5 (C1QTNF5/CTRP5) (Chavali et al., 2011). CTRP5 mutation is known to cause a type of macular degeneration called late-onset retinal degeneration (L-ORD) (Ayyagari et al., 2005). Both these proteins have been shown to physically interact at the cell membrane, but the consequence of this interaction for cell function and its effect on disease pathology is unknown. RPE cells derived from MFRP mutant iPSCs have disorganized cytoskeleton, lack epithelial organization, are not pigmented, and have reduced transepithelial resistance (Li et al., 2014). Previously, WNT signaling has been shown to be important for RPE development and differentiation (Burke, 2008; Fujimura et al., 2009; Westenskow et al., 2009). But in this particular case, it is not understood if RPE cells have an abnormal WNT pathway activity and how the mutation affects RPE cell development. It is, however, apparent that RPE cells from patients with MFRP mutations have severe structural abnormalities and provide a basis to explain the photoreceptor degeneration phenotype seen in RP patients. Lukovic et al. (2015) published an *in vitro* model for autosomal recessive RP caused by mutation affecting a cell surface protein MERTK. Similar to the classical RCS rat model with a mutation in MERTK gene and its RPE that is defective in the phagocytosis of photoreceptor outer segments (Edwards and Szamier, 1977; D'Cruz et al., 2000), RPE derived from iPSCs with mutant MERTK are also defective in the phagocytosis of photoreceptor outer segments. Among the syndromic RP type, there are three further subclasses of genes (Hartong et al., 2006): (a) genes involved in trafficking and cytoskeleton organization (USH1C, USH2A, USH3A, peripherin 2 [PRPH2]); (b) genes involved in mRNA splicing (RP9, PRPF31, PRPF8, PRPF3); and (c) genes involved in primary cilium formation and function (BBS1, BBS2, ARL6, BBS4, BBS5, MKKS, BBS7, TTC8). iPSCs have been made from patients with mutations in USH2A, male germ cell-associated kinase (MAK), PRPH2, RP9 (Jin et al., 2011; Tucker et al., 2011, 2013). However, in none of these cases the phenotype of RPE differentiated from iPSCs was investigated. Our recent unpublished work suggests that members of the cilia protein family are involved in apical–basal polarization of RPE cells in mouse models and in hiPSC-derived RPE. Mouse RPE cells with mutation in cilia proteins

BBS8 and BBS6 have reduced and defective apical processes and do not fully mature (unpublished data). Consistently, the pharmacological induction of the primary cilium in hiPSC-derived RPE leads to improved ciliogenesis and complete maturation of iPSC-derived RPE cultures *in vitro* (unpublished data). The resulting RPE cells have a higher expression of adult RPE-specific genes and improved ability to phagocytose photoreceptor outer segments, compared to primary human RPE (unpublished data). Based on these data, we hypothesize that RPE derived from iPSCs of patients with mutations in ciliogenesis proteins will display *in vitro* cellular endophenotypes related to cell maturation. We propose that this analysis will shed more insight into retinal degeneration phenotype seen in these patients with mutations in ciliogenesis proteins. This will provide a possibility of identifying an effective cure for ciliopathies.

10.4.3 Rare RPE-Associated Diseases

Gyrate atrophy is a disease caused by the progressive atrophy of the choroid and the retina (Kennaway et al., 1989). Although, classically, this disease is not considered a form of RP, it does lead to a progressive degeneration of rod photoreceptors. Because of rod cell death, gyrate atrophy patients initially present with night blindness symptoms, but by the second decade of their life, they are left with a tunnel vision due to progressive decline in visual function from the periphery of the eye. The disease is caused by a mutation in the enzyme ornithine aminotransferase (OAT), which helps convert ornithine into pyrroline-5-carboxylate (P5C), a precursor for amino acids glutamine and proline (Clayton, 2006). It is not clear how OAT mutations lead to gyrate atrophy symptoms, but it is thought that these symptoms are likely due to the accumulation of excess ornithine in the blood and the tissues of these patients or due to a deficiency of P5C or glutamine and proline. RPE cells generated from iPSCs reprogrammed from a gyrate atrophy patient present a specific mutation—A226V (Howden et al., 2011; Meyer et al., 2011). Although these authors measured the specific activity of the OAT enzyme and its ability to generate P5C from precursors ornithine and alpha-ketoglutarate, the degeneration of RPE in gyrate atrophy patients was not investigated (Meyer et al., 2011). As expected, RPE derived from gyrate atrophy patient's iPSCs with A226V showed almost no activity. The authors confirmed the specificity of their results using two independent controls: (a) OAT is a vitamin B6-dependent enzyme. This particular mutation A226V is known to interfere with vitamin B6 binding (Michaud et al., 1995; Clayton, 2006). The authors were able to rescue OAT activity by increasing the amounts of vitamin B6 in their assay; (b) authors also performed a genetic correction of the mutation using a bacterial artificial chromosome-based gene targeting, generated RPE from gene-corrected iPSCs and demonstrated a rescue of OAT activity (Howden et al., 2011). This study is an excellent example of the power of iPSC technology, because it allows the evaluation of physiological consequences of gene mutations that lead to severe human pathologies and, at the same time, provide the possibility of testing potential molecular and genetic therapeutic applications.

10.5 Macular Degenerations

The central portion of the human retina with an approximate diameter of 5.5 mm is called the *macula*. It is highly enriched in cone photoreceptors and is responsible for a large portion of the visual input to the brain. Diseases that affect the macula can severely compromise an individual's vision and the quality of life. Macular degeneration can be early onset where it affects children and young adults or late onset that mostly affects individuals beyond the fourth or fifth decade of their life. Three well-known examples of juvenile onset macular degeneration where RPE is thought to be the site of early damage are Best disease, Stargardt disease, and Sorsby's macular dystrophy (North et al., 2014). AMD and L-ORD are two main adult-onset forms of macular degeneration (Ayyagari et al., 2005; Swaroop et al., 2009).

10.5.1 Juvenile Onset Macular Degenerations

10.5.1.1 Best Disease

Best disease is an autosomal dominant form of macular degeneration mostly caused by point mutations in the gene Bestrophin (BEST1) (Boon et al., 2009). The disease is characterized by the accumulation of yellowish vitelliform deposits in the subretinal space (Zhang et al., 2011). It is thought that these deposits are caused by physiological changes in the RPE that lead to its reduced ability to digest and phagocytose photoreceptor outer segments and to transport water from the apical toward the basal sides of the RPE monolayer. Over time, the photoreceptor cells die in these patients likely due to the lack of functional support from the RPE, leading to severe vision loss. Approximately 120 different mutations in the gene BEST1 have been associated with the pathogenesis of Best disease; however, the exact mechanism of disease onset in RPE cells is unknown (Boon et al., 2009). A recent study by Singh et al. (2013) has provided some insight in cellular endophenotypes of Best disease. This group generated iPSC lines from two Best disease patients with specific point mutations A146K and N296H and compared the RPE derived from these iPSC lines to the RPE derived from corresponding healthy siblings. The authors showed *in vitro* long-term feeding of photoreceptor outer segments results in the reduced ability of patient-derived RPE cells to digest these outer segments as compared to cells from their healthy siblings. In a follow-up study, the authors were able to connect these defects in the ability of cells to phagocytose photoreceptor outer segments with general defects in the protein degradation machinery of the RPE cells (Singh et al., 2015). The authors were able to rescue some of these defects by using valproic acid, a U.S. Food and Drug Administration (FDA)-approved inhibitor of histone deacetylases. Consistent with the clinical findings, authors also showed reduced ability of Best disease patient's RPE cells to transport water from the apical to the basal sides. The reduced ability to digest photoreceptor outer segments may result in reduced ability to phagocytose new outer segments and lead to their accumulation in the subretinal space. Accumulated undigested outer segments combined with water that is not transported across the RPE may overtime lead to the accumulation of yellowish deposits seen in the eyes of Best disease patients. It is, however, not clear how these two mutations in BEST1 gene lead to reduced ability of RPE cells to digest photoreceptor outer segments or transport water. Further work is needed to address these molecular events and their direct association to disease-causing mutations. Another classical hallmark of Best disease patients is the abnormal electrophysiological responses of the RPE—mainly the light peak response, which is thought to originate from the basolateral chloride channels in RPE cells. There are no data in the current literature to address how BEST1 mutations might affect the activity of basolateral chloride channels. Nevertheless, this first thorough report on disease modeling for an eye disease provides hope that it is possible to model autosomal dominant diseases using the iPSC technology.

10.5.1.2 Sorsby's Macular Dystrophy

Sorsby's macular dystrophy is also a dominant form of macular degeneration caused by mutation in a metalloproteinase inhibitor gene tissue inhibitor of metalloproteinases-3 (TIMP3) (Weber et al., 1994; Li et al., 2005). Patients have dense drusen deposits underneath the RPE, abnormal growth of choroidal blood vessels, and atrophy of RPE cells. It is not clear how TIMP3 mutations lead to these pathological symptoms in these patients. Investigation of RPE derived from Sorsby's patients can provide some insight into the function of TIMP3, but an *in vitro* recapitulation of most cellular endophenotypes of disease will likely require a 3D tissue containing the RPE monolayer, a Bruch's membrane, and the choroicapillary-like network. Currently, there are no reports that have developed such a 3D tissue that resembles the back of human eye. iPSC technology can be used to build patient-specific 3D disease models that will allow the identification of disease-causing pathways.

10.5.1.3 Stargardt Disease

Stargardt disease provides a challenge similar to Sorsby's macular dystrophy of a disease that affects multiple tissues inside the eye. It is a recessive disease caused by mutation in an adenosine triphosphate (ATP)-binding transporter cassette subfamily A member 4 (ABCA4) that is predominantly localized in membranes of photoreceptor outer segments discs (Allikmets et al., 1997; Tsybovsky et al., 2010; Lambertus et al., 2014). It functions to transport N-retinyl-phosphatidylethanolamine (NR-PE) from the lumen of photoreceptor outer segments disks to their cytoplasm, where it is converted into all-trans retinol. All-trans retinol is transported to the RPE for conversion into the visual pigment 11-*cis* retinal. In patients with ABCA4 mutation, NR-PE accumulates inside disks of photoreceptor outer segments, where it gets photooxidized to make a toxic chemical called A2E. A2E combines with proteins to make lipofuscin. When RPE cells phagocytose ABCA4 mutant photoreceptor outer segments, A2E and lipofuscin start accumulating in RPE cells because they do not have a mechanism of metabolizing these by-products. The accumulation of these toxic products inside RPE cells leads to their eventual cell death, which, in turn, kills photoreceptors (Molday et al., 2009; Tsybovsky et al., 2010). Stargardt disease is an example of a complex disease that, for modeling, requires the *in vitro* development of multiple ocular tissues from iPSCs. In their recent work, Zhong et al. (2014) and Reichman et al. (2014) developed 3D optic vesicles that contain retina and at least a rudimentary RPE. More work will be needed to develop a system where both RPE and photoreceptors can simultaneously mature to represent a disease model for Stargardt disease.

10.5.2 Adult Onset Macular Degenerations

10.5.2.1 Late-Onset Retinal Degeneration

L-ORD, a monogenic autosomal dominant disease, is caused by a point mutation in a bicistronic gene that codes for two proteins MFRP and CTRP5 (Ayyagari et al., 2005; Chavali et al., 2010). The mutation changes one base in the 3′- UTR of MFRP that causes S163R amino acid substitution in the 281 amino acid-long CTRP5 protein. Even though L-ORD is a monogenic disease and AMD is polygenic, both diseases manifest drusen deposits underneath the RPE, show RPE atrophy and choroidal neovascularization in several cases (Ayyagari et al., 2005; Chavali et al., 2011; Zarbin et al., 2014). Because of this similarity, it is thought that a thorough analysis of the mechanism of L-ORD pathogenesis will also provide insight into the mechanism of AMD. Our lab has derived iPSC lines from two L-ORD patients and two of their healthy siblings. Preliminary analysis of passage matched RPE derived from L-ORD and healthy siblings iPSC lines shows that L-ORD RPE have metabolic homeostasis defects as compared to healthy RPE (unpublished data). Metabolic defects can be linked to the long-term health of RPE cells. Our aim is to establish L-ORD cellular endophenotypes using these patient RPE cells generated using the iPSC technology and use them to understand the mechanisms of initiation and progression of this disease.

10.5.2.2 Age-Related Macular Degeneration

AMD is the leading cause of irreversible blindness in developed countries. More than nine million individuals are affected with this disease in the Unites States (Rein et al., 2009). AMD is a complex disease that has been linked to a number of genetic loci and to different signaling and metabolic pathways, suggesting the possibility of an integration of multiple diseases into one clinical phenotype (Swaroop et al., 2009; van Lookeren Campagne et al., 2014). It is thought that AMD disease processes originate in the RPE–choroidal complex, and they severely compromise RPE health and functions (Ambati and Fowler, 2012). The disease has two main subtypes based on clinical presentation: (a) "dry" AMD is characterized by the accumulation of proteinaceous deposits called *drusen* underneath the RPE. The advanced stage of dry AMD is called *geographic atrophy* (GA), when RPE cells die off in the macular region of the eye leading to photoreceptor cell death and vision loss; (b) "wet" AMD or neovascular AMD is also an advanced disease stage when choroidal blood vessels start to abnormally proliferate, penetrate through the RPE blood–retina barrier, and leak fluid or blood in the subretinal space leading to vision loss (Swaroop et al., 2009; Ambati and Fowler, 2012). Although AMD is clinically well characterized, little is presently known about the molecular mechanisms that are responsible for its initiation (Zarbin and Rosenfeld, 2010). Several major roadblocks have precluded the study of disease initiation: (a) lack of any detectable clinical symptoms in patients until well past the middle age; (b) native RPE tissue supply from AMD cadavers is limited and is often disrupted and damaged by the end stage disease; (c) genetic complexity and aspects of disease pathology are not fully represented in animal models. iPSCs allow the possibility of addressing all these roadblocks. RPE derived from AMD patient iPSCs can be potentially used to discover disease-initiating events; iPSC-derived RPE are potentially available as an unlimited supply of fresh human material, and iPSCs derived from patients with different AMD risk alleles allow the possibility of linking patients' genetics with cellular endophenotypes of disease. Thus, iPSC technology is ideally suited to address the challenges of discovering potential drugs for AMD.

Epidemiologically, AMD is most strongly linked to aging, smoking, and high-fat diet, suggesting a major role for metabolic and oxidative stress (Swaroop et al., 2009). RPE cells are particularly vulnerable to stress-related damage because of their high metabolic activity (diurnal phagocytosis of photoreceptor outer segments) and their constant exposure to oxidative byproducts generated during the normal recycling of visual pigments (Sparrow and Boulton, 2005; Kevany and Palczewski, 2010). RPE's ability to manage this high metabolic and oxidative load decreases with age, causing increased ROS, protein aggregation, and unfolded protein response activation (Birol et al., 2007; Kaarniranta et al., 2009, 2010). It has been shown that chronically abnormal activation of intracellular energy-sensing pathways can mediate neurodegenerative and other aging-associated metabolic diseases. These diseases are triggered by the failure of proteostasis and genomic stability caused by high-calorie diet-induced activation of insulin/IGF-1 and mechanistic target of rapamycin signaling pathways and inhibition of sirtuins/SIRT (Wellen and Thompson, 2010; Haigis and Sinclair, 2010). The inhibition of SIRT1, for example, leads to the suppression of forkhead box O3 transcription factors, which upregulate stress response and repair genes (Miura and Endo, 2010). The inhibition of SIRT3 leads to a decreased availability of glutathione and the downregulation of the cell's defense mechanisms (Sebastian and Mostoslavsky, 2010; Someya et al., 2010). Thus, a metabolic stress-induced decline in antioxidant defense can further enhance the damage caused by repeated exposure to ROS leading to DNA breakage, oxidatively damaged proteins, and mitochondrial dysfunction. Similarly, in AMD, the cumulative effects of a high-fat diet, oxidative insults (smoking, sunlight), and genetic risk factors significantly increase the risk for disease onset and progression (Crabb et al., 2002; Nordgaard et al., 2006; Hollyfield et al., 2008; Wang et al., 2010). An ideal *in vitro* model for AMD pathogenesis will utilize a metabolic stressor to age iPSC-derived RPE cells to recapitulate some of the cellular endophenotypes of AMD. This aging platform can then be used to make robust genotype–phenotype associations. This will allow the possibility of identifying drugs that could selectively affect a given pathway and provide a personalized approach to AMD treatment.

A recent manuscript by Tsang lab is the first report on the use of AMD patient-specific iPSC-derived RPE as potential *in vitro* disease model for some aspects of AMD pathogenesis (Yang et al., 2014). The authors focused their attention on three AMD risk alleles, rs10490924 (homozygous protective—G/G; homozygous risk—T/T), InsDel (homozygous protective—WT/WT; homozygous risk—del443ins54/del443ins54), and rs11200638 (homozygous protective—G/G; homozygous risk—A/A); all of which are located on chromosome 10q26 in genes ARMS2 and HTRA1 and are in strong linkage disequilibrium. The function of these two genes is currently unknown; therefore, it is not clear how mutations in these genes affect their biological activity or change the susceptibility of RPE cells toward metabolic or oxidative stressors. Because of this reason, the authors followed an unbiased proteomic screening approach by comparing RPE cells with protective alleles to RPE cells with risk alleles after treating the cells with A2E. As discussed earlier, A2E is a toxic metabolite of visual cycle and is mostly generated in photoreceptors with ABCA4 mutation. But sufficient evidence is lacking that A2E as a stressor can really mimic AMD-like phenotypes in cellular models. In any case, the *in vitro* treatment of cells with A2E should elicit oxidative damage-induced stress response in RPE cells. In fact, the authors note dampened oxidative stress response in RPE cells with high-risk alleles for AMD compared to cells with low-risk alleles for AMD. It is, however, not clear how this difference in stress response is linked to these genetic loci. In any case, this report is the first of its kind and provides a unique opportunity to analyze AMD-causing disease pathways *in vitro* and a possibility to identify potential drugs for AMD treatment.

10.6 Use of iPSC-Derived RPE for Drug Screening

One major advantage of iPSC technology is the possibility to generate bulk-batches of phenotypically controlled derivatives for large-scale drug screenings. This has revolutionized the drug screening industry, which was so far dependent on immortalized cell lines that often do not truly represent the phenotype of a primary human tissue. iPSCs also provide the possibility of testing identified potential drug candidates on fully genotyped patient cells to determine genetic influence on drug toxicity or efficacy. This is certainly true for AMD, which, as stated earlier, is a multigenic disease where different disease-causing pathways merge toward a common disease end point. It is thought that drugs that are effective against a given AMD-causing pathway may not work well on other AMD-causing pathways. A similar observation has been made with certain anticancer drugs (Arnedos et al., 2014). *In vitro* drug testing on fully genotyped iPSC-derived RPE will guide future clinical trials for AMD.

There are three key requisites for using iPSC-derived RPE for large-scale drug screening: (a) ability to consistently produce cells in large scale; (b) ability to authenticate RPE quality before screening; (c) ability to culture and assay RPE cells in high throughput screening format. We recently published a protocol to easily differentiate iPSCs in to RPE in large quality and culture functionally authenticated RPE cells in high throughput format (Ferrer et al., 2014). Our differentiation protocol uses dual SMAD inhibition to generate neuroectoderm lineage from iPSCs followed by recombinant activin A and nicotinamide treatment to induce RPE fate from the neuroectoderm lineage (Chambers et al., 2009; Idelson et al., 2009). In this case, pigmented cell clusters were expanded first to obtain a sufficient number of RPE cells. These RPE cultures on transwell filters were functionally authenticated in parallel to using RPE cells for high throughput screening assays. We described a novel high-content multiplex screening assay that measures endogenous levels of genes in iPSC-derived RPE. We propose that this assay can be used to investigate the state of RPE differentiation or to discover specific modulators of disease relevant genes (Ferrer et al., 2014).

Similar approaches have been used to expand cultures of iPSC-derived RPE (Maruotti et al., 2013; Singh et al., 2013; Croze et al., 2014). In one case, researchers have started with confluent monolayer cultures of iPSCs, differentiated them into RPE, and obtained relatively pure RPE cultures by serial passaging of RPE cells under special culture conditions (Maruotti et al., 2013). Others have performed serial passaging and obtained up to

1000-fold expansion of RPE cell number. It is worth noting that these RPE cells beyond passage three did not maintain their RPE character and lost morphological and functional features of an epithelial cell (Singh et al., 2013). This likely happens because under repeated passaging, RPE cells undergo epithelial-to-mesenchymal transition (EMT). RPE EMT process could be inhibited by the use of an inhibitor (Y-27632) that blocks the activity of Rho-associated, coiled-coil protein kinase (ROCK) (Croze et al., 2014). There are two known ROCK genes in vertebrates, and they are involved in actin filament depolarization-induced cell contractility (Riento and Ridley, 2003). ROCK inhibition is widely used in ESC and iPSC cultures during dissociation to suppress loss-of-contact-induced apoptosis of cells (Watanabe et al., 2007). It is not clear if the effect of ROCK inhibition in RPE cells is similar to its effect on PSCs, but it allows culturing of PSC-derived RPE for 13 passages without cells losing RPE phenotype (Croze et al., 2014). In conclusion, iPSC-derived RPE can be easily expanded to sufficient numbers for large-scale drug screens. Going forward, these advances should encourage several high throughput drug screens focused on various aspects of RPE cell development, function, and pathology.

10.7 Use of iPSC-Derived RPE for *Ex Vivo* Gene Therapy

The first successful gene therapy trial was done recently for LCA, a disease affecting the RPE and caused by mutation in a visual cycle gene RPE65 (den Hollander et al., 2008; Maguire et al., 2008, 2009; Cideciyan, 2010). One main reason for the success of this gene therapy trial was the availability of pre-clinical animal models that allowed testing and establishing efficacy and safety of the gene delivery vector (Amado et al., 2010). However, the availability of appropriate animal models to allow the possibility of testing the efficacy of a human gene therapy vectors for all possible human diseases is limited. Two main limitations of animal models for testing gene therapy are: (a) some tissues are immune to transduction by certain serotypes of adeno-associated vector and require testing many different serotypes; (b) often for certain genes, the development of a clinical dose will require optimizing the expression of the transgene. This will require testing several different promoters and 5′- and/or 3′UTRs (reviewed by Burnight et al. [2014]). Overcoming both these limitations in animal models might require several hundred animals, thus increasing the cost and the time to achieve this goal. iPSC technology can easily overcome these two limitations. Gene therapy delivery constructs, cell type-specific promoters, and other elements of the vector can easily be tested in appropriate somatic cell types derived from hiPSCs. In many cases, gene therapy-mediated rescue can also be used to confirm the function of gene and its association with a disease cellular endophenotype. For example, in a first of its kind approach, Li et al. (2014) rescued iPSCs from patients with a form of RP caused by mutation in gene MFRP (see Section 10.4.2). The authors were able to rescue cellular endophenotypes of MFRP mutant RPE cells by a gene therapy approach. They used a specific AAV8 viral vector-expressing wild-type MFRP under a constitutive cytomegalovirus promoter. RPE derived from gene therapy-corrected iPSCs regained epithelial morphology, transepithelial resistance, and pigmentation. This approach confirms the role played by MFRP in RPE cells, links a specific mutation with a cellular endophenotype, and provides a proof-of-principle data that RP caused by MFRP mutation can be potentially treated by gene therapy. It is hoped that in the near future, FDA would start accepting *ex vivo* gene therapy data using patient-specific iPSCs as efficacy data in place of data from animal models. Furthermore, it is highly appealing to speculate that in the not too distant future, gene therapy can be combined with an autologous iPSC-based therapy to provide treatments for diseases where the target tissue is damaged and also needs cell replacement.

10.8 Use of iPSC-Derived RPE in Molecular Diagnostics

Molecular diagnostic is a process of detecting disease-causing risk factors or biomarkers at the genome, transcriptome, or the proteome level. It is often used to diagnose a disease or monitor a treatment, or predict outcomes of a therapy. In the case of RPE-associated congenital diseases, molecular diagnostic is routinely used for genetic counseling in families that carry disease-causing or risk alleles. Because this analysis is mostly done at the genomic level, an easy and noninvasive tissue of choice is blood. This approach works in cases where the mutation is exomic and leads to an alteration in the open reading frame (ORF). However, there can also be nonsense or missense mutations, or insertion or deletion of nucleotides. In cases where the mutation is either intronic or a synonymous nucleotide substitution in the coding region, the consequences of nucleotide alteration are hard to predict by sequence analysis at the genome level. In many cases, these mutations lead to splice alterations at the premessenger RNA level and can only be confirmed at the transcriptome level (Taneri et al., 2012; Lewandowska, 2013). Several of the RPE-specific genes are not at all expressed in blood, and therefore, a confirmatory test for a potential mRNA splice-altering mutation cannot be performed using blood samples.

iPSCs provide a platform to perform confirmatory molecular diagnostic analysis of mutations that can potentially lead to RPE-associated blinding eye diseases. By using RPE derived from patient-specific iPSCs, mutations that can potentially lead to mRNA splice alteration and mutations that lead to ORF change without a splice alteration can be confirmed at the transcriptome level. A number of different examples have been described that use iPSCs and their derivatives for molecular diagnostics. In two such cases, splice-altering mutations in genes USH2A and MAK were analyzed using iPSC-derived photoreceptors from these patients (Tucker et al., 2011, 2013). Of particular interest is a recent publication that analyzed a nonsense mutation (Arg120stop) in gene RP2 (Schwarz et al., 2014). By using iPSC-derived RPE, these authors confirmed that the nonsense mutation led to a complete loss of RP2 protein in cells. They showed that RPE cells have mislocalization of intraflagellar transport protein 20 (IFT-20), defects in heterotrimeric G protein subunit (Gβ1) trafficking, and defects in golgi cohesion. To confirm that this loss of protein is the likely cause of the cellular endophenotype seen in patient RPE cells, the authors used two different rescue strategies. In one case, they overexpressed RP2 gene, and in the other case, they used translational read-through-inducing drugs G418 and PTC124 to restore up to 20% of full-length RP2 protein in mutant cells. This strategy confirmed molecular diagnosis of RP2 mutation and provided a potential personalized therapeutic approach for patients with such nonsense mutations.

10.9 Use of iPSC-Derived RPE as a Cell-Based Therapy

It is well established that photoreceptor cell death in degenerative eye diseases is due to the lack of functional support from dying and dysfunctional RPE cells (Ambati and Fowler, 2012). Therefore, it is thought that in macular degenerative disease, such as AMD, L-ORD, and Stargardt's where RPE cells atrophy or die off, a potential therapy can be developed by transplanting healthy RPE cells that replace dying RPE in that region of the eye (Bharti et al., 2014). Several proof-of-principle studies have been previously done involving xenografts, allografts, and autologous grafts of RPE cells as potential cell therapy for AMD with varying degrees of success (Table 10.1) (Gouras et al., 1985; Peyman et al., 1991; Algvere et al., 1994; Weisz et al., 1999; Binder et al., 2002; van Meurs et al., 2004). Not surprisingly, most success was seen in cases where an autologous RPE/choroid graft from the periphery of the eye was translocated into the macular region using a procedure called *macular translocation*. Although the macular translocation procedure provides surgical and postsurgical challenges, and the newly derived peripheral RPE cells are still old and likely have a similar epigenetic stage as the diseased cells, these grafts are immunologically most compatible. It is likely because of this immunological compatibility that in a small number of cases where the surgery was successful, the grafts survived, integrated into the surrounding tissue, and provided visual rescue for the patients (van Meurs et al., 2004).

Based on these early surgical innovations, hESC-derived RPE cell suspension was injected into the subretinal space of 9 AMD and 9 Stargardt patients

TABLE 10.1

RPE Transplantation Strategies Used in AMD Patients

	Reference	Starting Cells	AMD Disease Stage	Transplantation Vehicle	Clinical Stage
Allografts					
1	Algvere et al. (1994)	pfRPE	CNV	RPE monolayer patch	Five-patient clinical study
2	Weisz et al. (1999)	pfRPE	GA	RPE cell suspension	One-patient clinical study
3	Schwartz et al. (2012, 2015)	ESCs	GA	Cell suspension	Phase I/IIa completed
4	Idelson et al. (2009)	ESCs	GA	Cell suspension	IND approved, three patients
5	Carr et al. (2009)	ESCs	Choroidal rupture and CNV	Polyester scaffold	IND approved, first patient transplanted
6	van Zeeburg et al. (2012), Hu et al. (2012), Rowland et al. (2013)	ESCs	GA	Paralene scaffold	IND approved, recruiting patients
Autografts					
1	Maaijwee et al. (2007), van Zeeburg et al. (2012)	paRPE	CNV	RPE/choroid patch	83-patient prospective study
2	Joussen et al. (2007)	paRPE	GA	RPE/choroid patch	12-patient prospective study
3	Kamao et al. (2014), Osakada et al. (2009a,b)	Fibroblast-derived iPSCs	CNV	RPE sheet, no scaffold	First patient transplanted (study halted)
4	Bharti et al. (2006, 2014)	CD34+ cell-derived iPSCs	GA	Biodegradable scaffold	Performing IND-enabling studies

Note: CNV: choroidal neovascularization; GA: geographic atrophy; IND: investigational new drug; pfRPE: primary fetal RPE; paRPE: primary adult RPE.

in a phase I trial (Table 10.1) (Schwartz et al., 2012, 2015). The authors report no major adverse events over a median follow-up on 22 months across these 18 patients (Schwartz et al., 2015). This first in-a-human trial of an ESC derivative provides hope for PSC-based therapies. However, unlike autologous macular translocation, these ESC-derived RPE are clearly an allogeneic product and risk the possibility of immune rejection in the long term. In contrast, iPSCs provide the possibility of developing an autologous RPE tissue, thus increasing the possibility of their survival when transplanted in the subretinal space of AMD patients. Even when using patient-specific autologous iPSCs, two additional impediments can hinder the success of derived RPE as a successful cell therapy: (a) the ability to generate functionally authentic RPE from iPSC lines derived from multiple different patients; and (b) the ability to transplant an electrically intact RPE tissue layer instead of cell suspension. Recent work suggests that RPE monolayer transplanted in subretinal space survives much longer compared to RPE cell suspension (Diniz et al., 2013). We have adapted previously published protocols for neuronal and RPE/retina differentiation to optimize specific iPSC to RPE differentiation (Lamba et al., 2006; Chambers et al., 2009; Idelson et al., 2009; Meyer et al., 2009; Osakada et al., 2009a,b). Our protocol involves dual SMAD inhibition combined with a canonical WNT inhibition at the embryoid body stage to efficiently generate neuroectoderm. By combining SMAD and canonical WNT inhibition with FGF pathway inhibition, these cells are efficiently differentiated into RPE precursors. The differentiation of iPSCs toward RPE fate follows a dramatic increase in the expression of transcription factors PAX6 and MITF that have been previously shown to induce RPE fate (Bharti et al., 2012; Raviv et al., 2014). RPE precursors are induced to committed RPE cells by a combination of activin A (TGF superfamily ligand) and WNT3a (canonical WNT ligand). At the end of a nine-week protocol, we can easily obtain 60–70% pure RPE cells (unpublished data). These cells are further purified using a differential trypsinization step to obtain more than 98% pure RPE cells (unpublished data). This differentiation protocol reproducibly makes RPE at high efficiency from iPSC lines of different origin (unpublished data). At this stage, RPE cells are transferred to a scaffold for maturation as an RPE tissue. Our ongoing results show that electrospun poly-(D-lactic-co-glycolic acid) (PLGA) scaffolds for iPSC-derived RPE form a polarized, confluent, and electrically tight monolayer that maintains transepithelial fluid flow from the apical to the basal side, a key functional property of RPE (unpublished data). By using electron microscopy and functional assays, we have determined that PLGA fibers with 400–500 nm diameter are optimal for RPE cells to form a native-like monolayer (PLGA properties: 50/50 DL-lactide/glycolide copolymer; molecular weight: 1.0 dl/g; degradation time: six to eight weeks) (unpublished data). Our data provide a stepwise protocol to reproducibly generate functional RPE tissue from multiple different iPSCs and provide a streamlined process toward developing an autologous cell therapy for AMD.

iPSC-derived RPE have been tested in preclinical animal models for their ability to rescue vision. However, in most cases, the preclinical animal of choice is RCS rat. In an RCS rat, photoreceptor cell death is linked to dysfunctional RPE due to mutation in the gene MERTK that does not allow RPE cells to phagocytose photoreceptor outer segments (Dowling and Sidman, 1962; Edwards and Szamier, 1977; D'Cruz et al., 2000). It is critical to note that unlike the GA stage of AMD, the RPE cells in RCS rat do not degenerate themselves. Thus, RCS rat does not provide an ideal model of choice for determining the engraftment of transplanted RPE cells into the host Bruch's membrane. It has nevertheless provided an important model to test the efficacy of RPE cells in preliminary preclinical work. Another important consideration for preclinical work using iPSC-derived RPE is that most efficacy models have tested cell suspension, and currently, there is no evidence that cell suspension when injected in the subretinal space can form a polarized functional monolayer integrated with the host Bruch's membrane. Recent work done by others and in our lab suggests that it is possible to transplant RPE monolayer tissue into the subretinal space (Diniz et al., 2013; Assawachananont et al., 2014; Kamao et al., 2014; Lu et al., 2014). In fact, in a first of its kind effort, Kamao et al. (2014) recently tested survival of autologous and allogeneic monkey iPSC-derived RPE tissues in monkey eyes and noticed long-term successful engraftment of autologous cells. However, in these wild-type monkeys, the efficacy of the transplant could not be investigated. Going forward, there is a critical need to develop a large-animal model to test the transplantation of iPSC-derived RPE tissue. These advances will certainly help push forward the clinical use of iPSC-derived RPE tissues.

The biggest advantage of using iPSCs compared to ESCs is the ability to use autologous cells, thereby hoping to prevent immune rejection of transplanted RPE cells. Because of this, the current effort in iPSC-derived RPE tissue is focused on autologous cells. In fact, the first AMD patient with advanced CNV has already been successfully transplanted in Japan using autologous iPSC-derived RPE tissue (Reardon and Cyranoski, 2014). Autologous cell therapy, however, poses unique logistic and financial challenges that could slow down the commercialization of this technology. Human leukocyte antigen (HLA)-matched iPSC banks provide an intermediate approach between autologous and allogeneic cell therapies. A handful of HLA-matched iPSC banks can theoretically treat up to 50–60% of the U.S. population and perhaps a higher number in certain less-diverse populations (Turner et al., 2013). Going forward, a universal donor iPSC line where all the HLA-antigens have been deleted by genetic engineering offers the most attractive possibility (Riolobos et al., 2013). Although this technology is currently only feasible for research purposes, a clinical application combining genetic manipulation and iPSC technology is likely a few years out in the future. This technology will help create an "off-the-shelf" iPSC-derived RPE tissue as a potential therapy for the entire AMD population.

10.10 Conclusions

The iPSC-RPE field has reached an important milestone. Robust and reproducible protocols have been generated to differentiate RPE from multiple healthy and patient-specific iPSC lines. Initial studies have generated proof-of-principle data for *in vitro* disease modeling of both early-onset developmental and late-onset degenerative diseases. Some studies have also provided assays for drug screens and molecular diagnostics using iPSC-derived RPE. There is at least one report of an autologous iPSC-derived RPE tissue transplantation in one AMD patient. Going forward, it is expected that this field will provide new discoveries to identify molecular pathways for disease onset, potential new drugs for patients with retinal degenerations, and a potential cell therapy for degenerative eye diseases. It is advised that appropriate universal control iPSC lines, differentiation protocols, and cell authentication assays be used so that the data generated across laboratories can be directly compared. This will reduce redundancy in efforts and help move the field forward for the greater good.

Acknowledgments

This work was supported by the National Eye Institute Intramural Funds and NIH Common Fund Therapeutic Challenge Award.

References

Adijanto, J., T. Banzon, S. Jalickee, N. S. Wang, and S. S. Miller. 2009. "CO_2-induced ion and fluid transport in human retinal pigment epithelium." *J Gen Physiol* no. 133 (6):603–22.

Algvere, P. V., L. Berglin, P. Gouras, and Y. Sheng. 1994. "Transplantation of fetal retinal pigment epithelium in age-related macular degeneration with subfoveal neovascularization." *Graefes Arch Clin Exp Ophthalmol* no. 232 (12):707–16.

Allikmets, R., N. F. Shroyer, N. Singh, J. M. Seddon, R. A. Lewis, P. S. Bernstein, A. Peiffer, N. A. Zabriskie, Y. Li, A. Hutchinson, M. Dean, J. R. Lupski, and M. Leppert. 1997. "Mutation of the Stargardt disease gene (ABCR) in age-related macular degeneration." *Science* no. 277 (5333):1805–7.

Amado, D., F. Mingozzi, D. Hui, J. L. Bennicelli, Z. Wei, Y. Chen, E. Bote et al. 2010. "Safety and efficacy of subretinal readministration of a viral vector in large animals to treat congenital blindness." *Sci Transl Med* no. 2 (21):21ra16.

Ambati, J., and B. J. Fowler. 2012. "Mechanisms of age-related macular degeneration." *Neuron* no. 75 (1):26–39.

Arnedos, M., J. C. Soria, F. Andre, and T. Tursz. 2014. "Personalized treatments of cancer patients: A reality in daily practice, a costly dream or a shared vision of the future from the oncology community?" *Cancer Treat Rev* no. 40 (10):1192–1198.

Assawachananont, J., M. Mandai, S. Okamoto, C. Yamada, M. Eiraku, S. Yonemura, Y. Sasai, and M. Takahashi. 2014. "Transplantation of embryonic and induced pluripotent stem cell-derived 3D retinal sheets into retinal degenerative mice." *Stem Cell Reports* no. 2 (5):662–74.

Ayala-Ramirez, R., F. Graue-Wiechers, V. Robredo, M. Amato-Almanza, I. Horta-Diez, and J. C. Zenteno. 2006. "A new autosomal recessive syndrome consisting of posterior microphthalmos, retinitis pigmentosa, foveoschisis, and optic disc drusen is caused by a MFRP gene mutation." *Mol Vis* no. 12:1483–9.

Ayyagari, R., M. N. Mandal, A. J. Karoukis, L. Chen, N. C. McLaren, M. Lichter, D. T. Wong et al. 2005. "Late-onset macular degeneration and long anterior lens zonules result from a CTRP5 gene mutation." *Invest Ophthalmol Vis Sci* no. 46 (9):3363–71.

Bharti, K., M. Gasper, J. Ou, M. Brucato, K. Clore-Gronenborn, J. Pickel, and H. Arnheiter. 2012. "A regulatory loop involving PAX6, MITF, and WNT signaling controls retinal pigment epithelium development." *PLoS Genetics* no. 8 (7):e1002757.

Bharti, K., W. Liu, T. Csermely, S. Bertuzzi, and H. Arnheiter. 2008. "Alternative promoter use in eye development: The complex role and regulation of the transcription factor MITF." *Development* no. 135 (6):1169–78.

Bharti, K., M. T. Nguyen, S. Skuntz, S. Bertuzzi, and H. Arnheiter. 2006. "The other pigment cell: Specification and development of the pigmented epithelium of the vertebrate eye." *Pigment Cell Res* no. 19 (5):380–94.

Bharti, K., M. Rao, S. C. Hull, D. Stroncek, B. P. Brooks, E. Feigal, J. C. van Meurs, C. A. Huang, and S. S. Miller. 2014. "Developing cellular therapies for retinal degenerative diseases." *Invest Ophthalmol Vis Sci* no. 55 (2):1191–202.

Binder, S., U. Stolba, I. Krebs, L. Kellner, C. Jahn, H. Feichtinger, M. Povelka et al. 2002. "Transplantation of autologous retinal pigment epithelium in eyes with foveal neovascularization resulting from age-related macular degeneration: A pilot study." *Am J Ophthalmol* no. 133 (2):215–25.

Birol, G., S. Wang, E. Budzynski, N. D. Wangsa-Wirawan, and R. A. Linsenmeier. 2007. "Oxygen distribution and consumption in the macaque retina." *Am J Physiol Heart Circ Physiol* no. 293 (3):H1696–704.

Boon, C. J., B. J. Klevering, B. P. Leroy, C. B. Hoyng, J. E. Keunen, and A. I. den Hollander. 2009. "The spectrum of ocular phenotypes caused by mutations in the BEST1 gene." *Prog Retin Eye Res* no. 28 (3):187–205.

Buchholz, D. E., S. T. Hikita, T. J. Rowland, A. M. Friedrich, C. R. Hinman, L. V. Johnson, and D. O. Clegg. 2009. "Derivation of functional retinal pigmented epithelium from induced pluripotent stem cells." *Stem Cells* no. 27 (10):2427–34.

Bumsted, K. M., L. J. Rizzolo, and C. J. Barnstable. 2001. "Defects in the MITF(mi/mi) apical surface are associated with a failure of outer segment elongation." *Exp Eye Res* no. 73 (3):383–92.

Burke, J. M. 2008. "Epithelial phenotype and the RPE: Is the answer blowing in the Wnt?" *Prog Retin Eye Res* no. 27 (6):579–95.

Burnight, E. R., Wiley, L. A., Mullins, R. F., Stone, E. M., and Tucker, B. A. 2014. "Gene therapy using stem cells." *Cold Spring Harb Perspect Med.* Nov. 12; 5(4). pii: a017434. doi: 10.1101/cshperspect.a017434.

Capowski, E. E., J. M. Simonett, E. M. Clark, L. S. Wright, S. E. Howden, K. A. Wallace, A. M. Petelinsek et al. 2014. "Loss of MITF expression during human embryonic stem cell differentiation disrupts retinal pigment epithelium development and optic vesicle cell proliferation." *Hum Mol Genet* no. 23 (23):6332–44.

Carr, A. J., A. A.Vugler, S. T. Hikita, J. M. Lawrence, C. Gias, L. L. Chen, D. E. Buchholz et al. 2009. "Protective effects of human iPS-derived retinal pigment epithelium cell transplantation in the retinal dystrophic rat." *PLoS One* no. 4 (12):e8152.

Cavodeassi, F., F. Carreira-Barbosa, R. M. Young, M. L. Concha, M. L. Allende, C. Houart, M. Tada, and S. W. Wilson. 2005. "Early stages of zebrafish eye formation require the coordinated activity of Wnt11, Fz5, and the Wnt/beta-catenin pathway." *Neuron* no. 47 (1):43–56.

Chambers, S. M., C. A. Fasano, E. P. Papapetrou, M. Tomishima, M. Sadelain, and L. Studer. 2009. "Highly efficient neural conversion of human ES and iPS cells by dual inhibition of SMAD signaling." *Nat Biotechnol* no. 27 (3):275–80.

Chavali, V. R., N. W. Khan, C. A. Cukras, D. U. Bartsch, M. M. Jablonski, and R. Ayyagari. 2011. "A CTRP5 gene S163R mutation knock-in mouse model for late-onset retinal degeneration." *Hum Mol Genet* no. 20 (10):2000–14.

Chavali, V. R., J. R. Sommer, R. M. Petters, and R. Ayyagari. 2010. "Identification of a promoter for the human C1Q-tumor necrosis factor-related protein-5 gene associated with late-onset retinal degeneration." *Invest Ophthalmol Vis Sci* no. 51 (11):5499–507.

Chen, C. M., and C. L. Cepko. 2000. "Expression of Chx10 and Chx100-1 in the developing chicken retina." *Mech Dev* no. 90 (2):293–7.

Cideciyan, A. V. 2010. "Leber congenital amaurosis due to RPE65 mutations and its treatment with gene therapy." *Prog Retin Eye Res* no. 29 (5):398–427.

Clayton, P. T. 2006. "B6–responsive disorders: A model of vitamin dependency." *J Inherit Metab Dis* no. 29 (2–3):317–26.

Crabb, J. W., M. Miyagi, X. Gu, K. Shadrach, K. A. West, H. Sakaguchi, M. Kamei et al. 2002. "Drusen proteome analysis: An approach to the etiology of age-related macular degeneration." *Proc Natl Acad Sci USA* no. 99 (23):14682–7.

Croze, R. H., D. E. Buchholz, M. J. Radeke, W. J. Thi, Q. Hu, P. J. Coffey, and D. O. Clegg. 2014. "ROCK inhibition extends passage of pluripotent stem cell-derived retinal pigmented epithelium." *Stem Cells Transl Med* no. 3 (9):1066–78.

Cunha-Vaz, J., R. Bernardes, and C. Lobo. 2011. "Blood-retinal barrier." *Eur J Ophthalmol* no. 21 Suppl 6:S3–9.

D'Cruz, P. M., D. Yasumura, J. Weir, M. T. Matthes, H. Abderrahim, M. M. LaVail, and D. Vollrath. 2000. "Mutation of the receptor tyrosine kinase gene Mertk in the retinal dystrophic RCS rat." *Hum Mol Genet* no. 9 (4):645–51.

del Barco Barrantes, I., G. Davidson, H. J. Grone, H. Westphal, and C. Niehrs. 2003. "Dkk1 and noggin cooperate in mammalian head induction." *Genes Dev* no. 17 (18):2239–44.

den Hollander, A. I., R. Roepman, R. K. Koenekoop, and F. P. Cremers. 2008. "Leber congenital amaurosis: Genes, proteins and disease mechanisms." *Prog Retin Eye Res* no. 27 (4):391–419.

Diniz, B., P. Thomas, B. Thomas, R. Ribeiro, Y. Hu, R. Brant, A. Ahuja et al. 2013. "Subretinal implantation of retinal pigment epithelial cells derived from human embryonic stem cells: Improved survival when implanted as a monolayer." *Invest Ophthalmol Vis Sci* no. 54 (7):5087–96.

Dowling, J. E., and R. L. Sidman. 1962. "Inherited retinal dystrophy in the rat." *J Cell Biol* no. 14:73–109.

Edwards, R. B., and R. B. Szamier. 1977. "Defective phagocytosis of isolated rod outer segments by RCS rat retinal pigment epithelium in culture." *Science* no. 197 (4307):1001–3.

Eiraku, M., N. Takata, H. Ishibashi, M. Kawada, E. Sakakura, S. Okuda, K. Sekiguchi T. Adachi, and Y. Sasai. 2011. "Self-organizing optic-cup morphogenesis in three-dimensional culture." *Nature* no. 472 (7341):51–6.

Ferda Percin, E., L. A. Ploder, J. J. Yu, K. Arici, D. J. Horsford, A. Rutherford, B. Bapat et al. 2000. "Human microphthalmia associated with mutations in the retinal homeobox gene CHX10." *Nat Genet* no. 25 (4):397–401.

Ferrer, M., B. Corneo, J. Davis, Q. Wan, K. J. Miyagishima, R. King, A. Maminishkis et al. 2014. "A multiplex high-throughput gene expression assay to simultaneously detect disease and functional markers in induced pluripotent stem cell-derived retinal pigment epithelium." *Stem Cells Transl Med* no. 3 (8):911–22.

Fitzpatrick, D. R., and V. van Heyningen. 2005. "Developmental eye disorders." *Curr Opin Genet Dev* no. 15 (3):348–53.

Fuhrmann, S., E. M. Levine, and T. A. Reh. 2000. "Extraocular mesenchyme patterns the optic vesicle during early eye development in the embryonic chick." *Development* no. 127 (21):4599–609.

Fujimura, N., M. M. Taketo, M. Mori, V. Korinek, and Z. Kozmik. 2009. "Spatial and temporal regulation of Wnt/beta-catenin signaling is essential for development of the retinal pigment epithelium." *Dev Biol* no. 334 (1):31–45.

Gouras, P., M. T. Flood, H. Kjedbye, M. K. Bilek, and H. Eggers. 1985. "Transplantation of cultured human retinal epithelium to Bruch's membrane of the owl monkey's eye." *Curr Eye Res* no. 4 (3):253–65.

Haigis, M. C., and D. A. Sinclair. 2010. "Mammalian sirtuins: Biological insights and disease relevance." *Annu Rev Pathol* no. 5:253–95.

Hartong, D. T., E. L. Berson, and T. P. Dryja. 2006. "Retinitis pigmentosa." *Lancet* no. 368 (9549):1795–809.

Haruta, M., Y. Sasai, H. Kawasaki, K. Amemiya, S. Ooto, M. Kitada, H. Suemori, N. Nakatsuji, C. Ide, Y. Honda, and M. Takahashi. 2004. "In vitro and in vivo characterization of pigment epithelial cells differentiated from primate embryonic stem cells." *Invest Ophthalmol Vis Sci* no. 45 (3):1020–5.

Hemesath, T. J., E. Steingrimsson, G. McGill, M. J. Hansen, J. Vaught, C. A. Hodgkinson, H. Arnheiter, N. G. Copeland, N. A. Jenkins, and D. E. Fisher. 1994. "microphthalmia, a critical factor in melanocyte development, defines a discrete transcription factor family." *Genes Dev* no. 8 (22):2770–80.

Hodgkinson, C. A., K. J. Moore, A. Nakayama, E. Steingrimsson, N. G. Copeland, N. A. Jenkins, and H. Arnheiter. 1993. "Mutations at the mouse microphthalmia locus are associated with defects in a gene encoding a novel basic-helix-loop-helix-zipper protein." *Cell* no. 74 (2):395–404.

Hollyfield, J. G., V. L. Bonilha, M. E. Rayborn, X. Yang, K. G. Shadrach, L. Lu, R. L. Ufret, R. G. Salomon, and V. L. Perez. 2008. "Oxidative damage-induced inflammation initiates age-related macular degeneration." *Nat Med* no. 14 (2):194–8.

Horsford, D. J., M. T. Nguyen, G. C. Sellar, R. Kothary, H. Arnheiter, and R. R. McInnes. 2005. "Chx10 repression of Mitf is required for the maintenance of mammalian neuroretinal identity." *Development* no. 132 (1):177–87.

Howden, S. E., A. Gore, Z. Li, H. L. Fung, B. S. Nisler, J. Nie, G. Chen et al. 2011. "Genetic correction and analysis of induced pluripotent stem cells from a patient with gyrate atrophy." *Proc Natl Acad Sci USA* no. 108 (16):6537–42.

Idelson, M., R. Alper, A. Obolensky, E. Ben-Shushan, I. Hemo, N. Yachimovich-Cohen, H. Khaner et al. 2009. "Directed differentiation of human embryonic stem cells into functional retinal pigment epithelium cells." *Cell Stem Cell* no. 5 (4):396–408.

Jin, Z. B., S. Okamoto, F. Osakada, K. Homma, J. Assawachananont, Y. Hirami, T. Iwata, and M. Takahashi. 2011. "Modeling retinal degeneration using patient-specific induced pluripotent stem cells." *PLoS One* no. 6 (2):e17084.

Joussen, A. M., S. Joeres, N. Fawzy, F. M. Heussen, H. Llacer, J. C. van Meurs, and B. Kirchhof. 2007. "Autologous translocation of the choroid and retinal pigment epithelium in patients with geographic atrophy." *Ophthalmology* no. 114 (3):551–60.

Kaarniranta, K., J. Hyttinen, T. Ryhanen, J. Viiri, T. Paimela, E. Toropainen, I. Sorri, and A. Salminen. 2010. "Mechanisms of protein aggregation in the retinal pigment epithelial cells." *Frontiers Bioscio* no. 2:1374–84.

Kaarniranta, K., A. Salminen, E. L. Eskelinen, and J. Kopitz. 2009. "Heat shock proteins as gatekeepers of proteolytic pathways-Implications for age-related macular degeneration (AMD)." *Ageing Res Rev* no. 8 (2):128–39.

Kamao, H., M. Mandai, S. Okamoto, N. Sakai, A. Suga, S. Sugita, J. Kiryu, and M. Takahashi. 2014. "Characterization of human induced pluripotent stem cell-derived retinal pigment epithelium cell sheets aiming for clinical application." *Stem Cell Reports* no. 2 (2):205–18.

Kameya, S., N. L. Hawes, B. Chang, J. R. Heckenlively, J. K. Naggert, and P. M. Nishina. 2002. "Mfrp, a gene encoding a frizzled related protein, is mutated in the mouse retinal degeneration 6." *Hum Mol Genet* no. 11 (16):1879–86.

Katoh, M. 2001. "Molecular cloning and characterization of MFRP, a novel gene encoding a membrane-type Frizzled-related protein." *Biochem Biophys Res Commun* no. 282 (1):116–23.

Kawasaki, H., H. Suemori, K. Mizuseki, K. Watanabe, F. Urano, H. Ichinose, M. Haruta et al. 2002. "Generation of dopaminergic neurons and pigmented epithelia from primate ES cells by stromal cell-derived inducing activity." *Proc Natl Acad Sci USA* no. 99 (3):1580–5.

Kennaway, N. G., L. Stankova, M. K. Wirtz, and R. G. Weleber. 1989. "Gyrate atrophy of the choroid and retina: Characterization of mutant ornithine aminotransferase and mechanism of response to vitamin B6." *Am J Hum Genet* no. 44 (3):344–52.

Kevany, B. M., and K. Palczewski. 2010. "Phagocytosis of retinal rod and cone photoreceptors." *Physiology (Bethesda)* no. 25 (1):8–15.

Kiser, P. D., M. Golczak, A. Maeda, and K. Palczewski. 2012. "Key enzymes of the retinoid (visual) cycle in vertebrate retina." *Biochim Biophys Acta* no. 1821 (1):137–51.

Klimanskaya, I., J. Hipp, K. A. Rezai, M. West, A. Atala, and R. Lanza. 2004. "Derivation and comparative assessment of retinal pigment epithelium from human embryonic stem cells using transcriptomics." *Cloning Stem Cells* no. 6 (3):217–45.

Krohne, T. U., P. D. Westenskow, T. Kurihara, D. F. Friedlander, M. Lehmann, A. L. Dorsey, W. Li et al. 2012. "Generation of retinal pigment epithelial cells from small molecules and OCT4 reprogrammed human induced pluripotent stem cells." *Stem Cells Transl Med* no. 1 (2):96–109.

Lamba, D. A., M. O. Karl, C. B. Ware, and T. A. Reh. 2006. "Efficient generation of retinal progenitor cells from human embryonic stem cells." *Proc Natl Acad Sci USA* no. 103 (34):12769–74.

Lambertus, S., R. A. van Huet, N. M. Bax, L. H. Hoefsloot, F. P. Cremers, C. J. Boon, B. J. Klevering, and C. B. Hoyng. 2014. "Early-onset Stargardt disease: Phenotypic and genotypic characteristics." *Ophthalmology* no. 122 (2):335–44.

Lehmann, G. L., I. Benedicto, N. J. Philp, and E. Rodriguez-Boulan. 2014. "Plasma membrane protein polarity and trafficking in RPE cells: Past, present and future." *Exp Eye Res* no. 126:5–15.

Lewandowska, M. A. 2013. "The missing puzzle piece: Splicing mutations." *Int J Clin Exp Pathol* no. 6 (12):2675–82.

Li, R., A. Maminishkis, T. Banzon, Q. Wan, S. Jalickee, S. Chen, and S. S. Miller. 2009. "IFN{gamma} regulates retinal pigment epithelial fluid transport." *Am J Physiol Cell Physiol* no. 297 (6):C1452–65.

Li, Y., W. H. Wu, C. W. Hsu, H. V. Nguyen, Y. T. Tsai, L. Chan, T. Nagasaki et al. 2014. "Gene therapy in patient-specific stem cell lines and a preclinical model of retinitis pigmentosa with membrane frizzled-related protein defects." *Mol Ther* no. 22 (9):1688–97.

Li, Z., M. P. Clarke, M. D. Barker, and N. McKie. 2005. "TIMP3 mutation in Sorsby's fundus dystrophy: Molecular insights." *Expert Rev Mol Med* no. 7 (24):1–15.

Liang, L., and J. H. Sandell. 2008. "Focus on molecules: Homeobox protein Chx10." *Exp Eye Res* no. 86 (4):541–2.

Lu, B., Y. C. Tai, and M. S. Humayun. 2014. "Microdevice-based cell therapy for age-related macular degeneration." *Dev Ophthalmol* no. 53:155–66.

Lukovic, D., A. Artero Castro, A. B. Delgado, L. Bernal Mde, N. Luna Pelaez, A. Diez Lloret et al. 2015. "Human iPSC derived disease model of MERTK-associated retinitis pigmentosa." *Sci Rep* no. 5:12910.

Maaijwee, K., H. Heimann, T. Missotten, P. Mulder, A. Joussen, and J. van Meurs. 2007. "Retinal pigment epithelium and choroid translocation in patients with exudative age-related macular degeneration: Long-term results." *Graefes Arch Clin Exp Ophthalmol.* no. 245 (11):1681–9.

Maeda, T., M. J. Lee, G. Palczewska, S. Marsili, P. J. Tesar, K. Palczewski, M. Takahashi, and A. Maeda. 2013. "Retinal pigmented epithelial cells obtained from human induced pluripotent stem cells possess functional visual cycle enzymes in vitro and in vivo." *J Biol Chem* no. 288 (48):34484–93.

Maguire, A. M., K. A. High, A. Auricchio, J. F. Wright, E. A. Pierce, F. Testa, F. Mingozzi et al. 2009. "Age-dependent effects of RPE65 gene therapy for Leber's congenital amaurosis: A phase 1 dose-escalation trial." *Lancet* no. 374 (9701):1597–605.

Maguire, A. M., F. Simonelli, E. A. Pierce, E. N. Pugh, Jr., F. Mingozzi, J. Bennicelli, S. Banfi et al. 2008. "Safety and efficacy of gene transfer for Leber's congenital amaurosis." *N Engl J Med* no. 358 (21):2240–8.

Marneros, A. G. 2013. "NLRP3 inflammasome blockade inhibits VEGF-A-induced age-related macular degeneration." *Cell Rep* no. 4 (5):945–58.

Martinez-Morales, J. R., V. Dolez, I. Rodrigo, R. Zaccarini, L. Leconte, P. Bovolenta, and S. Saule. 2003. "OTX2 activates the molecular network underlying retina pigment epithelium differentiation." *J Biol Chem* no. 278 (24):21721–31.

Martinez-Morales, J. R., I. Rodrigo, and P. Bovolenta. 2004. "Eye development: A view from the retina pigmented epithelium." *Bioessays* no. 26 (7):766–77.

Maruotti, J., K. Wahlin, D. Gorrell, I. Bhutto, G. Lutty, and D. J. Zack. 2013. "A simple and scalable process for the differentiation of retinal pigment epithelium from human pluripotent stem cells." *Stem Cells Transl Med* no. 2 (5):341–54.

Mazzoni, F., H. Safa, and S. C. Finnemann. 2014. "Understanding photoreceptor outer segment phagocytosis: Use and utility of RPE cells in culture." *Exp Eye Res* no. 126:51–60.

Meyer, J. S., S. E. Howden, K. A. Wallace, A. D. Verhoeven, L. S. Wright, E. E. Capowski, I. Pinilla et al. 2011. "Optic vesicle-like structures derived from human pluripotent stem cells facilitate a customized approach to retinal disease treatment." *Stem Cells* no. 29 (8):1206–18.

Meyer, J. S., R. L. Shearer, E. E. Capowski, L. S. Wright, K. A. Wallace, E. L. McMillan, S. C. Zhang, and D. M. Gamm. 2009. "Modeling early retinal development with human embryonic and induced pluripotent stem cells." *Proc Natl Acad Sci USA* no. 106 (39):16698–703.

Michaud, J., G. N. Thompson, L. C. Brody, G. Steel, C. Obie, G. Fontaine, K. Schappert, C. G. Keith, D. Valle, and G. A. Mitchell. 1995. "Pyridoxine-responsive gyrate atrophy of the choroid and retina: Clinical and biochemical correlates of the mutation A226V." *Am J Hum Genet* no. 56 (3):616–22.

Miura, Y., and T. Endo. 2010. "Survival responses to oxidative stress and aging." *Geriatr Gerontol Int* no. 10 Suppl 1:S1–9.

Molday, R. S., M. Zhong, and F. Quazi. 2009. "The role of the photoreceptor ABC transporter ABCA4 in lipid transport and Stargardt macular degeneration." *Biochim Biophys Acta* no. 1791 (7):573–83.

Muniz, A., W. A. Greene, M. L. Plamper, J. H. Choi, A. J. Johnson, A. T. Tsin, and H. C. Wang. 2014. "Retinoid uptake, processing, and secretion in human iPS-RPE support the visual cycle." *Invest Ophthalmol Vis Sci* no. 55 (1):198–209.

Nakayama, A., M. T. Nguyen, C. C. Chen, K. Opdecamp, C. A. Hodgkinson, and H. Arnheiter. 1998. "Mutations in microphthalmia, the mouse homolog of the human deafness gene MITF, affect neuroepithelial and neural crest-derived melanocytes differently." *Mech Dev* no. 70 (1–2):155–66.

Nasonkin, I. O., S. L. Merbs, K. Lazo, V. F. Oliver, M. Brooks, K. Patel, R. A. Enke et al. 2013. "Conditional knockdown of DNA methyltransferase 1 reveals a key role of retinal pigment epithelium integrity in photoreceptor outer segment morphogenesis." *Development* no. 140 (6):1330–41.

Nguyen, M., and H. Arnheiter. 2000. "Signaling and transcriptional regulation in early mammalian eye development: A link between FGF and MITF." *Development* no. 127 (16):3581–91.

Nobukuni, Y., A. Watanabe, K. Takeda, H. Skarka, and M. Tachibana. 1996. "Analyses of loss-of-function mutations of the MITF gene suggest that haploinsufficiency is a cause of Waardenburg syndrome type 2A." *Am J Hum Genet* no. 59 (1):76–83.

Nordgaard, C. L., K. M. Berg, R. J. Kapphahn, C. Reilly, X. Feng, T. W. Olsen, and D. A. Ferrington. 2006. "Proteomics of the retinal pigment epithelium reveals altered protein expression at progressive stages of age-related macular degeneration." *Invest Ophthalmol Vis Sci* no. 47 (3):815–22.

North, V., R. Gelman, and S. H. Tsang. 2014. "Juvenile-onset macular degeneration and allied disorders." *Dev Ophthalmol* no. 53:44–52.

Osakada, F., H. Ikeda, Y. Sasai, and M. Takahashi. 2009a. "Stepwise differentiation of pluripotent stem cells into retinal cells." *Nat Protoc* no. 4 (6):811–24.

Osakada, F., Z. B. Jin, Y. Hirami, H. Ikeda, T. Danjyo, K. Watanabe, Y. Sasai, and M. Takahashi. 2009b. "In vitro differentiation of retinal cells from human pluripotent stem cells by small-molecule induction." *J Cell Sci* no. 122 (Pt 17):3169–79.

Peyman, G. A., K. J. Blinder, C. L. Paris, W. Alturki, N. C. Nelson, Jr., and U. Desai. 1991. "A technique for retinal pigment epithelium transplantation for age-related macular degeneration secondary to extensive subfoveal scarring." *Ophthalmic Surg* no. 22 (2):102–8.

Phillips, M. J., E. T. Perez, J. M. Martin, S. T. Reshel, K. A. Wallace, E. E. Capowski, R. Singh et al. 2014. "Modeling human retinal development with patient-specific induced pluripotent stem cells reveals multiple roles for visual system homeobox 2." *Stem Cells* no. 32 (6):1480–92.

Pingault, V., D. Ente, F. Dastot-Le Moal, M. Goossens, S. Marlin, and N. Bondurand. 2010. "Review and update of mutations causing Waardenburg syndrome." *Hum Mutat* no. 31 (4):391–406.

Raviv, S., K. Bharti, S. Rencus-Lazar, Y. Cohen-Tayar, R. Schyr, N. Evantal, E. Meshorer et al. 2014. "PAX6 regulates melanogenesis in the retinal pigmented epithelium through feed-forward regulatory interactions with MITF." *PLoS Genet* no. 10 (5):e1004360.

Raymond, S. M., and I. J. Jackson. 1995. "The retinal pigmented epithelium is required for development and maintenance of the mouse neural retina." *Curr Biol* no. 5 (11):1286–95.

Reardon, S., and D. Cyranoski. 2014. "Japan stem-cell trial stirs envy." *Nature* no. 513 (7518):287–8.

Reichman, S., A. Terray, A. Slembrouck, C. Nanteau, G. Orieux, W. Habeler, E. F. Nandrot, J. A. Sahel, C. Monville, and O. Goureau. 2014. "From confluent human iPS cells to self-forming neural retina and retinal pigmented epithelium." *Proc Natl Acad Sci USA* no. 111 (23):8518–23.

Rein, D. B., J. S. Wittenborn, X. Zhang, A. A. Honeycutt, S. B. Lesesne, and J. Saaddine. 2009. "Forecasting age-related macular degeneration through the year 2050: The potential impact of new treatments." *Archives of Ophthalmology* no. 127 (4):533–40.

Riento, K., and A. J. Ridley. 2003. "Rocks: Multifunctional kinases in cell behaviour." *Nat Rev Mol Cell Biol* no. 4 (6):446–56.

Riolobos, L., R. K. Hirata, C. J. Turtle, P. R. Wang, G. G. Gornalusse, M. Zavajlevski, S. R. Riddell, and D. W. Russell. 2013. "HLA engineering of human pluripotent stem cells." *Mol Ther* no. 21 (6):1232–41.

Rivolta, C., D. Sharon, M. M. DeAngelis, and T. P. Dryja. 2002. "Retinitis pigmentosa and allied diseases: Numerous diseases, genes, and inheritance patterns." *Hum Mol Genet* no. 11 (10):1219–27.

Rodriguez-Boulan, E., and I. G. Macara. 2014. "Organization and execution of the epithelial polarity programme." *Nat Rev Mol Cell Biol* no. 15 (4):225–42.

Rowan, S., C. M. Chen, T. L. Young, D. E. Fisher, and C. L. Cepko. 2004. "Transdifferentiation of the retina into pigmented cells in ocular retardation mice defines a new function of the homeodomain gene Chx10." *Development* no. 131 (20):5139–52.

Schwartz, S. D., J. P. Hubschman, G. Heilwell, V. Franco-Cardenas, C. K. Pan, R. M. Ostrick, E. Mickunas, R. Gay, I. Klimanskaya, and R. Lanza. 2012. "Embryonic stem cell trials for macular degeneration: A preliminary report." *Lancet* no. 379 (9817):713–20.

Schwartz, S. D., C. D. Regillo, B. L. Lam, D. Eliott, P. J. Rosenfeld, N. Z. Gregori, J. P. Hubschman et al. 2015. "Human embryonic stem cell-derived retinal pigment epithelium in patients with age-related macular degeneration and Stargardt's macular dystrophy: Follow-up of two open-label phase 1/2 studies." *Lancet* no. 385 (9967):509–16.

Schwarz, N., A. J. Carr, A. Lane, F. Moeller, L. L. Chen, M. Aguila, B. Nommiste et al. 2014. "Translational read-through of the RP2 Arg120stop mutation in patient iPSC-derived retinal pigment epithelium cells." *Hum Mol Genet.*

Sebastian, C., and R. Mostoslavsky. 2010. "SIRT3 in calorie restriction: Can you hear me now?" *Cell* no. 143 (5):667–8.

Shi, G., A. Maminishkis, T. Banzon, S. Jalickee, R. Li, J. Hammer, and S. S. Miller. 2008. "Control of chemokine gradients by the retinal pigment epithelium." *Invest Ophthalmol Vis Sci* no. 49 (10):4620–30.

Singh, R., D. Kuai, K. E. Guziewicz, J. Meyer, M. Wilson, J. Lu, M. Smith et al. 2015. "Pharmacological modulation of photoreceptor outer segment degradation in a human iPS cell model of inherited macular degeneration." *Mol Ther.* no. 23 (11):1700–11.

Singh, R., W. Shen, D. Kuai, J. M. Martin, X. Guo, M. A. Smith, E. T. Perez et al. 2013. "iPS cell modeling of Best disease: Insights into the pathophysiology of an inherited macular degeneration." *Hum Mol Genet* no. 22 (3):593–607.

Slavotinek, A. M. 2011. "Eye development genes and known syndromes." *Mol Genet Metab* no. 104 (4):448–56.

Someya, S., W. Yu, W. C. Hallows, J. Xu, J. M. Vann, C. Leeuwenburgh, M. Tanokura, J. M. Denu, and T. A. Prolla. 2010. "Sirt3 mediates reduction of oxidative damage and Prevention of age-Related hearing loss under caloric restriction." *Cell* no. 143 (5):802–12.

Sparrow, J. R., and M. Boulton. 2005. "RPE lipofuscin and its role in retinal pathobiology." *Exp Eye Res* no. 80 (5):595–606.

Steinfeld, J., I. Steinfeld, N. Coronato, M. L. Hampel, P. G. Layer, M. Araki, and A. Vogel-Hopker. 2013. "RPE specification in the chick is mediated by surface ectoderm-derived BMP and Wnt signalling." *Development* no. 140 (24):4959–69.

Swaroop, A., E. Y. Chew, C. B. Rickman, and G. R. Abecasis. 2009. "Unraveling a multifactorial late-onset disease: From genetic susceptibility to disease mechanisms for age-related macular degeneration." *Annu Rev Genomics Hum Genet* no. 10:19–43.

Tachibana, M., L. A. Perez-Jurado, A. Nakayama, C. A. Hodgkinson, X. Li, M. Schneider, T. Miki, J. Fex, U. Francke, and H. Arnheiter. 1994. "Cloning of MITF, the human homolog of the mouse microphthalmia gene and assignment to chromosome 3p14.1–p12.3." *Hum Mol Genet* no. 3 (4):553–7.

Taneri, B., E. Asilmaz, and T. Gaasterland. 2012. "Biomedical impact of splicing mutations revealed through exome sequencing." *Mol Med* no. 18:314–9.

Tassabehji, M., V. E. Newton, and A. P. Read. 1994. "Waardenburg syndrome type 2 caused by mutations in the human microphthalmia (MITF) gene." *Nat Genet* no. 8 (3):251–5.

Tsybovsky, Y., R. S. Molday, and K. Palczewski. 2010. "The ATP-binding cassette transporter ABCA4: Structural and functional properties and role in retinal disease." *Adv Exp Med Biol* no. 703:105–25.

Tucker, B. A., R. F. Mullins, L. M. Streb, K. Anfinson, M. E. Eyestone, E. Kaalberg, M. J. Riker, A. V. Drack, T. A. Braun, and E. M. Stone. 2013. "Patient-specific iPSC-derived photoreceptor precursor cells as a means to investigate retinitis pigmentosa." *Elife* no. 2:e00824.

Tucker, B. A., T. E. Scheetz, R. F. Mullins, A. P. DeLuca, J. M. Hoffmann, R. M. Johnston, S. G. Jacobson, V. C. Sheffield, and E. M. Stone. 2011. "Exome sequencing and analysis of induced pluripotent stem cells identify the cilia-related gene male germ cell-associated kinase (MAK) as a cause of retinitis pigmentosa." *Proc Natl Acad Sci USA* no. 108 (34):E569–76.

Turner, M., S. Leslie, N. G. Martin, M. Peschanski, M. Rao, C. J. Taylor, A. Trounson, D. Turner, S. Yamanaka, and I. Wilmut. 2013. "Toward the development of a global induced pluripotent stem cell library." *Cell Stem Cell* no. 13 (4):382–4.

van Lookeren Campagne, M., J. LeCouter, B. L. Yaspan, and W. Ye. 2014. "Mechanisms of age-related macular degeneration and therapeutic opportunities." *J Pathol* no. 232 (2):151–64.

van Meurs, J. C., E. ter Averst, L. J. Hofland, P. M. van Hagen, C. M. Mooy, G. S. Baarsma, R. W. Kuijpers, T. Boks, and P. Stalmans. 2004. "Autologous peripheral retinal pigment epithelium translocation in patients with subfoveal neovascular membranes." *Br J Ophthalmol* no. 88 (1):110–3.

van Soest, S., A. Westerveld, P. T. de Jong, E. M. Bleeker-Wagemakers, and A. A. Bergen. 1999. "Retinitis pigmentosa: Defined from a molecular point of view." *Surv Ophthalmol* no. 43 (4):321–34.

van Zeeburg, E. J., K. J. Maaijwee, T. O. Missotten, H. Heimann, and J. C. van Meurs. 2012. "A free retinal pigment epithelium-choroid graft in patients with exudative age-related macular degeneration: Results up to 7 years." *Am J Ophthalmol.* no. 153 (1):120–7.

von Lintig, J., P. D. Kiser, M. Golczak, and K. Palczewski. 2010. "The biochemical and structural basis for trans-to-cis isomerization of retinoids in the chemistry of vision." *Trends Biochem Sci.*

Wang, S., G. Birol, E. Budzynski, R. Flynn, and R. A. Linsenmeier. 2010. "Metabolic responses to light in monkey photoreceptors." *Curr Eye Res* no. 35 (6):510–8.

Watanabe, K., M. Ueno, D. Kamiya, A. Nishiyama, M. Matsumura, T. Wataya, J. B. Takahashi et al. 2007. "A ROCK inhibitor permits survival of dissociated human embryonic stem cells." *Nat Biotechnol* no. 25 (6):681–6.

Weber, B. H., G. Vogt, R. C. Pruett, H. Stohr, and U. Felbor. 1994. "Mutations in the tissue inhibitor of metalloproteinases-3 (TIMP3) in patients with Sorsby's fundus dystrophy." *Nat Genet* no. 8 (4):352–6.

Weisz, J. M., M. S. Humayun, E. De Juan, Jr., M. Del Cerro, J. S. Sunness, G. Dagnelie, M. Soylu, L. Rizzo, and R. B. Nussenblatt. 1999. "Allogenic fetal retinal pigment epithelial cell transplant in a patient with geographic atrophy." *Retina* no. 19 (6):540–5.

Wellen, K. E., and C. B. Thompson. 2010. "Cellular metabolic stress: Considering how cells respond to nutrient excess." *Mol Cell* no. 40 (2):323–32.

Westenskow, P., S. Piccolo, and S. Fuhrmann. 2009. "Beta-catenin controls differentiation of the retinal pigment epithelium in the mouse optic cup by regulating Mitf and Otx2 expression." *Development* no. 136 (15):2505–10.

Yang, J., Y. Li, L. Chan, Y. T. Tsai, W. H. Wu, H. V. Nguyen, C. W. Hsu et al. 2014. "Validation of genome-wide association study (GWAS)-identified disease risk alleles with patient-specific stem cell lines." *Hum Mol Genet* no. 23 (13):3445–55.

Yang, Z., B. V. Alvarez, C. Chakarova, L. Jiang, G. Karan, J. M. Frederick, Y. Zhao et al. 2005. "Mutant carbonic anhydrase 4 impairs pH regulation and causes retinal photoreceptor degeneration." *Hum Mol Genet* no. 14 (2):255–65.

Yasumoto, K., S. Amae, T. Udono, N. Fuse, K. Takeda, and S. Shibahara. 1998. "A big gene linked to small eyes encodes multiple Mitf isoforms: Many promoters make light work." *Pigment Cell Res* no. 11 (6):329–36.

Zarbin, M. A., R. P. Casaroli-Marano, and P. J. Rosenfeld. 2014. "Age-related macular degeneration: Clinical findings, histopathology and imaging techniques." *Dev Ophthalmol* no. 53:1–32.

Zarbin, M. A., and P. J. Rosenfeld. 2010. "Pathway-based therapies for age-related macular degeneration: An integrated survey of emerging treatment alternatives." *Retina* no. 30 (9):1350–67.

Zhang, Q., K. W. Small, and H. E. Grossniklaus. 2011. "Clinicopathologic findings in Best vitelliform macular dystrophy." *Graefes Arch Clin Exp Ophthalmol* no. 249 (5):745–51.

Zhong, X., C. Gutierrez, T. Xue, C. Hampton, M. N. Vergara, L. H. Cao, A. Peters et al. 2014. "Generation of three-dimensional retinal tissue with functional photoreceptors from human iPSCs." *Nat Commun* no. 5:4047.

Zuber, M. E., G. Gestri, A. S. Viczian, G. Barsacchi, and W. A. Harris. 2003. "Specification of the vertebrate eye by a network of eye field transcription factors." *Development* no. 130 (21):5155–67.

11

Modeling Neuroretinal Development and Disease in Stem Cells

Deepak A. Lamba

CONTENTS

11.1 Introduction

11.1.1 Retina and Its Development

The retina is the main light-sensing region of the eye. The light sensing is carried out by a group of cells lining the back of the retina called the *photoreceptors*. The cells are involved in the conversion of the light signal into a chemical signal by the process of phototransduction, which is passed onto the downstream inner retinal neurons. The retina has three main cellular layers: the outer nuclear layer where the photoreceptors reside; the inner nuclear layers, which contain the main excitatory inner neurons called the *bipolar cells* along with two types of inhibitory interneurons called the *horizontal* and *amacrine cells*; and finally, the ganglion cell layer, which contains the retinal ganglion cells (RGCs) whose axons project to the central visual cortical areas (Figure 11.1). As stated previously, photoreceptors are the light-sensing cells of the retina. There are two main types of photoreceptors in the vertebrate retina, the rod and the cone. Cone photoreceptors are born earlier than the rods and are critical for high acuity vision and crowd the central part of the retina. One of the earliest photoreceptor genes expressed

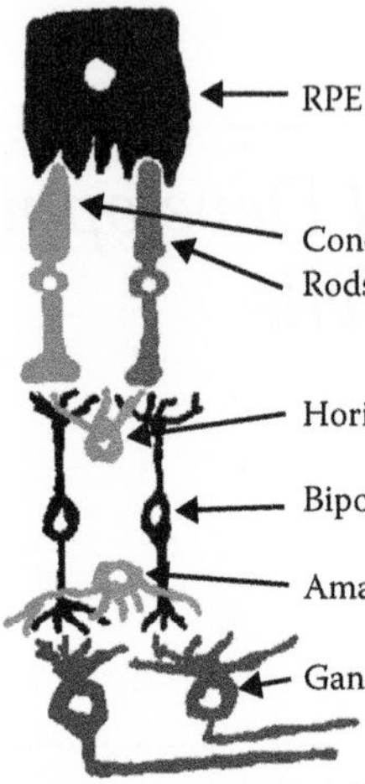

FIGURE 11.1
Cartoon showing key cell types in the mammalian retina and disorders associated with some of them. AMD, age-related macular degeneration; LCA, Leber congenital amaurosis; RP, retinitis pigmentosa.

is cone–rod homeobox gene (*Crx*) (Furukawa et al., 1997). Following specification, the photoreceptors take up either a rod or a cone fate. The cone photoreceptors subsequently decide between short and medium wavelengths (and long wavelength in Old World primates like humans) (Hunt et al., 1998).

The eye is essentially a highly specialized extension of the forebrain. In vertebrates, the eyes appear as bilateral evaginations from the diencephalon. In mammals, this appears first in the form of the optic pit. The continued evagination results in the optic vesicles, which come into close contact with the overlying ectoderm. Interaction between the two tissues induces the formation of the lens and the cornea in the overlying ectoderm. The head of the optic vesicle forms a two-layered optic cup following the indentation of the tip. The outer layer of the cup gives rise to the retinal pigment epithelium, while the inner layer undergoes proliferation to form the multilayered neural retina, which consists of the various retinal neuronal cells and the Müller glia. The order of the generation of the retinal neurons has been shown to be conserved across species (Sidman, 1961). Ganglion cells are the first to differentiate. This is followed by the cone photoreceptors and the horizontal cells and then the amacrine cells. The final wave of differentiation consists of the rod photoreceptors, the bipolar cells, and, finally, the Müller glia.

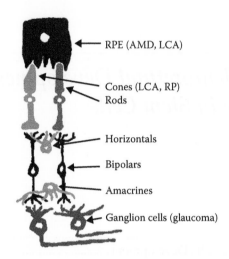

The presumptive eye field is specified prior to the development of the optic vesicles. This eye field specification in the neural plate is related to a group of transcription factors called the *eye field transcription factors* (EFTFs). These include paired box 6 (*Pax6*) or eyeless, retina and anterior neural fold homeobox (*Rx*; *Rax*), LIM homeobox 2 (*Lhx2*), SIX homeobox 3 (*Six3*), and SIX homeobox 6 (*Six6*) (Zuber et al., 2003). These EFTFs very likely induce or cross-regulate the expression of each other promoting an eye fate. While we know much about EFTFs and early eye development, we know less about the role of extracellular signaling molecules and how they regulate EFTFs. Various groups have looked into the role of Wnt signaling in the initiation and the regulation of the eye fields (Rasmussen et al., 2001; Cavodeassi et al., 2005). Wnt signaling can be divided into a canonical, β-catenin-dependent pathway and a noncanonical, β-catenin-independent pathway, and ligands and receptors for both pathways are expressed at the site of the prospective eye field. Wnt1 or Wnt8b activates the canonical Wnt-β-catenin pathway and can lead to a reduction in the size of the eye fields when overexpressed in *Xenopus* embryos. This seems to occur due to the suppression of Rx and Six3 expression. On the other hand, Wnt11 activates the noncanonical pathway and results in larger eyes in *Xenopus* when overexpressed (Cavodeassi et al., 2005). The overexpression of the Wnt receptor Frizzled-3 (Fz3) in *Xenopus* results in the formation of multiple ectopic eyes. Here, Fz3 is believed to preferentially activate the noncanonical Wnt pathway (Rasmussen et al., 2001). Wnt4 is also required for *Xenopus* eye formation and likely acts via Fz3 to induce the expression of ELL-associated factor 2 (EAF2), which in turn regulates Rx expression in *Xenopus* (Maurus et al., 2005). Loss of EAF2 function results in loss of the eyes, while loss of Wnt-4 function is rescued by EAF2 overexpression.

Antagonizing both Wnt and BMP signaling also seems important for eye development. Dickkopf1 (Dkk1) is a potent antagonist of canonical Wnt signaling, and mice carrying null deletions of Dkk1 lose all cranial structures anterior to the midbrain, including the eyes (Mukhopadhyay et al., 2001). Noggin (*Nog*), a known inhibitor of BMP signaling, is believed to play an important role in neural induction and eye field formation (Lamb et al., 1993; Zuber et al., 2003). In animal cap assays, Nog causes increased expression of many EFTFs including Pax6, Six3, Rx, Lhx2, and Optx2 (Zuber et al., 2003). In mice, Dkk1 and Nog pathways synergize to induce head formation during gastrulation by dually antagonizing Wnt and BMP signaling, acting as a head organizer (del Barco Barrantes et al., 2003); loss of one copy each of Dkk1 and Nog also results in total loss of the anterior head. Insulin-like growth factors (IGF-1) are believed to play an important role in head and eye formations (Pera et al., 2001), and it is achieved by the inhibition of canonical Wnt signaling (Richard-Parpaillon et al., 2002) via kermit2 (*Gipc1*), which is both an IGF receptor- and a Frizzled receptor-interacting protein (Wu et al., 2006). Recently, work using mouse embryonic stem cells (mESCs) has confirmed the importance of the PDZ domain-containing protein GIPC PDZ domain-containing family member 1 (GIPC1) (La Torre et al., 2015). They showed that the overexpression of a dominant negative form of GIPC PDZ domain-containing family member 1 (dnGIPC1), as well as the downregulation of endogenous GIPC1, is sufficient to inhibit the development of eye field cells from mESCs. Here, GIPC1 interacts directly with IGF receptor and participates in Akt1 activation. This was confirmed by experiments where the pharmacological inhibition of Akt1 phosphorylation mimicked the dnGIPC1 phenotype. After the initial formation of the eye field, Sonic hedgehog (Shh) signaling from the midline splits the eye field to form two eyes (Chiang et al., 1996). Shh has been shown to repress Pax6 expression via ventral anterior homeobox 2 (Vax-2), a homeodomain transcription factor (Li et al., 1997; Kim and Lemke, 2006). Loss of Shh in the midline results in cyclopia, the development of a single midline eye (Chiang et al., 1996). Thus, a balance largely between Wnt and BMP signaling activations/inhibitions by localized extracellular signaling centers is crucial in regulating eye field induction and morphogenesis during early stages of development.

Even less is known about cell intrinsic factors regulating EFTF activity. CCCTC-binding factor (CTCF), a transcriptional regulator, has been shown to control Pax6 transcription by interacting with a repressor element located in the 5′-flanking region upstream of the Pax6 P0 promoter. The overexpression of CTCF results in the suppression of *PAX6* gene expression, which in turn leads to a small eye phenotype similar to those in mice with Pax6 mutation (Li et al., 2004). Also, members of the three amino acid loop extension (TALE) family of homeobox proteins (Meis homeobox 1 [Meis1], Meis 2 [*Mrg1*], Prep1 [*Pknox1*]) have been implicated in the regulation of Pax6 expression. Loss of Meis1/2 homeobox activity in the prospective lens ectoderm represses Pax6 expression. This in turn affects lens placodal development (Zhang et al., 2002). Meis1 and Prep1 are also expressed in the retina (Zhang et al., 2002; Ferretti et al., 2006). Loss of Prep1 activity concomitantly leads to loss of Meis1 and pre-B-cell leukemia homeobox (Pbx) 1/2 (TALE activation partners) expression, resulting in *Pax6* gene expression levels (Ferretti et al., 2006). Recent studies using explant culture models show that LHX2 integrates extrinsic cues, such as BMPs including BMP4 and BMP7 with intrinsic transcription factors such as visual system homeobox 2 (*VSX2*) and SRY-box 2 genes, serving an important role in the regulatory network for optic morphogenesis and retinal differentiation (Yun et al., 2009; Gordon et al., 2013).

Another recently emerging signaling pathway with roles in the specification of neural tissues is the *Notch* pathway. The overexpression of a constitutively active Notch internal cytoplasmic domain in *Xenopus* embryos induced the expression of Pax6 and the formation of ectopic eyes (Onuma et al., 2002). Mouse genetic studies of Notch pathway components also suggest an early role for some in eye development. Hes family bHLH transcription factor 1 (Hes1), a key component of the Notch signal transduction pathway, is expressed at E8.5 in the anterior neural plate and subsequent stages of optic cup formation (Lee et al., 2005). Loss of Hes1 alone results in microphthalmia (Tomita et al., 1996), but the combined loss of Hes1 and Pax6 leads to a complete loss of optic cup formation, although Rx expression is not affected (Lee et al., 2005). Loss of Hes5, another component of the Notch pathway, which is expressed later in neuroretinal progenitors, leads to a reduction in the number of Müller glia cells without any obvious microphthalmia (Hojo et al., 2000). On the other hand, a combined loss of Hes1 and Hes5, which leads to loss of most of the Notch pathway output, results in the complete failure of optic cup formation (Hatakeyama et al., 2004). The likely ligand for the Notch pathway at these early stages of eye development is Jagged1, which is expressed in the early optic cup and becomes restricted to the peripheral retina later (Lindsell et al., 1996; Bao and Cepko, 1997). Recent work suggests that the different Notch pathway components regulate downstream retinal progenitor genes differentially (Maurer et al., 2014). They showed that Notch1 and recombination signal binding protein for immunoglobulin kappa J region (Rbpj) block atonal bHLH transcription factor 7 (*Atoh7*) and *Neurog2* gene expression, while Notch1, Notch3, and Rbpj regulate Neurog2 especially in the distal retina. Also, the effectors of Notch regulate genes differentially such that Hes1 repressor mediates Atoh7 suppression while not affecting Neurog2 expression.

The preceding analysis highlights how the eye field is defined through the expression of a set of transcription factors, which are constrained to this domain by multiple signaling factors.

11.1.2 RGC Development

RGCs are one of the first neurons to arrive from the retinal stem cells during mammalian development. These cells are normally generated by the eye in excess and get pruned out as appropriate connectivity develops with the other generated retinal neurons. A number of key transcription factors are suggested to play a role in their generation and fate specification. One of the main RGC fate specifier genes is *ATOH7* (Math5, Cath5). ATOH7 is the ortholog of the *Drosophila* atonal gene, which is a basic helix-loop-helix transcription factor. It is expressed transiently in the mammalian retina around the time of RGC genesis with overlapping expression pattern (Brown et al., 1998). Studies in mice lacking the *Math5* gene confirmed a severe reduction in ganglion cell formation with a very tiny optic nerve (Brown et al., 2001; Le et al., 2006). Another transcription fact playing a critical role in RGC fate is POU4F2 (POU class 4 homeobox 2) (Brn3b). The knockout of Brn3b results in up to a 70% loss of ganglion cell numbers in adult mice even though they are born normally, suggesting that this gene is important for RGC survival (Wang et al., 2002; Mu et al., 2004; Pan et al., 2005). Interestingly, Math5 knockout mice have a complete loss of Brn3b-expressing cells, suggesting that Math5 lies upstream of Brn3b in RGCs (Liu et al., 2001). Another transcription factor, ISL LIM homeobox 1 (ISL1), is similarly expressed in ganglion cells, and mice lacking Isl1 have normal ganglion cell birth, but they soon die thereafter by apoptosis (Elshatory et al., 2007; Pan et al., 2008). Mice lacking two other related transcription factors, Sox4 (SRY-box 4) and Sox11 (SRY-box 11), show similar phenotype to Isl1 knockout mice (Jiang et al., 2013).

11.1.3 Photoreceptor Development

Photoreceptors are the main light-sensing cells of the retina. There are two main types of photoreceptors in the vertebrate retina, the rod and the cone. Following specification, the photoreceptors take up either a rod or a cone fate. Cone photoreceptors are born earlier than the rods. The cone photoreceptors subsequently decide between short and medium wavelengths (and long wavelength in Old World primates like humans) (Hunt et al., 1998).

Photoreceptor development is controlled by several different types of transcription factors. The homeodomain transcription factor orthodenticle homeobox 2 (Otx2) is an early factor that biases progenitor cells to become photoreceptors: conditional deletion of Otx2 leads to loss of photoreceptors; retroviral gene transfer of Otx2 into retinal progenitors promotes the photoreceptor fate (Nishida et al., 2003). Otx2 also activates the transcription of CRX (Furukawa et al., 1997), which is also required for the expression of photoreceptor-specific genes, including the opsins. In mice lacking Crx, photoreceptor cells do differentiate but completely fail to mature and eventually undergo apoptosis. Blimp1, a third critical and newly identified factor (Brzezinski et al., 2010) in mice, is shown to critically bias photoreceptor fate over bipolar cell fate during development. In mice with conditional Blimp1 deletion, there is a reduction in photoreceptor formation with a corresponding increase in differentiation of bipolar cells in the inner retina. Blimp1 has been shown to specifically repress VSX2 expression, which is required for the bipolar cell fate by binding to the enhancer region of VSX2 and blocking its activity (Katoh et al., 2010). A cis-regulatory element, B108, has recently been described as critical for Blimp1 expression and function (Wang et al., 2014). Otx2 and retinoic acid-related orphan receptor β (RORβ) regulate Blimp1 expression via B108. Additionally, the studies show that Blimp1 and Otx2 formed a negative feedback loop that regulates the level of Otx2, which in turn regulates the production of the correct ratio of rods and bipolar cells.

The Otx2/Crx/Blimp1-expressing photoreceptors develop as either rods or cones depending on their expression of at least two key rod fate specifying transcription factors: neural retina leucine zipper (Nrl) and nuclear receptor subfamily 2 group E member 3 (Nr2e3). Nrl (neural leucine zipper1; Swaroop et al., 1992) is a member of the Maf family of transcriptional activators (Friedman et al., 2004). Nrl binds to and activates the promoters of a number of key rod-specific genes (Kumar et al., 1996; Rehemtulla et al., 1996) and is required for rod photoreceptor development; deletion of this gene results in a complete loss of rod cell fate. In this mouse, the photoreceptors instead adopt a cone-like fate with predominantly S-opsin expression (Mears et al., 2001; Daniele et al., 2005). Downstream of Nrl lies another transcription factor, Nr2e3, a member of the nuclear hormone receptor family of transcription factors. Patients with mutations in *Nr2e3* gene result in cases of the so-called enhanced S-cone syndrome in humans (Haigh et al., 2003), and a similar phenotype has been identified in mice (Haider et al., 2001). Like Nrl mutant mice, retinas from mice lacking *Nr2e3* develop excess cone-like photoreceptors at the expense of rods (Haider et al., 2001; Chen et al., 2005; Corbo and Cepko, 2005). Nr2e3 also acts as a transcriptional activator of a number of rod-specific genes including rhodopsin and likely occupies these regions in association with both Nrl and Crx (Chen et al., 2005). Interestingly, Nrl and Nr2e3 have unique roles to play in rod maturation as the replacement of Nrl with Nr2e3 results in the activation of some, but not all, of the rod-specific gene profile (Cheng et al., 2006). It has also been confirmed that Nr2e3 is a strong repressor of the cone fate. In mice, the overexpression of Nr2e3 under the control of the S-opsin promoter causes the developing cones to switch their fate and develop into rod photoreceptor-like cells (Cheng et al., 2006).

11.2 Retinal Degenerative Disorders

The two main cell types involved in retinal degenerations include the RGCs and the photoreceptor cells. Ganglion cell loss is a common complication of glaucoma, while photoreceptor loss occurs secondary to inherited mutations as observed in retinitis pigmentosa (RP) and Leber congenital amaurosis (LCA) or secondary to age-associated changes as observed in age-related macular degeneration (AMD). In the following sections, we will discuss both these broad degenerations.

11.2.1 Glaucoma

Glaucoma is a progressive degenerative retinal disorder characterized by the death of RGCs and is the second leading cause of blindness in the world. Glaucoma currently affects over 60 million people worldwide and is expected to affect over 74 million people by 2020. The ganglion cells, which receive signal from the light-sensing cells in the eye, send the signal to the brain through the optic nerve. In the human retina, there are approximately one million RGCs, which extend their axons through the optic nerve. Once lost, these ganglion cells are not regenerated. Unfortunately, the degenerative process is slow with the patients not appreciating any significant vision loss until much later in the disorder. Thus, the resultant blindness in late stages of the disease is severe and permanent.

The cause of glaucoma is not very well understood. The disorder is multifactorial, although many cases are linked to increased pressure within the eye and a subsequent loss of blood supply, leading to ganglion cell death (Qu et al., 2010).

There are treatments available for glaucoma, administered via eye drops, which lower the pressure in the eye. These are particularly effective at the early stages of the disease. However, a number of patients do not respond well to these drops and often require combination therapy. Additionally, in later stages of the disease when a number of the ganglion cells are lost, the drops, at the most, help prevent further deterioration. There are no effective treatments to restore lost vision. Developing novel approaches is urgently required to restore RGC loss in patients with advanced glaucoma in order to recover lost visual function.

11.2.2 Photoreceptor Degenerative Disorders

RP and LCA encompass a group of inherited retinal degenerations with a prevalence of 1 in 4000–6000. They are progressive degenerative disorders. RP usually begins in adolescents and young adults with night blindness followed by decreasing visual fields, leading to tunnel vision and eventually legal blindness or, in many cases, complete blindness. LCA, on the other hand, appears at birth or in the first few years of life and accounts for more than 20% of children attending schools for the blind. Both RP and LCA occur due to mutations in either the photoreceptor genes or the retinal pigment epithelial (*RPE*) genes.

AMD is the leading cause of worldwide blindness in the elderly affecting almost 15 million people in the United States. AMD changes are present in approximately 10% of people over 65 and as many as one in three people over the age of 80. The pathogenesis of AMD is unclear, and multiple factors may be involved in the progression of the disorder including chronic oxidative stress, genetic factors, small RNA processing, choriocapillary atrophy, lipofuscin accumulation and inflammation, and complement activation. The disease starts in the macula, the region of central retina with highest visual acuity. There are two main forms of the disorder: dry AMD with choriocapillary atrophy, and wet form which is associated with exudative changes and neovascularization. Dysfunction in the RPE seems to be the primary cause of photoreceptor cell loss. A number of studies have associated oxidative stress as the key driver of AMD (Cai et al., 2000; Jarrett and Boulton, 2012). The retina has one of the highest oxygen consumption of any tissue in the body. This results in significant reactive oxygen species (ROS) production. Increasing age results in the loss of ability to deal with excessive ROS, leading to oxidative damage. Evidence pointing to this includes (a) decrease in microsomal glutathione *S*-transferase-1, (b) increased lipid peroxidation, and (c) age-associated decrease in catalase activity in the human RPE. However, there is increasing evidence of the involvement of choroidal vasculature and cross-interaction with RPE in the pathogenesis of both wet (neovascular) and dry (capillary atrophy) forms of AMD (Lutty et al., 1999; McLeod et al., 2009; Mullins et al., 2011; Biesemeier et al., 2014). Despite the loss of photoreceptors, the inner retinal circuitry is intact for many years, and this has led to the possibility of photoreceptor replacement as a potential therapy.

11.3 Pluripotent Stem Cell Differentiation to Generate Retinal Cells

Mouse and human PSCs have been shown to be induced to generate retinal stem cells and, subsequently, various retinal neurons by a number of groups to date. The earliest demonstration of a directed stepwise differentiation protocol using mESCs was demonstrated by Sasai's group in 2005 (Ikeda et al., 2005). By using Wnt and nodal signaling antagonists under serum-free culture conditions followed by serum and activin A, the group could promote retinal stem cell fate in about 16% of cells. These cells could subsequently be differentiated in the various retinal neurons including photoreceptors and ganglion cells. RGCs differentiated within 10 days of differentiation and expressed markers including ELAV-like protein (Hu), class III beta-tubulin (Tuj1), and POU4F2/3 (Brn3). Subsequently, protocols were generated using either coculture (Aoki et al., 2006), *Pax6* gene transfection (Kayama et al., 2010), or timing-specific growth factors such as Shh and bFGF (Jagatha et al., 2009).

The work from our lab was the first to describe a defined directed differentiation protocol to generate retinal cells from hESCs (Figure 11.2). Similar to what was observed in mice, Wnt inhibition combined with inhibition of the BMP pathway but activation of IGF-1 signaling promoted retinal differentiation in 70–80% of all the cells (Lamba et al., 2006). These cells could then be expanded over two to three months to generate both RGCs and different photoreceptors. The same protocol could also be applied to generate retinal neurons from hiPSCs (Lamba et al., 2010). It has also been used and modified by a number of other labs working on generating retinal neurons (Bae et al., 2012; Hambright et al., 2012; Mellough et al., 2012; Lidgerwood et al., 2015). We next tested the efficiency of the photoreceptor generated *in vitro* to integrate into a host retina (Lamba et al., 2009) (Figure 11.3). Remarkably, in a wild-type eye, photoreceptors integrated in the right layer and were able to extend inner and outer segments. Finally, we tested whether they had the potential to restore light responsiveness in a congenitally blind mouse. Crx mutant mice lack a key transcription factor rendering them blind due to the absence of key phototransduction cascade components. Upon transplantation, we observed that the transplanted cells integrated and made synaptic contact with host photoreceptors. Upon testing their vision using an electroretinogram, the transplanted eyes had an electrophysiological response to light exposure. The b-wave observed is a response of the host inner retinal neurons, suggesting that they are functionally connected to transplanted human photoreceptors.

A seminal work came out of the Sasai lab using mESCs to generate self-organizing optic cups (Eiraku et al., 2011; Eiraku and Sasai, 2012) (Figure 11.4). The authors described a methodology to differentiate ESCs

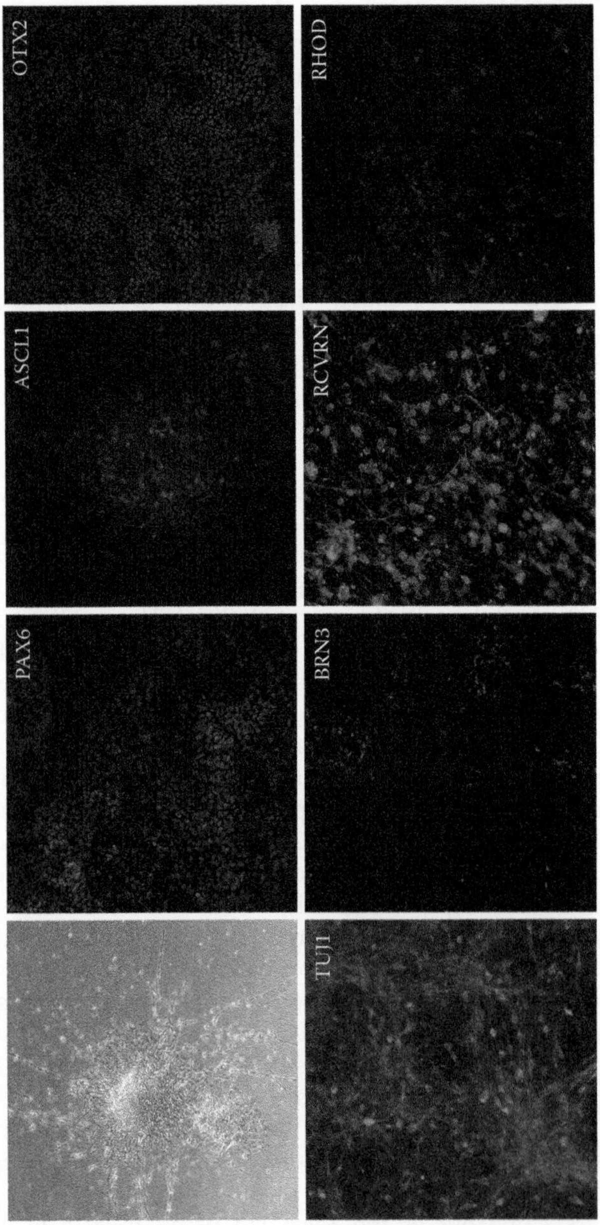

FIGURE 11.2
(See color insert.) Characterization of 2D differentiated retinal cells from PSCs.

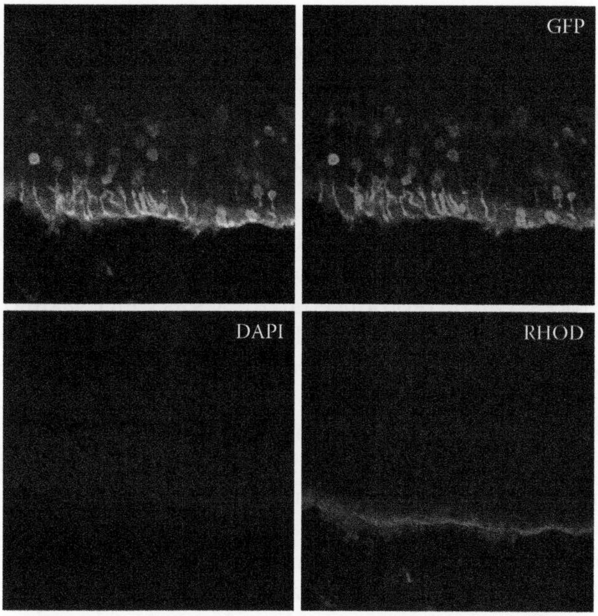

FIGURE 11.3
(See color insert.) Characterization of 3D differentiated retinal cells from PSCs.

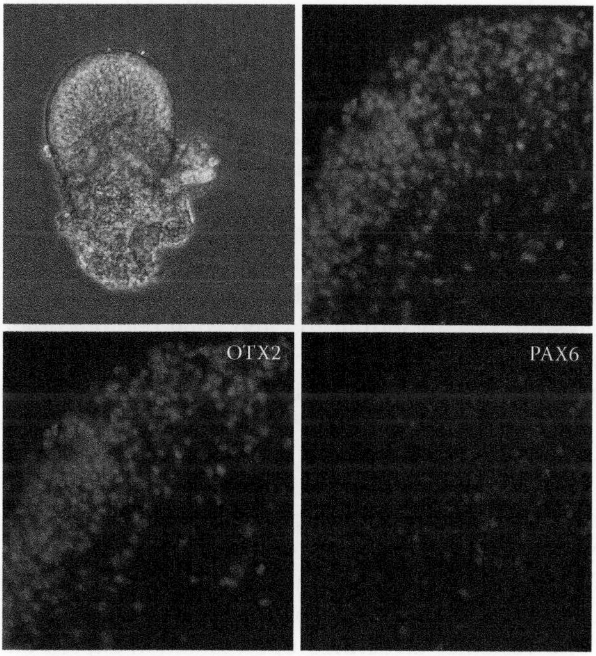

FIGURE 11.4
(See color insert.) Integration of stem cell-derived retinal photoreceptors in a mouse retina.

in low-attachment plates in order to allow them to self-organize. Within a matter of a week, vesicle-like extensions protruded from the sides of the 3D cultures. The authors used an Rx-green fluorescent protein (GFP) line to easily identify these structures, which could then be pinched off. The optic vesicles then matured to form bilayered optic cups as in the embryo such that the outer layer formed MITF+ RPE and the inner layer differentiated into VSX2+ neural retina. The neural retina subsequently matured to form a multilaminated retina with appropriately oriented photoreceptor layer and inner retina. This was soon followed by another report by the group showing that the process can be reproduced in hESCs as well (Nakano et al., 2012). The authors used ROCK pathway inhibitor to allow single-cell survival and Wnt pathway inhibitor, IWR-1-endo, promote optic vesicle formation. To further promote neural retinal formation, the 3D clusters were treated with 10% fetal bovine serum and Hedgehog pathway component smoothened agonist. Interestingly, the addition of a small-molecule Wnt agonist at this stage of the protocol promoted RPE over the neuroretinal fate in the optic cups. Upon further differentiation for over 120+ days in 3D low-attachment conditions, the tissue formed a multi-laminated retina with the presence of all major cell types including rods, cones, retinal interneurons, and RGCs in mostly the right layers. The addition of a small-molecule Notch pathway inhibitor promoted stage-specific synchronized differentiation. The group recently published a report showing that the addition of BMP4 from days 6–15 of the protocol enhanced neuroretinal fate in these cells and prevented Foxg1+ telencephalic neuroepithelium induction (Kuwahara et al., 2015). The addition of glycogen synthase kinase 3 inhibitors along with FGF receptor inhibitors from days 18–24 promoted RPE fates and could be reversed by the change of media back to that containing retinoic acid and taurine. A very recent report analyzed the ability of these 3D spheres to integrate in immune-deficient nude rats and in two primate models of retinal degeneration (Shirai et al., 2016). The primate models were generated either by subretinal injection of cobalt chloride or by laser photocoagulation. Although the authors reported that the grafted hESC retina was observed differentiating into various retinal neuron subtypes, including rod and cone photoreceptors following transplantation, the integration appeared to be limited at the sites of apposition only. The photoreceptors tended to stay in the sphere and not migrate into the host retina.

Similar 3D differentiation has been described by David Gamm's group (Meyer et al., 2011; Phillips et al., 2012). The group used a slightly different approach on inducing generalized neural differentiation using media containing N2 and B27 supplements. This was followed by allowing the aggregates to self-assemble. They found that the clusters had either a translucent vesicular morphology or a more opaque appearance in brightfield microscopy. The vesicular structures upon isolation expressed markers of optic vesicles including Pax6, Rax, Lhx2, and Six6, while the nonvesicular structures had other central nervous system markers including distal-less homeobox 1/2 and NKX2.1. The vesicular structures went to mature into multilaminated retina with all known retinal cell types. Further modification of the protocol by others have allowed them to further mature the iPSC-derived retinal photoreceptors by late addition of retinoic acid to the media at week 10 of differentiation in these cultures, leading to the formation of outer segment-like extensions with RHO expression (Zhong et al., 2014). This was confirmed by electron microscopy analysis to show organized disks in the outer segment region and expression of phototransduction proteins including phosphodiesterases, cyclic nucleotide-gated channels, and transducin. Others have used the protocol to generate RGCs at high efficiency using brain derived neurotrophic factor (BDNF) (Tanaka et al., 2015). The ganglion cell axons generated by this protocol following plating on coated plates exhibited anterograde axonal transport and sodium-dependent action potentials.

The retinas generated from ESCs and iPSCs have a huge potential to be used for better understanding of the steps involved in retinal development. This, however, requires a confirmation that retinal cells generated *in vitro* faithfully follow normal developmental stages and timelines. Toward this, an initial study was carried out directly comparing *in vitro* retinal differentiation and human fetal retina by using microarray technology (Lamba and Reh, 2011). This study confirmed that the *in vitro* two-dimensional (2D) retinal cultures were at similar developmental stages as the corresponding human fetal retinas and varied very little (less than 1% genes) from them. A more recent study was recently published detailing the transcriptional changes in GFP-expressing photoreceptors with age by using RNA-sequencing technology (Kaewkhaw et al., 2015). They reported that a number of photoreceptor-specific genes exhibited a gradual increase in expression with time. Additionally, as expected during human fetal retinal development, cone genes including GNAT2, GNB3, RXRG, ARR3, GNGT2, and CNGA3 showed higher expression by day 67 prior to the expression of rod-specific genes including NRL, NR2E3, and GNAT1 by day 90, further validating the genesis of cones before rods. The analysis also identified a number of putative cell surface molecules such as prominin, SLC6A17, SLC40A1, KCNH2, RTN4RL1, ST3GAL5, GNGT2, and EPHA10 that could be used for cell sorting.

11.4 Modeling Photoreceptor Degenerations *In Vitro*

Patient-derived iPSCs have the potential of being ideal models to better understand the pathophysiology of the degeneration *in vitro*. Thus, they may provide the opportunity to design or identify drugs for specific forms of degeneration. Toward this, a few groups have looked at photoreceptor degeneration in specific patient-derived lines. A report out of the Masayo

Takahashi lab compared iPSC lines from five different patients with distinct mutations in the *RP1, RP9, PRPH2,* or *RHO* genes (Jin et al., 2011). They showed that all the different lines had the ability to generate both RPE and neural retina. However, the different patient lines lacked the ability to either generate or maintain the expression of a key rod photoreceptor marker, RHO, at 150 days in culture. A number of these lines had markers of either oxidative stress accumulation, DNA oxidation, apoptosis, or endoplasmic reticulum (ER) stress-associated markers, binding immunoglobulin protein (BIP), and CCAAT enhancer-binding protein homologous protein (CHOP). Finally, they tested certain antioxidant vitamins that had a neuroprotective role in preserving rod photoreceptors. They found that, in their lines, only α-tocopherol was protective in only one of the mutant lines (RP9 mutations), while the other vitamins were ineffective in rod preservation.

Another study looked at iPSC line generated from a defined (E181K) mutation in the RHO gene (Yoshida et al., 2014). As a control for the experiments, the authors used genome editing to correct the mutation using an adenoviral vector system. This allowed the group to compare changes more directly due to the isogenic background. The iPSC lines were differentiated to neuroretinal lineage, and rod photoreceptors identified by the infection of the culture with an adenoviral vector-containing Nrl promoter-driven enhanced GFP. These cells were then purified using flow cytometry after five weeks of culture. The study found a reduced survival rate in the rod photoreceptor cells generated from iPSC lines with the E181K mutation. This correlated with the increased expression of both ER stress and apoptotic markers. Finally, the cell lines were used to screen potential therapeutic agents for their ability to promote photoreceptor survival. This allowed them to identify a number of agents including rapamycin as potential rod photoreceptor-protective compounds, which resulted in the reduction in ER stress markers along with a concomitant reduction in autophagy. This study provides a proof-of-principle support to the advantage of the use of patient-derived iPSCs for disease understanding and drug discovery.

The modeling of a form of sporadic RP was demonstrated using iPSC lines for a patient with splicing in the male germ cell-associated kinase (*MAK*) gene (Tucker et al., 2011). This was sequenced and shown to be due to the presence of an Alu insertion in the ninth exon of the gene. The gene splicing leads to a loss of the MAK transcript, presumably via nonsense-mediated decay, resulting in impaired photoreceptor maturation and eventual degeneration. The study also identified a novel splice variant of the *MAK* gene that contains an extra 75 base pairs in frame, which was dependent on the expression of the exon 9-containing splice form. The group used a similar strategy in another publication to look at a novel intronic USH2A variant in a patient with nonsyndromic RP (Tucker et al., 2013). Analysis of the patient cell-derived retinal cell gene expression showed that the mutation resulted in the conversion of a portion of the intron into an exon, leading to a frameshift and a premature stop codon in the *USH2A* gene. This was combined with a second mutation in the other allele (Arg4192His). They reported that this leads to the elevation of ER stress markers heat shock protein family A (Hsp70) member 5 (Hspa5) and heat shock protein 90 kDa beta family member 1 (HSP90B1). Another study looked at modeling mutations in the RP GTPase regulator (*RPGR*) gene, which cause 20% of all cases (Megaw et al., 2015). iPSC lines were generated from patients with RPGR mutations and their unaffected relatives and differentiated using the 3D differentiation protocols for up to 100 days. The group reported that photoreceptor cultures from RPGR-mutated iPSCs had increased actin polymerization compared with control lines. Using both Western blot and unbiased protein arrays, there was evidence of reduction in both Src and extracellular signal–regulated kinases (ERK) phosphorylations in RPGR-mutated photoreceptor culture, suggesting that RPGR mutations lead to actin dysregulation.

iPSCs have also been used to study the efficacy of gene therapy *in vitro* in human cells. In a study by the Tucker lab, the group generated iPSCs from patients with a form of LCA due to mutation in the centrosomal protein 290 (*CEP290*) gene, which leads to an autosomal recessive form of the disease (Burnight et al., 2014). In this study, the authors attempted to test the feasibility of gene therapy *in vitro* using lentiviral vectors carrying cytomegalovirus-driven full-length human CEP290. The authors showed that the patient-derived lines had a defect in ciliogenesis upon analysis of their fibroblasts. The cilia were less numerous and shorter compared to those of the controls, and this defect could be rescued following transduction with lentiviruses containing normal full-length CEP290. This study highlights the advantage of the patient iPSCs as a tool to test the therapeutic potential of gene therapy. Others have attempted the repair of RP mutations using CRISPR/Cas9 technology. The Mahajan lab tested whether CRISPR/Cas9 could be used in patient-specific iPSCs to precisely repair an RPGR point mutation that causes X-linked RP (Bassuk et al., 2016). The iPSCs were transduced with CRISPR guide RNAs, Cas9 endonuclease, and donor homology template, and this resulted in mutation correction in 13% of *RPGR* gene copies and conversion to the wild-type allele. The group, however, did not test if this resulted in improvement in photoreceptor survival *in vitro* between the disease and gene-corrected lines.

One recent study looked at ganglion cell disease modeling *in vitro* using iPSC lines (Chen et al., 2016). The authors used iPSCs to model autosomal dominant optic atrophy, which is the most common hereditary optic atrophy, characterized by the degeneration of RGCs, and has been strongly associated with mutations in optic atrophy 1 (*OPA1*) gene. Upon retinal differentiation of iPSC lines with the mutant, OPA1 exhibited a significant increase in apoptosis at the retinal progenitor stage already with loss of RGC differentiation. Interestingly, BMP pathway inhibitor, Nog, and 17β-estradiol rescued ganglion cell differentiation.

11.5 Conclusion and Future of Disease Modeling and Repair

Much progress has been made in developing protocols for the differentiation of retinal neurons from undifferentiated cells. The focus has now moved to validating the use of these cells for replacement therapies and using them to better model retinal degenerations. The biggest hurdle for developing accurate models is identifying the means to induce the phenotype. Most retinal degenerations affecting the photoreceptors or the RGCs are adult-onset degenerations. Some forms of LCA do appear soon after birth, but they too would require very long-term cultures. Thus, there is a need to modify culture conditions that will force premature appearance of the degenerative phenotype. One could postulate what such conditions could include such as light-induced stress, ER-specific stressor, or mitochondria-specific stress including ROS for RP or pressure induction chambers for high-pressure glaucoma. These will likely speed up the pathophysiology, allowing researchers to model other aspects of degenerative process *in vitro*. This should then lead to more focused drug screening assays. The hope for much of these studies is a future of personalized medicine where treatments are focused to the specific mutation allowing for increased success in the clinic. The human PSC progeny will also allow for the easy identification of drugs that may have the potential to be toxic for patients, thereby minimizing the risk of failure at early stages of trials.

References

Aoki, H., Hara, A., Nakagawa, S., Motohashi, T., Hirano, M., Takahashi, Y., and Kunisada, T. (2006). Embryonic stem cells that differentiate into RPE cell precursors in vitro develop into RPE cell monolayers in vivo. *Exp Eye Res* 82, 265–274.

Bae, D., Mondragon-Teran, P., Hernandez, D., Ruban, L., Mason, C., Bhattacharya, S. S., and Veraitch, F. S. (2012). Hypoxia enhances the generation of retinal progenitor cells from human induced pluripotent and embryonic stem cells. *Stem Cells Dev* 21, 1344–1355.

Bao, Z. Z., and Cepko, C. L. (1997). The expression and function of Notch pathway genes in the developing rat eye. *J Neurosci* 17, 1425–1434.

Bassuk, A. G., Zheng, A., Li, Y., Tsang, S. H., and Mahajan, V. B. (2016). Precision medicine: Genetic repair of retinitis pigmentosa in patient-derived stem cells. *Sci Rep* 6, 19969.

Biesemeier, A., Taubitz, T., Julien, S., Yoeruek, E., and Schraermeyer, U. (2014). Choriocapillaris breakdown precedes retinal degeneration in age-related macular degeneration. *Neurobiol Aging* 35(11), 2562–2573.

Brown, N. L., Kanekar, S., Vetter, M. L., Tucker, P. K., Gemza, D. L., and Glaser, T. (1998). Math5 encodes a murine basic helix-loop-helix transcription factor expressed during early stages of retinal neurogenesis. *Development* 125, 4821–4833.

Brown, N. L., Patel, S., Brzezinski, J., and Glaser, T. (2001). Math5 is required for retinal ganglion cell and optic nerve formation. *Development* 128, 2497–2508.

Brzezinski, J. A. T., Lamba, D. A., and Reh, T. A. (2010). Blimp1 controls photoreceptor versus bipolar cell fate choice during retinal development. *Development* 137, 619–629.

Burnight, E. R., Wiley, L. A., Drack, A. V., Braun, T. A., Anfinson, K. R., Kaalberg, E. E., Halder, J. A. et al. (2014). CEP290 gene transfer rescues Leber congenital amaurosis cellular phenotype. *Gene Ther* 21, 662–672.

Cai, J., Nelson, K. C., Wu, M., Sternberg, P., Jr., and Jones, D. P. (2000). Oxidative damage and protection of the RPE. *Prog Retin Eye Res* 19, 205–221.

Cavodeassi, F., Carreira-Barbosa, F., Young, R. M., Concha, M. L., Allende, M. L., Houart, C., Tada, M., and Wilson, S. W. (2005). Early stages of zebrafish eye formation require the coordinated activity of Wnt11, Fz5, and the Wnt/beta-catenin pathway. *Neuron* 47, 43–56.

Chen, J., Rattner, A., and Nathans, J. (2005). The rod photoreceptor-specific nuclear receptor Nr2e3 represses transcription of multiple cone-specific genes. *J Neurosci* 25, 118–129.

Chen, J., Riazifar, H., Guan, M. X., and Huang, T. (2016). Modeling autosomal dominant optic atrophy using induced pluripotent stem cells and identifying potential therapeutic targets. *Stem Cell Res Ther* 7, 2.

Cheng, H., Aleman, T. S., Cideciyan, A. V., Khanna, R., Jacobson, S. G., and Swaroop, A. (2006). In vivo function of the orphan nuclear receptor NR2E3 in establishing photoreceptor identity during mammalian retinal development. *Hum Mol Genet* 15, 2588–2602.

Chiang, C., Litingtung, Y., Lee, E., Young, K. E., Corden, J. L., Westphal, H., and Beachy, P. A. (1996). Cyclopia and defective axial patterning in mice lacking Sonic hedgehog gene function. *Nature* 383, 407–413.

Corbo, J. C., and Cepko, C. L. (2005). A hybrid photoreceptor expressing both rod and cone genes in a mouse model of enhanced S-cone syndrome. *PLoS Genet* 1, e11.

Daniele, L. L., Lillo, C., Lyubarsky, A. L., Nikonov, S. S., Philp, N., Mears, A. J., Swaroop, A., Williams, D. S., and Pugh, E. N., Jr. (2005). Cone-like morphological, molecular, and electrophysiological features of the photoreceptors of the Nrl knockout mouse. *Invest Ophthalmol Vis Sci* 46, 2156–2167.

del Barco Barrantes, I., Davidson, G., Grone, H. J., Westphal, H., and Niehrs, C. (2003). Dkk1 and noggin cooperate in mammalian head induction. *Genes Dev* 17, 2239–2244.

Eiraku, M., and Sasai, Y. (2012). Mouse embryonic stem cell culture for generation of three-dimensional retinal and cortical tissues. *Nat Protoc* 7, 69–79.

Eiraku, M., Takata, N., Ishibashi, H., Kawada, M., Sakakura, E., Okuda, S., Sekiguchi, K., Adachi, T., and Sasai, Y. (2011). Self-organizing optic-cup morphogenesis in three-dimensional culture. *Nature* 472, 51–56.

Elshatory, Y., Deng, M., Xie, X., and Gan, L. (2007). Expression of the LIM-homeodomain protein Isl1 in the developing and mature mouse retina. *J Comp Neurol* 503, 182–197.

Ferretti, E., Villaescusa, J. C., Di Rosa, P., Fernandez-Diaz, L. C., Longobardi, E., Mazzieri, R., Miccio, A. et al. (2006). Hypomorphic mutation of the TALE gene Prep1 (pKnox1) causes a major reduction of Pbx and Meis proteins and a pleiotropic embryonic phenotype. *Mol Cell Biol* 26, 5650–5662.

Friedman, J. S., Khanna, H., Swain, P. K., Denicola, R., Cheng, H., Mitton, K. P., Weber, C. H., Hicks, D., and Swaroop, A. (2004). The minimal transactivation domain of the basic motif-leucine zipper transcription factor NRL interacts with TATA-binding protein. *J Biol Chem* 279, 47233–47241.

Furukawa, T., Morrow, E. M., and Cepko, C. L. (1997). Crx, a novel otx-like homeobox gene, shows photoreceptor-specific expression and regulates photoreceptor differentiation. *Cell* 91, 531–541.

Gordon, P. J., Yun, S., Clark, A. M., Monuki, E. S., Murtaugh, L. C., and Levine, E. M. (2013). Lhx2 balances progenitor maintenance with neurogenic output and promotes competence state progression in the developing retina. *J Neurosci* 33, 12197–12207.

Haider, N. B., Naggert, J. K., and Nishina, P. M. (2001). Excess cone cell proliferation due to lack of a functional NR2E3 causes retinal dysplasia and degeneration in rd7/rd7 mice. *Hum Mol Genet* 10, 1619–1626.

Haigh, J. J., Morelli, P. I., Gerhardt, H., Haigh, K., Tsien, J., Damert, A., Miquerol, L. et al. (2003). Cortical and retinal defects caused by dosage-dependent reductions in VEGF-A paracrine signaling. *Dev Biol* 262, 225–241.

Hambright, D., Park, K. Y., Brooks, M., McKay, R., Swaroop, A., and Nasonkin, I. O. (2012). Long-term survival and differentiation of retinal neurons derived from human embryonic stem cell lines in un-immunosuppressed mouse retina. *Mol Vis* 18, 920–936.

Hatakeyama, J., Bessho, Y., Katoh, K., Ookawara, S., Fujioka, M., Guillemot, F., and Kageyama, R. (2004). Hes genes regulate size, shape and histogenesis of the nervous system by control of the timing of neural stem cell differentiation. *Development* 131, 5539–5550.

Hojo, M., Ohtsuka, T., Hashimoto, N., Gradwohl, G., Guillemot, F., and Kageyama, R. (2000). Glial cell fate specification modulated by the bHLH gene Hes5 in mouse retina. *Development* 127, 2515–2522.

Hunt, D. M., Dulai, K. S., Cowing, J. A., Julliot, C., Mollon, J. D., Bowmaker, J. K., Li, W. H., and Hewett-Emmett, D. (1998). Molecular evolution of trichromacy in primates. *Vision Res* 38, 3299–3306.

Ikeda, H., Osakada, F., Watanabe, K., Mizuseki, K., Haraguchi, T., Miyoshi, H., Kamiya, D. et al. (2005). Generation of Rx+/Pax6+ neural retinal precursors from embryonic stem cells. *PNAS* 102, 11331–11336.

Jagatha, B., Divya, M. S., Sanalkumar, R., Indulekha, C. L., Vidyanand, S., Divya, T. S., Das, A. V., and James, J. (2009). In vitro differentiation of retinal ganglion-like cells from embryonic stem cell derived neural progenitors. *Biochem Biophys Res Commun* 380, 230–235.

Jarrett, S. G., and Boulton, M. E. (2012). Consequences of oxidative stress in age-related macular degeneration. *Mol Aspects Med* 33, 399–417.

Jiang, Y., Ding, Q., Xie, X., Libby, R. T., Lefebvre, V., and Gan, L. (2013). Transcription factors SOX4 and SOX11 function redundantly to regulate the development of mouse retinal ganglion cells. *J Biol Chem* 288, 18429–18438.

Jin, Z. B., Okamoto, S., Osakada, F., Homma, K., Assawachananont, J., Hirami, Y., Iwata, T., and Takahashi, M. (2011). Modeling retinal degeneration using patient-specific induced pluripotent stem cells. *PLoS One* 6, e17084.

Kaewkhaw, R., Kaya, K. D., Brooks, M., Homma, K., Zou, J., Chaitankar, V., Rao, M., and Swaroop, A. (2015). Transcriptome dynamics of developing photoreceptors in three-dimensional retina cultures recapitulates temporal sequence of human cone and rod differentiation revealing cell surface markers and gene networks. *Stem Cells* 33, 3504–3518.

Katoh, K., Omori, Y., Onishi, A., Sato, S., Kondo, M., and Furukawa, T. (2010). Blimp1 suppresses Chx10 expression in differentiating retinal photoreceptor precursors to ensure proper photoreceptor development. *J Neurosci* 30, 6515–6526.

Kayama, M., Kurokawa, M. S., Ueda, Y., Ueno, H., Kumagai, Y., Chiba, S., Takada, E., Ueno, S., Tadokoro, M., and Suzuki, N. (2010). Transfection with pax6 gene of mouse embryonic stem cells and subsequent cell cloning induced retinal neuron progenitors, including retinal ganglion cell-like cells, in vitro. *Ophthalmic Res* 43, 79–91.

Kim, J. W., and Lemke, G. (2006). Hedgehog-regulated localization of Vax2 controls eye development. *Genes Dev* 20, 2833–2847.

Kumar, R., Chen, S., Scheurer, D., Wang, Q. L., Duh, E., Sung, C. H., Rehemtulla, A., Swaroop, A., Adler, R., and Zack, D. J. (1996). The bZIP transcription factor Nrl stimulates rhodopsin promoter activity in primary retinal cell cultures. *J Biol Chem* 271, 29612–29618.

Kuwahara, A., Ozone, C., Nakano, T., Saito, K., Eiraku, M., and Sasai, Y. (2015). Generation of a ciliary margin-like stem cell niche from self-organizing human retinal tissue. *Nat Commun* 6, 6286.

La Torre, A., Hoshino, A., Cavanaugh, C., Ware, C. B., and Reh, T. A. (2015). The GIPC1-Akt1 pathway is required for the specification of the eye field in mouse embryonic stem cells. *Stem Cells* 33, 2674–2685.

Lamb, T. M., Knecht, A. K., Smith, W. C., Stachel, S. E., Economides, A. N., Stahl, N., Yancopolous, G. D., and Harland, R. M. (1993). Neural induction by the secreted polypeptide noggin. *Science* 262, 713–718.

Lamba, D. A., Gust, J., and Reh, T. A. (2009). Transplantation of human embryonic stem cell-derived photoreceptors restores some visual function in Crx-deficient mice. *Cell Stem Cell* 4, 73–79.

Lamba, D. A., Karl, M. O., Ware, C. B., and Reh, T. A. (2006). Efficient generation of retinal progenitor cells from human embryonic stem cells. *PNAS* 103, 12769–12774.

Lamba, D. A., McUsic, A., Hirata, R. K., Wang, P. R., Russell, D., and Reh, T. A. (2010). Generation, purification and transplantation of photoreceptors derived from human induced pluripotent stem cells. *PLoS One* 5, e8763.

Lamba, D. A., and Reh, T. A. (2011). Microarray characterization of human embryonic stem cell—Derived retinal cultures. *IOVS* 52, 4897–4906.

Le, T. T., Wroblewski, E., Patel, S., Riesenberg, A. N., and Brown, N. L. (2006). Math5 is required for both early retinal neuron differentiation and cell cycle progression. *Dev Biol* 295, 764–778.

Lee, H. Y., Wroblewski, E., Philips, G. T., Stair, C. N., Conley, K., Reedy, M., Mastick, G. S., and Brown, N. L. (2005). Multiple requirements for Hes 1 during early eye formation. *Dev Biol* 284, 464–478.

Li, H., Tierney, C., Wen, L., Wu, J. Y., and Rao, Y. (1997). A single morphogenetic field gives rise to two retina primordia under the influence of the prechordal plate. *Development* 124, 603–615.

Li, T., Lu, Z., and Lu, L. (2004). Regulation of eye development by transcription control of CCCTC binding factor (CTCF). *J Biol Chem* 279, 27575–27583.

Lidgerwood, G. E., Lim, S. Y., Crombie, D. E., Ali, R., Gill, K. P., Hernandez, D., Kie, J. et al. (2015). Defined medium conditions for the induction and expansion of human pluripotent stem cell-derived retinal pigment epithelium. *Stem Cell Rev* 12, 179–188.

Lindsell, C. E., Boulter, J., diSibio, G., Gossler, A., and Weinmaster, G. (1996). Expression patterns of Jagged, Delta1, Notch1, Notch2, and Notch3 genes identify ligand-receptor pairs that may function in neural development. *Mol Cell Neurosci* 8, 14–27.

Liu, W., Mo, Z., and Xiang, M. (2001). The Ath5 proneural genes function upstream of Brn3 POU domain transcription factor genes to promote retinal ganglion cell development. *PNAS* 98, 1649–1654.

Lutty, G., Grunwald, J., Majji, A. B., Uyama, M., and Yoneya, S. (1999). Changes in choriocapillaris and retinal pigment epithelium in age-related macular degeneration. *Mol Vis* 5, 35.

Maurer, K. A., Riesenberg, A. N., and Brown, N. L. (2014). Notch signaling differentially regulates Atoh7 and Neurog2 in the distal mouse retina. *Development* 141, 3243–3254.

Maurus, D., Heligon, C., Burger-Schwarzler, A., Brandli, A. W., and Kuhl, M. (2005). Noncanonical Wnt-4 signaling and EAF2 are required for eye development in *Xenopus laevis*. *Embo J* 24, 1181–1191.

McLeod, D. S., Grebe, R., Bhutto, I., Merges, C., Baba, T., and Lutty, G. A. (2009). Relationship between RPE and choriocapillaris in age-related macular degeneration. *Invest Ophthalmol Vis Sci* 50, 4982–4991.

Mears, A. J., Kondo, M., Swain, P. K., Takada, Y., Bush, R. A., Saunders, T. L., Sieving, P. A., and Swaroop, A. (2001). Nrl is required for rod photoreceptor development. *Nat Genet* 29, 447–452.

Megaw, R., Mellough, C., Wright, A., Lako, M., and Ffrench-Constant, C. (2015). Use of induced pluripotent stem-cell technology to understand photoreceptor cytoskeletal dynamics in retinitis pigmentosa. *Lancet* 385 *Suppl* 1, S69.

Mellough, C. B., Sernagor, E., Moreno-Gimeno, I., Steel, D. H., and Lako, M. (2012). Efficient stage-specific differentiation of human pluripotent stem cells toward retinal photoreceptor cells. *Stem Cells* 30, 673–686.

Meyer, J. S., Howden, S. E., Wallace, K. A., Verhoeven, A. D., Wright, L. S., Capowski, E. E., Pinilla, I. et al. (2011). Optic vesicle-like structures derived from human pluripotent stem cells facilitate a customized approach to retinal disease treatment. *Stem Cells* 29, 1206–1218.

Mu, X., Beremand, P. D., Zhao, S., Pershad, R., Sun, H., Scarpa, A., Liang, S., Thomas, T. L., and Klein, W. H. (2004). Discrete gene sets depend on POU domain transcription factor Brn3b/Brn-3.2/POU4f2 for their expression in the mouse embryonic retina. *Development* 131, 1197–1210.

Mukhopadhyay, M., Shtrom, S., Rodriguez-Esteban, C., Chen, L., Tsukui, T., Gomer, L., Dorward, D. W. et al. (2001). Dickkopf1 is required for embryonic head induction and limb morphogenesis in the mouse. *Dev Cell* 1, 423–434.

Mullins, R. F., Johnson, M. N., Faidley, E. A., Skeie, J. M., and Huang, J. (2011). Choriocapillaris vascular dropout related to density of drusen in human eyes with early age-related macular degeneration. *Invest Ophthalmol Vis Sci* 52, 1606–1612.

Nakano, T., Ando, S., Takata, N., Kawada, M., Muguruma, K., Sekiguchi, K., Saito, K., Yonemura, S., Eiraku, M., and Sasai, Y. (2012). Self-formation of optic cups and storable stratified neural retina from human ESCs. *Cell Stem Cell* 10, 771–785.

Nishida, A., Furukawa, A., Koike, C., Tano, Y., Aizawa, S., Matsuo, I., and Furukawa, T. (2003). Otx2 homeobox gene controls retinal photoreceptor cell fate and pineal gland development. *Nat Neurosci* 6, 1255–1263.

Onuma, Y., Takahashi, S., Asashima, M., Kurata, S., and Gehring, W. J. (2002). Conservation of Pax 6 function and upstream activation by Notch signaling in eye development of frogs and flies. *PNAS* 99, 2020–2025.

Pan, L., Deng, M., Xie, X., and Gan, L. (2008). ISL1 and BRN3B co-regulate the differentiation of murine retinal ganglion cells. *Development* 135, 1981–1990.

Pan, L., Yang, Z., Feng, L., and Gan, L. (2005). Functional equivalence of Brn3 POU-domain transcription factors in mouse retinal neurogenesis. *Development* 132, 703–712.

Pera, E. M., Wessely, O., Li, S. Y., and De Robertis, E. M. (2001). Neural and head induction by insulin-like growth factor signals. *Dev Cell* 1, 655–665.

Phillips, M. J., Wallace, K. A., Dickerson, S. J., Miller, M. J., Verhoeven, A. D., Martin, J. M., Wright, L. S. et al. (2012). Blood-derived human iPS cells generate optic vesicle-like structures with the capacity to form retinal laminae and develop synapses. *IOVS* 53, 2007–2019.

Qu, J., Wang, D., and Grosskreutz, C. L. (2010). Mechanisms of retinal ganglion cell injury and defense in glaucoma. *Exp Eye Res* 91, 48–53.

Rasmussen, J. T., Deardorff, M. A., Tan, C., Rao, M. S., Klein, P. S., and Vetter, M. L. (2001). Regulation of eye development by frizzled signaling in *Xenopus*. *PNAS* 98, 3861–3866.

Rehemtulla, A., Warwar, R., Kumar, R., Ji, X., Zack, D. J., and Swaroop, A. (1996). The basic motif-leucine zipper transcription factor Nrl can positively regulate rhodopsin gene expression. *PNAS* 93, 191–195.

Richard-Parpaillon, L., Heligon, C., Chesnel, F., Boujard, D., and Philpott, A. (2002). The IGF pathway regulates head formation by inhibiting Wnt signaling in *Xenopus*. *Dev Biol* 244, 407–417.

Shirai, H., Mandai, M., Matsushita, K., Kuwahara, A., Yonemura, S., Nakano, T., Assawachananont, J. et al. (2016). Transplantation of human embryonic stem cell-derived retinal tissue in two primate models of retinal degeneration. *PNAS* 113, E81–90.

Sidman, R. L. (1961). Histogenesis of the mouse retina. Studies with [3H] thymidine. *In the Structure of the Eye* (Academic Press, New York), 487–506.

Swaroop, A., Xu, J. Z., Pawar, H., Jackson, A., Skolnick, C., and Agarwal, N. (1992). A conserved retina-specific gene encodes a basic motif/leucine zipper domain. *PNAS* 89, 266–270.

Tanaka, T., Yokoi, T., Tamalu, F., Watanabe, S., Nishina, S., and Azuma, N. (2015). Generation of retinal ganglion cells with functional axons from human induced pluripotent stem cells. *Sci Rep* 5, 8344.

Tomita, K., Ishibashi, M., Nakahara, K., Ang, S. L., Nakanishi, S., Guillemot, F., and Kageyama, R. (1996). Mammalian hairy and enhancer of split homolog 1 regulates differentiation of retinal neurons and is essential for eye morphogenesis. *Neuron* 16, 723–734.

Tucker, B. A., Mullins, R. F., Streb, L. M., Anfinson, K., Eyestone, M. E., Kaalberg, E., Riker, M. J., Drack, A. V., Braun, T. A., and Stone, E. M. (2013). Patient-specific iPSC-derived photoreceptor precursor cells as a means to investigate retinitis pigmentosa. *Elife* 2, e00824.

Tucker, B. A., Scheetz, T. E., Mullins, R. F., DeLuca, A. P., Hoffmann, J. M., Johnston, R. M., Jacobson, S. G., Sheffield, V. C., and Stone, E. M. (2011). Exome sequencing and analysis of induced pluripotent stem cells identify the cilia-related gene male germ cell-associated kinase (MAK) as a cause of retinitis pigmentosa. *PNAS* 108, E569–576.

Wang, S., Sengel, C., Emerson, M. M., and Cepko, C. L. (2014). A gene regulatory network controls the binary fate decision of rod and bipolar cells in the vertebrate retina. *Dev Cell* 30, 513–527.

Wang, S. W., Mu, X., Bowers, W. J., Kim, D. S., Plas, D. J., Crair, M. C., Federoff, H. J., Gan, L., and Klein, W. H. (2002). Brn3b/Brn3c double knockout mice reveal an unsuspected role for Brn3c in retinal ganglion cell axon outgrowth. *Development* 129, 467–477.

Wu, J., O'Donnell, M., Gitler, A. D., and Klein, P. S. (2006). Kermit 2/XGIPC, an IGF1 receptor interacting protein, is required for IGF signaling in *Xenopus* eye development. *Development* 133, 3651–3660.

Yoshida, T., Ozawa, Y., Suzuki, K., Yuki, K., Ohyama, M., Akamatsu, W., Matsuzaki, Y. et al. (2014). The use of induced pluripotent stem cells to reveal pathogenic gene mutations and explore treatments for retinitis pigmentosa. *Mol Brain* 7, 45.

Yun, S., Saijoh, Y., Hirokawa, K. E., Kopinke, D., Murtaugh, L. C., Monuki, E. S., and Levine, E. M. (2009). Lhx2 links the intrinsic and extrinsic factors that control optic cup formation. *Development* 136, 3895–3906.

Zhang, X., Friedman, A., Heaney, S., Purcell, P., and Maas, R. L. (2002). Meis homeoproteins directly regulate Pax6 during vertebrate lens morphogenesis. *Genes Dev* 16, 2097–2107.

Zhong, X., Gutierrez, C., Xue, T., Hampton, C., Vergara, M. N., Cao, L. H., Peters, A. et al. (2014). Generation of three-dimensional retinal tissue with functional photoreceptors from human iPSCs. *Nat Commun* 5, 4047.

Zuber, M. E., Gestri, G., Viczian, A. S., Barsacchi, G., and Harris, W. A. (2003). Specification of the vertebrate eye by a network of eye field transcription factors. *Development* 130, 5155–5167.

Index

Page numbers followed by f and t indicate figures and tables, respectively.

H

J

K

U

V

W

X

Printed and bound by CPI Group (UK) Ltd, Croydon, CR0 4YY
01/11/2024
01782617-0008